T0139829

Security of Cyber-Physical Systems

Hadis Karimipour • Pirathayini Srikantha
Hany Farag • Jin Wei-Kocsis
Editors

Security of Cyber-Physical Systems

Vulnerability and Impact

Editors
Hadis Karimipour
School of Engineering
University of Guelph
Guelph, ON, Canada

Pirathayini Srikantha
Lassonde School of Engineering
York University, Toronto, ON, Canada

Hany Farag
Lassonde School of Engineering
York University
Toronto, ON, Canada

Jin Wei-Kocsis
College of Computer and Information
Technology
Purdue University
West Lafayette, IN, USA

ISBN 978-3-030-45543-9 ISBN 978-3-030-45541-5 (eBook)
https://doi.org/10.1007/978-3-030-45541-5

This Springer imprint is published by the registered company Springer Nature Switzerland AG.
The registered company address is: Gewerbestrasse 11, 6330 Cham, Switzerland

Contents

About the Editors

Hadis Karimipour is the director of Smart Grid Lab in the School of Engineering, University of Guelph, Ontario, Canada. She received a Ph.D. degree in Energy System from the Department of Electrical and Computer Engineering in the University of Alberta in February 2016. Before joining the University of Guelph, she was a postdoctoral fellow in University of Calgary working on cybersecurity of the smart power grids. She is currently an Assistant Professor at the School of Engineering, Engineering Systems and Computing Group at the University of Guelph, Ontario, Canada. Her research interests include large-scale power system state estimation, cyber-physical modeling, cybersecurity of the smart grids, and parallel and distributed computing. She is a member of IEEE and IEEE Computer Society. She serves as the Chair of the IEEE Women in Engineering (WIE) and Chapter Chair of IEEE Information Theory in the Kitchener-Waterloo section.

Pirathayini Srikantha is currently an Assistant Professor in the Lassonde School of Engineering at York University. She obtained her Ph.D. degree from The Edward S. Rogers Sr. Department of Electrical and Computer Engineering at the University of Toronto in 2017. She is a certified Professional Engineer (P.Eng.) in Ontario. Her main research interests are in the areas of large-scale optimization and distributed control for enabling adaptive, sustainable, and resilient power grid operations. Her work has been published in premier smart grid journal and conference venues. Her research efforts have received recognitions that include the best paper award (IEEE Smart Grid Communications) and runner-up best poster award (ACM Women in Computing). She is also actively involved in professional and social activities. She has served as the Workshop Chair, Session Chair, and Technical Program Committee member in IEEE conferences. She is a reviewer in numerous IEEE transactions journals.

Hany E. Farag (M'13–SM'18) received the B.Sc. (with honors) and the M.Sc. degrees in electrical engineering from Assiut University, Assiut, Egypt, in 2004 and 2007, respectively, and the Ph.D. degree from in Electrical and Computer Engineering the University of Waterloo, Waterloo, ON, Canada, in 2013. Since July 2013, he has been with the Department of Electrical Engineering and Computer

Science, Lassonde School of Engineering, York University, Toronto, ON, Canada, where he is currently working as an Associate Professor. He is a York Research Chair in Integrated Smart Energy Grids, the Principal Investigator with the Smart Grid Research Laboratory, York University, and the Lead Developer of curriculum and over two million dollars worth laboratories for the power system specialization in the new electrical engineering program at York University. He is also the Executive Secretary of CIGRE International Workgroup C6.28 for standardization of off-supply microgrids. His research interests include the areas of power distribution networks, integration of distributed and renewable energy resources, electric mobility, modeling, analysis, and design of microgrids, and applications of multiagent technologies in smart grids. He is a Registered Professional Engineer in Ontario. He was the recipient of the prestigious Early Research Award (ERA) from Ontario Government.

Jin Wei-Kocsis (M'09) received the Ph.D. degree in Electrical and Computer Engineering from the University of Toronto, ON, Canada, in 2014, the M.S. degree in Electrical Engineering from the University of Hawaii at Manoa, Honolulu, HI, USA, in 2008, and the B.E. degree in Electronic Information Engineering from the Beijing University of Aeronautics and Astronautics, Beijing, China, in 2004. She is currently an Assistant Professor of Electrical and Computer Engineering at the University of Akron. She worked as a Postdoctoral Fellow in National Renewable Energy Laboratory from April 2014 to July 2014. Her research interests include developing low-complexity and hybrid weakly supervised deep learning techniques, developing data-driven cognitive computing and networking mechanisms, developing a blockchain-enabled decentralized and privacy-preserving computing platform, developing blockchain-powered software-defined network-based secure and disaster-resilient networks, designing a blockchain-powered crowdsourced situational-awareness system, and developing data-driven cyber-physical-social security and privacy solutions for critical infrastructures. She received NASA Early Career Faculty Grant, the DoE/SuNLaMP Award, the BIRD Foundation Award, UA NSF I-Corps Grant, and Firestone Research Initiative Fellowship Award based on the research achievements. She has also achieved multiple best paper awards for our journal and conference publications.

Contributors

Mohammad Amiri-Zarandi School of Computer Science, University of Guelph, Canada

Tomomi Aoyama Nagoya Institute of Technology, Nagoya, Aichi, Japan

Haruna Asai Nagoya Institute of Technology, Nagoya, Aichi, Japan

Mohammad Ashrafnejad University of Tabriz, Tabriz, Iran

Ali Dehghantanha University of Guelph, Guelph, ON, Canada

Farnaz Derakhshan University of Tabriz, Tabriz, Iran

Zahra Faraji University of Guelph, Guelph, ON, Canada

Shahrzad Hadayeghparast University of Guelph, Guelph, ON, Canada

Yoshihiro Hashimoto Nagoya Institute of Technology, Nagoya, Aichi, Japan

Bri-Mathias Hodge University of Colorado Boulder, Boulder, CO, USA

Hadis Karimipour School of Engineering, University of Guelph, Guelph, ON, Canada

Faiq Khalid Technische Universität Wien, Vienna, Austria

Ichiro Koshijima Nagoya Institute of Technology, Nagoya, Aichi, Japan

Deepa Kundur Department of Electrical and Computer Engineering, University of Toronto, Toronto, ON, Canada

Weixian Liao Towson University, Towson, MD, USA

Pan Li Case Western Reserve University, Cleveland, OH, USA

Jingyuan Liu York University, Toronto, ON, Canada

Farnaz Seyyed Mozaffari Department of Engineering, University of Guelph, Guelph, ON, Canada

Sanaz Nakhodchi School of Computer Science, University of Guelph, Guelph, ON, Canada

Yuitaka Ota Nagoya Institute of Technology, Nagoya, Aichi, Japan

Reza M. Parizi College of Computing and Software Engineering, Kennesaw State University, Kennesaw, GA, USA

Semeen Rehman Technische Universität Wien, Vienna, Austria

Hossein Mohammadi Rouzbahani University of Guelph, Guelph, ON, Canada

Jacob Sakhnini University of Guelph, Guelph, ON, Canada

Muhammad Shafique Technische Universität Wien, Vienna, Austria

Anna Shi Department of Electrical and Computer Engineering, University of Toronto, Toronto, ON, Canada

Pirathayini Srikantha Lassonde School of Engineering, York University, Toronto, ON, Canada

Mucun Sun The University of Texas at Dallas, Richardson, TX, USA

Chee-Wooi Ten Department of Electrical and Computer Engineering, Michigan Technological University, Houghton, MI, USA

Aaruni Upadhyay School of Computer Science, University of Guelph, Guelph, ON, Canada

Lingfeng Wang Department of Electrical Engineering, University of Wisconsin–Milwaukee, Milwaukee, WI, USA

Jin Wei Purdue University, West Lafayette, IN, USA

Yifu Wu Purdue University, West Lafayette, IN, USA

Koji Yamashita Department of Electrical and Computer Engineering, Michigan Technological University, Houghton, MI, USA

Bokun Zhang Department of Electrical and Computer Engineering, University of Toronto, Toronto, ON, Canada

Jie Zhang The University of Texas at Dallas, Richardson, TX, USA

AI and Security of Cyber Physical Systems: Opportunities and Challenges

Jacob Sakhnini and Hadis Karimipour

1 Introduction

Today's technology is continually evolving towards inter-connectivity among devices. The interconnection of devices through the internet allows for various types of data to be collected to optimize the functionality of the system. This inter-connectivity phenomenon is often referred to as Internet of Things (IoT). IoT devices refer to devices that are connected to the internet; such devices are pivotal to many applications including health-care, transportation, and power generation [1].

IoT technology is used to enhance the performance of systems in many applications. IoT medical devices collect patient data to aid health-care professionals in providing better care for patients. Similarly, power systems collect consumption data to generate and distribute data more efficiently. This integration of physical and cyber components within a system is associated with many benefits; these systems are often referred to as Cyber Physical Systems (CPS).

CPS operate on physical and cyber levels simultaneously for enhanced performance. The physical components of CPS comprise of sensors and actuators. The data from sensors are collected to the cyber component or the network layer, which sends the appropriate response to the actuators of the system. As such, the physical characteristics of the system are controlled through the network based on readings collected from the sensors.

The CPS and IoT technologies are used in many industries critical to our daily lives. In health-care, CPS and IoT devices provide the ability to observe patients' conditions for enhanced care [2]. CPS have the potential to reduce costs,

J. Sakhnini · H. Karimipour (✉)
School of Engineering, University of Guelph, Guelph, ON, Canada
e-mail: jsakhnin@uoguelph.ca; hkarimi@uoguelph.ca

H. Karimipour et al. (eds.), *Security of Cyber-Physical Systems*,
https://doi.org/10.1007/978-3-030-45541-5_1

enhance mobility and independence of patients, and reach the body using minimally invasive techniques. In transportation, CPS are used for tracking vehicles and passengers for improved management and more efficient scheduling. The European Rail Traffic Management System (ERTMS) is an excellent example of CPS in transportation. The ERTMS uses communications among IoT integrated vehicles and devices to manage train schedules [3]. In manufacturing, CPS are often used in life-cycle analysis which involves collecting and analyzing system data for future improvements. CPS also offer increased sustainability through continuous monitoring of the physical components.

Although this inter-connectivity of devices can pave the road for immense advancement in technology and automation, the integration of network components into any system increases its vulnerability to cyber threats. Using internet networks to connect devices together creates access points for adversaries. And considering the critical applications of some of these devices, adversaries have the potential of exploiting sensitive data and interrupting the functionality of critical infrastructure. Additionally, many Advanced Persistent Threat (APT) actors and hacking teams are targeting critical infrastructure and services [4] such as health-care [5] and safety critical systems [6].

2 Book Outline

This book presents an overview of security in CPS by analyzing issues and vulnerabilities in CPS and examining state of the art security measures. Furthermore, the book proposes various defense strategies including intelligent attack and anomaly detection algorithms.

This book is comprised of 15 chapters. The next chapter, chapter "Overview of Security for Smart Cyber-Physical Systems", presents a brief overview of security threats in CPS [7]. Chapter "Design and Operation Framework for Industrial Control System Security Exercise" demonstrates a design framework for training exercises that can be undertaken by businesses to educate their staff in regards to cyber security [8]. Chapter "Cascading Failure Attacks in the Power System" demonstrates the dangers for cyber attacks on power systems and discusses high-impact cascading failure attacks [9]. Chapter "The Risk of Botnets in Cyber Physical Systems" analyzes Botnets in CPS and state of the art countermeasures [10].

The book then considers anomaly detection in CPS. This begins with chapter "Learning Based Anomaly Detection in Critical Cyber-Physical Systems" which discusses learning based anomaly detection methods in CPS presenting a case study using machine learning algorithms [11]. This is followed by data-driven anomaly detection in modern power systems discussed in chapter "Data-driven Anomaly Detection in Modern" [12].

Next, the book delves into intelligent security solutions for CPS. This section begin by introducing AI-enabled security monitoring in CPS in chapter "AI-enabled Security Monitoring in Smart Cyber Physical Grids" [13]. Chapter "Application of

Machine Learning in State Estimation of Smart Cyber-Physical Grid" covers the use of machine learning in state estimation of power systems; in which machine learning is used to enhance the computational efficiency of smart power systems [14]. Chapter "A Comparison Between Different Machine Learning Models for IoT Malware Detection" compares between different machine learning algorithms for malware detection [15]. This chapter also tests various machine learning algorithms for detection of malware in IoT datasets. Chapter "A Bibliometric Analysis on the Application of Deep Learning in Cybersecurity" ends the discussion of intelligent algorithms with a bibliometric survey of state of the art of deep learning in cybersecurity [16].

The last section of this book examines more advanced and specific topics in industrial CPS. Chapter "Dynamical Analysis of Cyber-Related Contingencies Initiated from Substations" analyzes cyber-related contingencies in smart power systems tackling stability concerns and the impact of physical operations on power systems [17]. Chapter "Distributed Attack and Mitigation Strategies for Active Power Distribution Networks" introduces a stealthy attack strategy capable of compromising distribution network operations in power systems [18]. Chapter "Privacy-Preserving Homomorphic Masking for Smart Grid Data Analytics in the Cloud" proposes a lightweight homomorphic masking technique that allows operators to perform comprehensive data analytic [19]. This technique also includes privacy-preserving strategies so it can be used in sensitive applications. Chapter "A Distributed Middleware Architecture for Attack-Resilient Communications in Smart Grids" concludes this book with a novel intelligent communication middle-ware architecture intended to enhance data acquisition and communication infrastructure in smart power systems [20].

References

1. D. Miorandi, S. Sicari, F. De Pellegrini, I. Chlamtac, Internet of things: vision, applications and research challenges. Ad Hoc Netw. **10**, 1497–1516 (2012)
2. A. Milenković, C. Otto, E. Jovanov, Wireless sensor networks for personal health monitoring: issues and an implementation. Comput. Commun. **29**, 2521–2533 (2006)
3. A. Ferlin, R. Ben-Ayed, P. Sun, S. Collart-Dutilleul, P. Bon, Implementation of ERTMS: a methodology based on formal methods and simulation with respect to French national rules. Transp. Res. Procedia **14**, 1957–1966 (2016). Transport Research Arena TRA2016
4. J.J.P.C. Rodrigues, D.B. De Rezende Segundo, H.A. Junqueira, M.H. Sabino, R.M. Prince, J. Al-Muhtadi, V.H.C. De Albuquerque, Enabling technologies for the Internet of health things. IEEE Access **6**, 13129–13141 (2018)
5. M. Hassanalieragh, A. Page, T. Soyata, G. Sharma, M. Aktas, G. Mateos, B. Kantarci, S. Andreescu, Health monitoring and management using Internet-of-Things (IoT) sensing with cloud-based processing: opportunities and challenges, in *2015 IEEE International Conference on Services Computing, New York City, NY, June* (IEEE, Piscataway, 2015), pp. 285–292
6. A. Luque-Ayala, S. Marvin, Developing a critical understanding of smart urbanism? Urban Stud. **52**, 2105–2116 (2015)

7. F. Khalid, S. Rehman, M. Shafique, Overview of security for smart cyber-physical systems, in *Security of Cyber-Physical System: Vulnerability and Impact*, ch. 2, pp. 5–25 (Springer International Publishing, Cham, 2019)
8. H. Asai, T. Aoyama, Y. Ota, T. Hashimoto, I. Koshijima, Design and operation framework for industrial control system security exercise, in *Security of Cyber-Physical System: Vulnerability and Impact*, ch. 3, pp. 26–62 (Springer International Publishing, Cham, 2019)
9. W. Liao, P. Li, Cascading failure attacks in the power system, in *Security of Cyber-Physical System: Vulnerability and Impact*, ch. 4, pp. 63–88 (Springer International Publishing, Cham, 2019)
10. F. Derakhshan, M. Ashrafnejad, The risk of botnets in cyber physical systems, in *Security of Cyber-Physical System: Vulnerability and Impact*, ch. 5, pp. 88–119 (Springer International Publishing, Cham, 2019)
11. F.S. Mozaffari, H. Karimipour, R.M. Parizi, Learning based anomaly detection in critical cyber-physical systems, in *Security of Cyber-Physical System: Vulnerability and Impact*, ch. 6, pp. 120–147 (Springer International Publishing, Cham, 2019)
12. M. Sun, J. Zhang, Data-driven anomaly detection in modern power systems, in *Security of Cyber-Physical System: Vulnerability and Impact*, ch. 7, pp. 148–161 (Springer International Publishing, Cham, 2019)
13. H.M. Rouzbahani, Z. Faraji, M. Amiri-Zarandi, H. Karimipour, Ai-enabled security monitoring in smart cyber physical grids, in *Security of Cyber-Physical System: Vulnerability and Impact*, ch. 8, pp. 162–194 (Springer International Publishing, Cham, 2019)
14. S. Hadayeghparast, H. Karimipour, Application of machine learning in state estimation of smart cyber-physical grid, in *Security of Cyber-Physical System: Vulnerability and Impact*, ch. 9, pp. 195–227 (Springer International Publishing, Cham, 2019)
15. S. Nakhodchi, A. Upadhyay, A. Dehghantanha, A comparison between different machine learning models for IoT malware detection, in *Security of Cyber-Physical System: Vulnerability and Impact*, ch. 10, pp. 228–242 (Springer International Publishing, Cham, 2019)
16. S. Nakhodchi, A. Dehghantanha, A bibliometric analysis on the application of deep learning in cybersecurity, in *Security of Cyber-Physical System: Vulnerability and Impact*, ch. 11, pp. 243–272 (Springer International Publishing, Cham, 2019)
17. K. Yamashita, C.-W. Ten, L. Wang, Dynamical analysis of cyber-related contingencies initiated from substations, in *Security of Cyber-Physical System: Vulnerability and Impact*, ch. 12, pp. 273–297 (Springer International Publishing, Cham, 2019)
18. J. Liu, P. Srikantha, Distributed attack and mitigation strategies for active power distribution networks, in *Security of Cyber-Physical System: Vulnerability and Impact*, ch. 13, pp. 298–315 (Springer International Publishing, Cham, 2019)
19. P. Srikantha, A. Shi, B. Zhang, D. Kundur, Privacy-preserving homomorphic masking for smart grid data analytics in the cloud, in *Security of Cyber-Physical System: Vulnerability and Impact*, ch. 14, pp. 316–342 (Springer International Publishing, Cham, 2019)
20. Y. Wu, J. Wei, B.-M. Hodge, A distributed middleware architecture for attack-resilient communications in smart grids, in *Security of Cyber-Physical System: Vulnerability and Impact*, ch. 16, pp. 365–394 (Springer International Publishing, Cham, 2019)

Overview of Security for Smart Cyber-Physical Systems

Faiq Khalid, Semeen Rehman, and Muhammad Shafique

1 Introduction

The technological advancements in several application domains, i.e., industry, healthcare, transportation, vehicles, etc., have increased the complexity of the interaction between the cyber domain and the physical domain [1]. This requirement of the complex interaction leads to Cyber-Physical Systems (CPS) for efficient integration of the cyber domain and the physical domain [2]. CPS is *a tightly-coupled communication network where several embedded computing devices, smart controllers, physical environments, and humans systematically interact with each other* [3, 4], as shown in Fig. 1. Typically, CPS devices interact with each other by sharing the information and communicating the appropriate control commands. CPS collects the data from the physical domain using sensors and controllers analyze it to issue the necessary control commands, which guide/control the physical domain via actuators.

The complex and massive integration of networked computing devices, sensors, and actuators with humans in CPS is playing a significant role in the rise of the Internet-of-Things (IoT), where different cyber and physical (sub-)systems are integrated over the Internet [5, 6]. Due to their ability to handle such complex interaction, IoT has revolutionized several application domains, e.g., smart traffic control, healthcare, transport systems, industrial automation, smart grids, autonomous vehicles, and smart homes/buildings (as shown in Fig. 2). The scope and usage/opportunities of IoT vary with respect to its application in the corresponding industries (as shown in Fig. 3). For instance, in 2017, several application domains, e.g., multimedia, manufacturing, financial corporations, etc., have used the

F. Khalid (✉) · S. Rehman · M. Shafique
Technische Universität Wien, Vienna, Austria
e-mail: faiq.khalid@tuwien.ac.at; semeen.rehman@tuwien.ac.at; muhammad.shafique@tuwien.ac.at

H. Karimipour et al. (eds.), *Security of Cyber-Physical Systems*,
https://doi.org/10.1007/978-3-030-45541-5_2

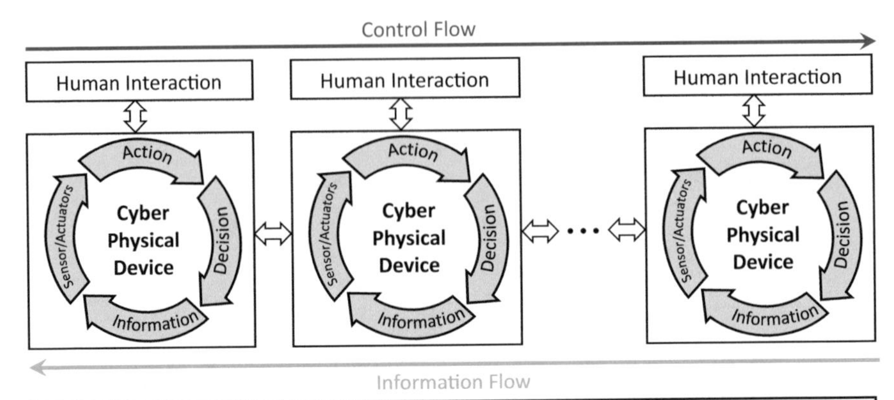

Fig. 1 Key features of a CPS: (1) collects the data from the sensors, (2) shares the data via communication networks, (3) analyzes the data, (4) generates the appropriate control commands, and (5) controls or guides the physical domain via actuators

IoT in many of their applications, primarily in security and surveillance, energy, and asset management [7] (as shown in Fig. 3). This extensive use of IoT in industries, especially in safety-critical applications comes with the following challenges:

1. One of the biggest challenges is to handle the enormous amounts of generated data that must be stored and processed. For example, an analytical survey by the company "Statista" shows that almost 75.42 billion connected devices will be utilized by 2025, which will approximately generate 180 zettabytes of measured data [8], as shown in Fig. 3.
2. The complex integration of the cyber domain (i.e., networked computing devices), physical domain (i.e., actuators), and humans make CPS vulnerable to various security threats [9, 10]. Therefore, several security incidents have been reported in the real-world. Some of the prominent incidents include blocking the city water pipeline [11–13], hacking the pacemaker [14, 15], anti-lock braking system [16], wheel speed sensor spoofing in smart cars [17, 18], relay attacks (to disable lights) [19] and several industrial attacks [20–22]. These incidents encourage the researchers to address the following key research questions:
   - How to ensure secure data acquisition from sensors?
   - How to ensure the security of the controllers and corresponding control signals over different CPS layers?
   - How to ensure secure inter-layer, intra-layer, and stack communication of data and control signals?

To address these research questions, researchers have proposed several security measures, i.e., online anomaly detection [23, 24], anonymization [5], trusted computing (attestation) [25, 26], encryption of the data and control signals [27], and

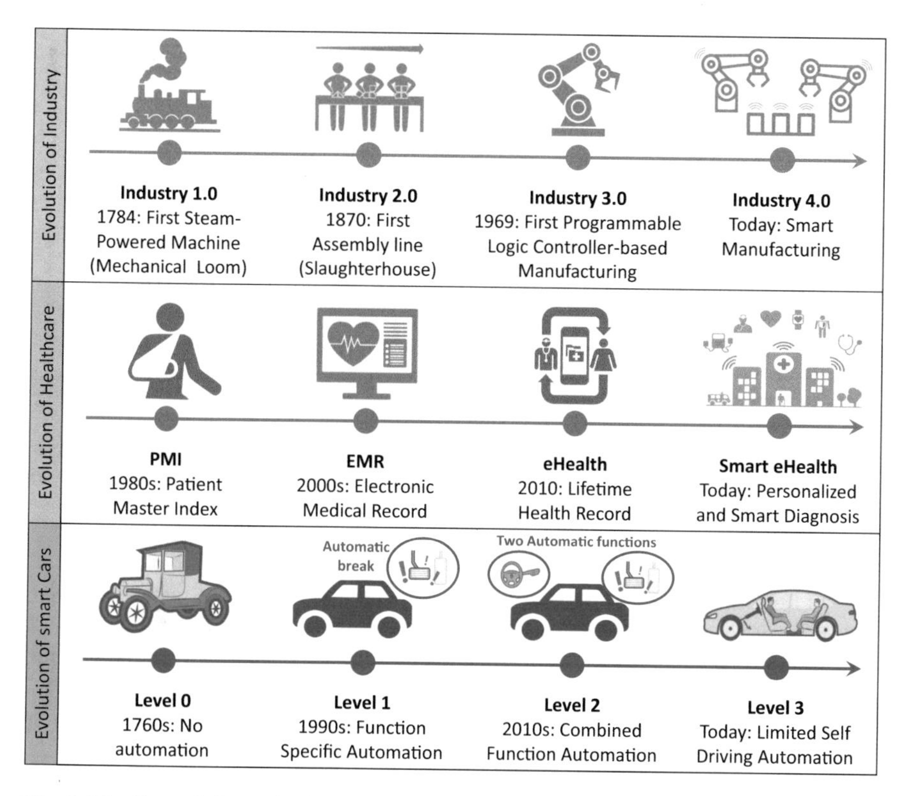

Fig. 2 Timeline of the technological advancements in different applications, e.g., industry, healthcare, and cars. The industry evolved from manual work to the powered machine (industry 1.0), assembly line (industry 2.0), mechanical automation (industry 3.0), and smart manufacturing based on cyber-physical systems (industry 2.0). The healthcare evolved from manual treatment, diagnosis, and health record maintenance to electronic lifetime health records, and to personalized and smart diagnosis. The evolution of cars includes single-function automation, multi-function automation to limited self-driving

verifiable computation. However, the ever-growing number of connected devices and corresponding data, and resource constraints make it very challenging to develop security measures, especially in battery-operated devices in CPS. Therefore, security measures in CPS need to be energy-efficient yet adaptive and sustainable for handling the unpredictable operating conditions over a lifetime. Moreover, for sustainable secure CPS, these security measures must be intelligent enough to adapt to the unforeseen attack surface.

In the following sections, we discuss:

1. A brief overview of *security attacks* for CPS and their respective *attack models* (Sect. 2).
2. The intelligent security measures for CPSs along with their associated research challenges (Sect. 3).

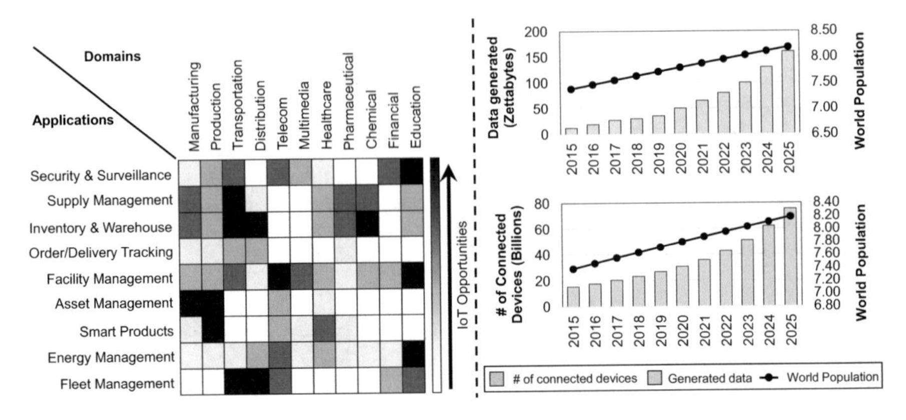

Fig. 3 On the left, a heat map shows the opportunities and scope of the IoT for different applications [7]. On the right, a growing trend of the connected devices (in billions) and the generated data (in zettabytes) for IoT from 2015 to 2025 [8]

3. Low power online anomaly detection technique and machine learning (ML)-based security measures for smart CPSs, and we present case studies of MC8051 and MC8051-based communication network along with the trust-Hub benchmark attacks [28, 29] on MC8051, i.e., T200 to T800 (Sect. 4).
4. The security issues for ML along with the summary of state-of-the-art security attacks and corresponding defenses (Sect. 5).

2 CPS Security

Security measures in traditional systems exploit communication behavior and other side-channel parameters. However, these techniques do not apply to CPS because of complex interdependencies on different layers, especially uncertainties in physical layers [3, 30]. For example, Table 1 shows some of the real-world CPS attacks that exploit the vulnerabilities at different CPS layers. Consequently, traditional definitions for threats and corresponding threat models cannot be used to analyze and develop security measures in CPS [31]. Therefore, in this section, we provide a brief overview of security vulnerabilities in CPS at different layers of CPS, respective payloads, and associated threat models.

2.1 Security Attacks in CPS

The security attacks in CPS vary from traditional cyber attacks to cyber-physical attacks that exploit the vulnerabilities at different CPS layer [3, 42, 43]. Typically,

Table 1 Examples of the real-world security attacks on different layers of CPS

CPS layer	Applications	Attacks	Description
Sensors/Actuators	Smart grids	False data injection	Interrupt the data acquisition by feeding false data to sensors [32]
Network	Smart grids	Cyber extortion	Hack and exploit the CPS component that can connected to Internet [33]
Sensors/Actuators	Control system	False data/Signal injection	Corrupting the sensor data (sensors) or control commands (actuators) [34, 35]
Network	Control system	Replay attacks	Hack the network to delay or corrupt control commands [36]
Network or physical	Smart grids	Aurora experiment	Maliciously interrupt brakes [37, 38]
Network	Smart healthcare	False data injection	Corrupt the patient record [39]
Physical	Smart healthcare	Unauthorized commands injection	Remotely send the false commands to insulin injection pump [39]
Network	Smart cars	Denial-of-service	Disable the communication with brakes [40, 41]

these vulnerabilities are categorized based on the CPS layers. Figure 4 shows different security attacks for CPS with respect to several CPS layers.

1. **Physical Layer:** This layer is vulnerable to the physical intervention of an attacker, which may lead to physical damage, taking control of the sensors (inaccurate measurements) [44], actuators (inappropriate decisions/actions), and controllers [30]. Moreover, the sensors and actuators are also vulnerable to several data sniffing, spoofing and leaking attacks, which raises concerns about its confidentiality [45, 46]. Based on the payload and attack methods, there are the following possible security threats for this layer:

   - Hacking the sensors/actuators to modify, manipulate or destroy the sensors/actuators (e.g., side channel-based brute force attacks) [22].
   - Hacking the power distribution network of CPS to steal the energy for denial-of-service attacks or to generate random fluctuations for malicious payloads [22].
   - Leaking the private key in case of encrypted measurement and control mechanisms, i.e., brute force, dictionary or monitoring-based key stealing attacks [21].

2. **Network Layer:** This layer is one of the most vulnerable layers in CPS because of several sophisticated security attacks aiming at intruding the communication channels for manipulating the communication (i.e., man-in-middle attack, information leakage, denial-of-service) [47]. Broadly, the attacks on the communication layer can be categorized as:

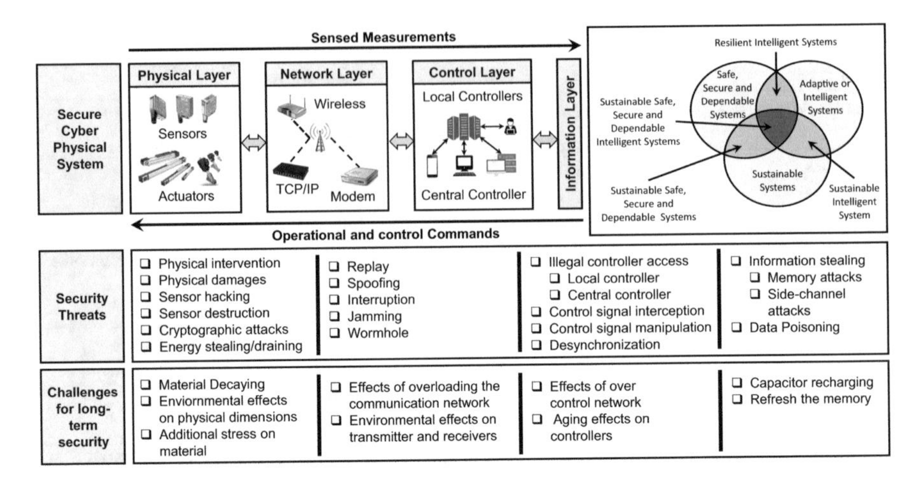

Fig. 4 Security attacks for CPS with respect to several CPS layers and associated challenges for long-term and sustainable security measures for CPS

- Replay Attack: In this attack, the attacker either transmits the data to an incorrect destination or introduces synchronization problems (i.e., timing violations) [48].
- Denial-of-service (DoS): In this attack, a communication channel is manipulated to perform several malicious payloads. For example, in the jamming attack, the attacker reduces the bandwidth or increases the data rate to jam the communication nodes [49, 50]). Similarly, there are several alternative methods to perform DoS attack, e.g., collision (violating the communication protocol to generate inconsistencies [51]), routing ill-directing (changing the route of messages to introduce synchronization issues [3, 4, 42]), flooding (inserting unnecessary request signals or data to overload the communication channel [52, 53]) and selective forwarding (controlling the transmitting data and diverting it to an illegal destination, e.g., Sybil [54]).

3. **Control Layer:** This layer is one of the critical layers because on gaining access to the control layer, small changes in the control signal can be triggered that may have a significant impact on the overall security and safety of CPS. Typically, the security attacks in control layers introduce synchronization problems [55] to disturb the timing properties of the CPS, which eventually lead to malfunctioning or physical destruction [56].

In CPS, the information or data measured from the physical world and control signals are one of the most critical parameters. However, these parameters are vulnerable to software layer information-stealing attacks (eavesdropping [57, 58] or information monitoring). Moreover, if an attacker can manipulate the information [59], then he can perform almost all types of attacks, i.e., replay, denial of services, information stealing, etc.

Table 2 Potential attack models for CPS with respect to its layers (M: Manufacturer, D: Designer, E: External Attacker)

Layers	Attackers	Attacking mechanism	Payloads
Physical	M, D, E	Physical intervention	DoS, reliability (e.g., aging)
Sensor/Actuator	M, D, E	Hacking, control access, data manipulations	Energy stealing, DoS, information leakage, desynchronization
Communication	M, D, E	Replay, Sybil, jamming, flooding, spoofing	Energy stealing, DoS, information leakage, desynchronization
Control	M, D, E	Control access, Eavesdropping	Information leakage, DoS, desynchronization
Information	M, D, E	Eavesdropping	Information leakage
Integration level	M, D, E	All possible control and communication attacks	Energy stealing, DoS, information leakage, desynchronization

2.2 Attack Models for CPS

Unlike the traditional embedded systems, the attack surface in CPS not only depends upon the attacker's access but also on the behavior of the CPS layers. Therefore, to develop the security measure for CPS, an appropriate attack model [3, 4] is required. Table 2 provides a summary of the possible threat models at each layer of the CPS. In CPS, the attack model depends upon the following factors:

1. *Attacker:* it is defined as any element that intentionally modifies the CPS behavior to fulfill a particular motive (e.g., criminal, spying, or cyberwar, etc.).
2. *Attacker's Access*: it is defined as the access to the different layers, e.g., physical access to the sensors or actuators, wireless or internet access to the network layer, etc.
3. *Attacking Mechanism:* it is defined as the vectors and methodologies to perform an attack, e.g., interception, interruption, and modification.
4. *Payloads:* it is defined as the targeted functionality that fulfills the attacker's motive, e.g., confidentiality, integrity, availability, privacy, or safety.

3 Designing a Secure CPS

The above-mentioned security vulnerabilities raise serious concerns over the usage of CPS in safety-critical applications, like autonomous vehicles, smart healthcare, industry 4.0. Therefore, there is a dire need to develop security measures that are adaptive to unforeseen attack surfaces and sustainable enough to handle the long-term impact of environmental changes and technological advancements [4]. In summary, to design sustainable and secure CPS, the following research challenges must be addressed:

1. **Inclusion of the security in design constraints:** Several security measures for CPS are being introduced, but security is not included as the design constraint in the CPS design cycle. Therefore, it is imperative to integrate security constraints into traditional design constraints.
2. **Resource-efficient adaptive design:** The complex interaction and integration of physical-domain and cyber-domain make the CPS highly vulnerable to unforeseen attack surfaces. Moreover, the limited resources (in battery-operated CPS) also limit the runtime security measures. Therefore, it is imperative to develop such security measures that are adaptive and also fulfill the resource constraints and energy budget.
3. **Secure communication:** To design a secure CPS, it is imperative to ensure the secure interaction and integration of several heterogeneous cyber-physical devices.
4. **Data confidentiality:** Information and control signals are an integral part of CPS. Therefore, it is imperative to ensure the secure communication and storage of the information and control signals.

3.1 Security Measures for CPS

To address the above-mentioned design challenges several security measures have been proposed to counter the security attacks (Sect. 2.1) while considering the attack models (Sect. 2.2). Depending upon the defender's goal, these techniques can be categorized as:

3.1.1 Authentication-Based Security Measures

One of the critical challenges in CPS is to protect sensitive information and measured data. Therefore, several security measures have been proposed to ensure information privacy, i.e., Physical Unclonable Functions (PUFs) [60, 61], True Random Number Generators (TRNGs) [62], and protection against physical tampering [63]. Though these techniques provide encrypted key/authentication IDs for information leakage and obfuscation against reverse engineering, they have the following limitations:

- Due to the requirement for precise calibration, these techniques cannot incorporate the effects of dormant attacks and uncertainties of the physical layer.
- The key-based encryption and authentication IDs require additional computational resources, which leads to energy overhead. Thus, it restricts their applicability to design and test time security measures for low power battery-operated CPS devices.

3.1.2 Runtime Security Measures

To capture the effects of dormant attacks at runtime, several runtime techniques have been proposed that can also incorporate uncertainties, i.e., sensor [64, 65] and context-aware [66] detection techniques. Similarly, several communication-based [67] and side-channel parameter-based profiling techniques [68, 69], roundtable discussion [33], and controller protection techniques have been proposed and are being deployed in several industries [34]. Even though these techniques address the runtime detection challenges, given the current exponential increase in connected devices of CPS, they are not adaptive enough to handle the growing runtime computational requirements.

3.1.3 Machine Learning-Based Security Measures for CPS

To address the requirement of handling the enormous data, several machine learning (ML)-based [70] techniques have been proposed. Moreover, the ML-based approaches extract some hidden features from sensed/measured data [38], which can help to identify the abnormal behavior efficiently. Though ML-based techniques address the problem of handling a large amount of data, these techniques inherently possess an energy overhead, and most of them are focused on data protection from leakage. However, with efficient implementation, ML-based techniques can be extended to side-channel parameters and communication patterns, for identifying other security threats in CPS, i.e., DoS, desynchronization, etc.

4 Design a Secure Internet-of-CPS

To address two key limitations of the state-of-the-art security measures, i.e., energy overhead and feature extraction and analysis of enormous measured data, researchers have investigated different side-channel parameters and communication behaviors to develop low power online anomaly detection techniques. In general, the security of the autonomous systems require the two phases, i.e., security vulnerability analysis and runtime anomaly detection (see Fig. 5):

4.1 Security Vulnerability Analysis

The first phase in the designing of the Internet-of-Cyber Physical System (IoCPS) is to analyze the CPS to find security vulnerabilities. Though several techniques have been proposed for security vulnerability analysis, most of them are based on traditional simulation and emulation. Therefore, they are inherently incomplete, which may compromise the security and safety in safety-critical applications, e.g.,

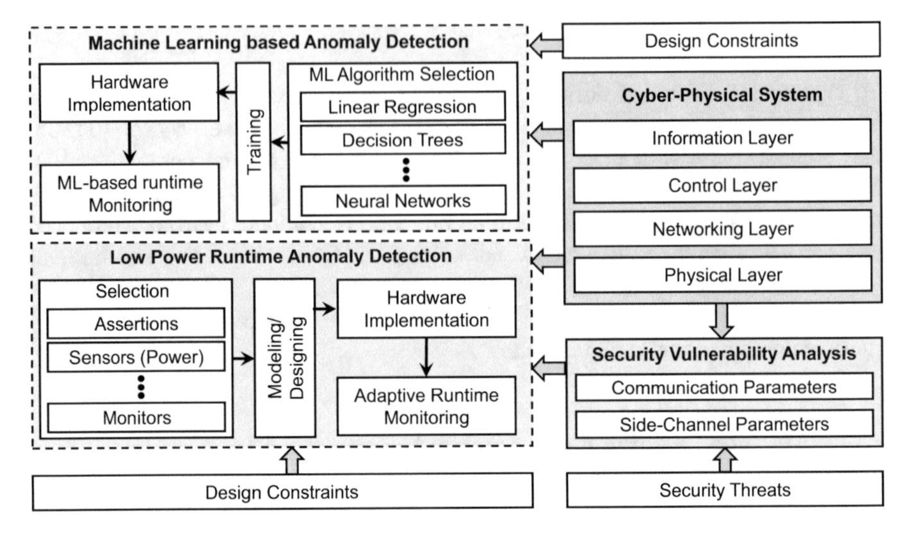

Fig. 5 Low power and ML-based runtime anomaly detection for secure Internet-of-Cyber Physical Systems (IoCPS)

autonomous driving. To address this completeness problem, in this paper, we discuss a formal verification-based security vulnerability analysis [71–74] that consists of the following steps:

- In the first step, the proposed methodology translates the functional and side-channel-based parametric behavior into the corresponding formal model.
- In the second step, functional verification is performed to ensure accurate and reliable formal modeling.
- In the final step, the safety and security properties to identify possible security vulnerabilities are identified, which may aid to the development of efficient security measures.

4.2 Runtime Anomaly Detection

After identifying the security vulnerability, the next step is to use this vulnerability information to develop online anomaly detection techniques using different parameters, e.g., communication behavior and side-channel parameters. Here, we present a brief overview of two anomaly detection techniques that leverage the communication behavior and power (dynamic and leakage) for designing low-power and ML-based anomaly detection techniques, respectively.

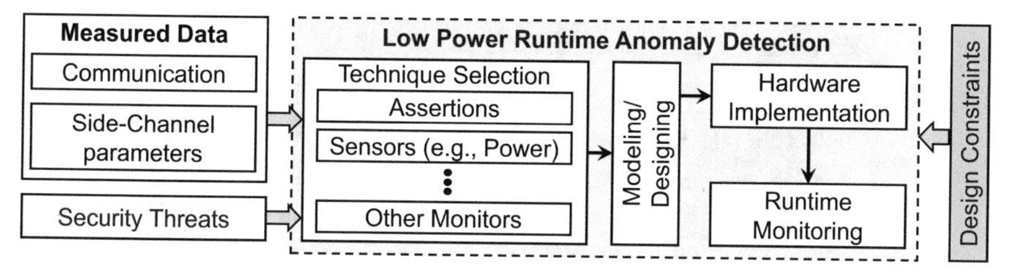

Fig. 6 Overview of the low power anomaly detection for secure IoCPS

4.2.1 Low Power Anomaly Detection

To reduce the power overhead in CPS, researchers have proposed a methodology that leverages the traditional techniques, i.e., assertions, sensor-based analysis, and runtime monitoring, for designing low-power online anomaly detection techniques [67] (as shown in Fig. 6). The first step of this methodology is to choose an appropriate detection scheme based on security threats, metrics and design constraints. In the next step, based on the selected detection scheme, the corresponding assertions or sensor-based runtime monitoring setup is developed and implemented. In this section, we briefly discuss a method that uses the communication behavior-based assertions to identify the online anomalies with significantly low power and area overhead. To illustrate the effectiveness of this technique, we present an experimental analysis that shows the effects of security attacks on the communication behavior of the MC8051-based communication network. The experimental analysis in Fig. 7 shows that in the case of a DoS attack, the output packets of the statistical parameter for the communication behavior is close to zero. However, in the case of communication attacks, like flooding, jamming, and information leakage, the value of the statistical parameter for communication behavior is larger than clean communication behavior. Hence, communication behavior can be used to identify anomalous behavior.

4.2.2 Machine Learning Based Anomaly Detection

The number of communication channels is increasing exponentially with an increase in a number of connected devices. Therefore, communication behavior-based assertions are not feasible for the larger and complex network of CPS. To address this issue, researchers have proposed to leverage the side-channel parameters, like power [68–70]. Though side-channel parameters effectively identify the abnormal behavior, these techniques posses a significant computational overhead. Therefore, researchers have leveraged the ML algorithms to handle this complex computational task. In this chapter, we briefly discuss a methodology that leverages the ML algorithm to detect abnormal behavior using power profiles, as shown in Fig. 8. The first step of this methodology is to select an appropriate ML algorithm and side-

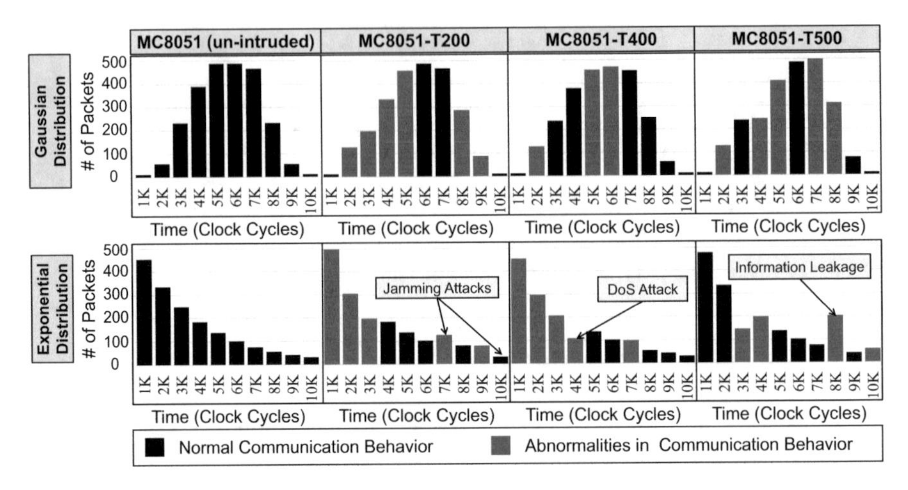

Fig. 7 The effects of trust-hub Trojan benchmarks (i.e., MC8051-T200, T300, T400, and T8500) on the communication behavior of MC8051 for Gaussian and Exponential input data distribution

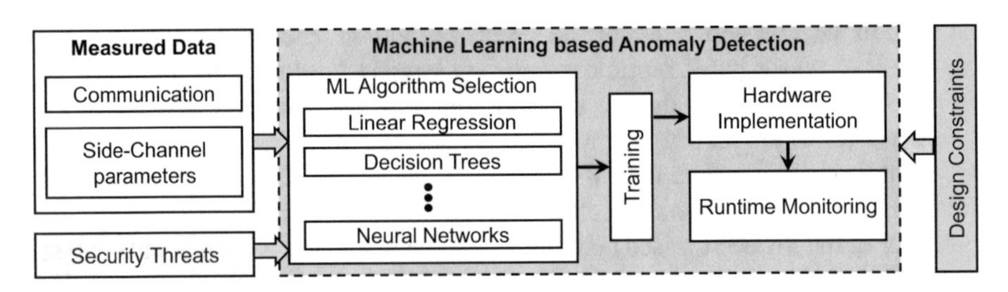

Fig. 8 Overview of ML-based anomaly detection in Secure IoCPS

channel parameter based on the design constraints, security threats, and complexity of the measured data. In the second step, the power profiles are extracted during the design time and then the selected ML algorithm is trained using the obtained power profiles. Finally, an efficient data acquisition block to measure the runtime power profiles is designed, which is then used to identify the abnormal behavior using the trained ML algorithm.

To illustrate the effect of security attacks on the power behavior of CPS controllers, we analyzed the MC8051 with and without trust-Hub benchmarks [28, 29], i.e., MC8051-T200, T400, T500, T600, T700, T800, in Xilinx power analyzer. The experimental analysis in Fig. 9 shows that intrusions in MC8051 have a significant impact on the power distribution for different pipeline stages, as depicted from labels 1, 2, 3, and 4. This concludes that the power profiling of the processing elements/controllers in CPS can be used to identify the abnormalities.

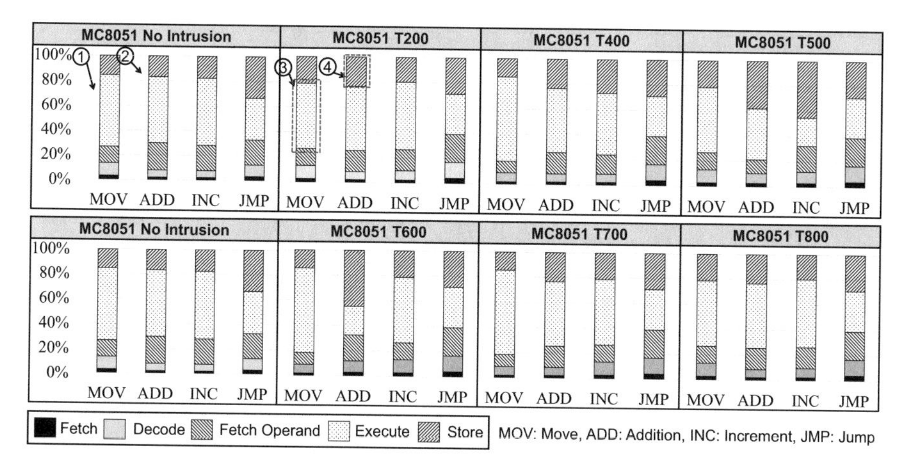

Fig. 9 Effects of trust-hub Trojan benchmarks (i.e., MC8051-T200, T300, T400, T500, T600, T700 and T800) on power correlation with respect to pipeline stages for different instructions, i.e., MOV, ADD, INC, JMP

5 Security for Machine Learning

ML is becoming an integral part of the CPS because of its ability to handle complex computations for an enormous amount of data efficiently. Moreover, recently, ML is being used for many security measures for CPS, i.e., online anomaly detection, ML-based online monitoring for controllers, etc. Though ML effectively handles the complex computations and significantly improves the security of the CPS, it possesses the inherent security vulnerabilities [9, 75, 76], i.e., adversarial examples, data poisoning, hardware attacks, etc. (as shown in Fig. 10). The security vulnerabilities add a new attack surface in CPS security [77]. Therefore, for designing a secure CPS, the ML systems and ML-based security measures must be protected from security attacks. In this section, we briefly discuss the possible attack models, state-of-the-art security attacks and corresponding defenses for ML systems.

5.1 Attack Models

The following factors define the attack surface, which leads to all possible attack models:

1. Attacker's Access: it referred to all the possible steps, tools, datasets, and other parameters in the design cycle that can be exploited to perform an attack. For example, in the data poisoning attacks [76, 78], the attacker manipulates the

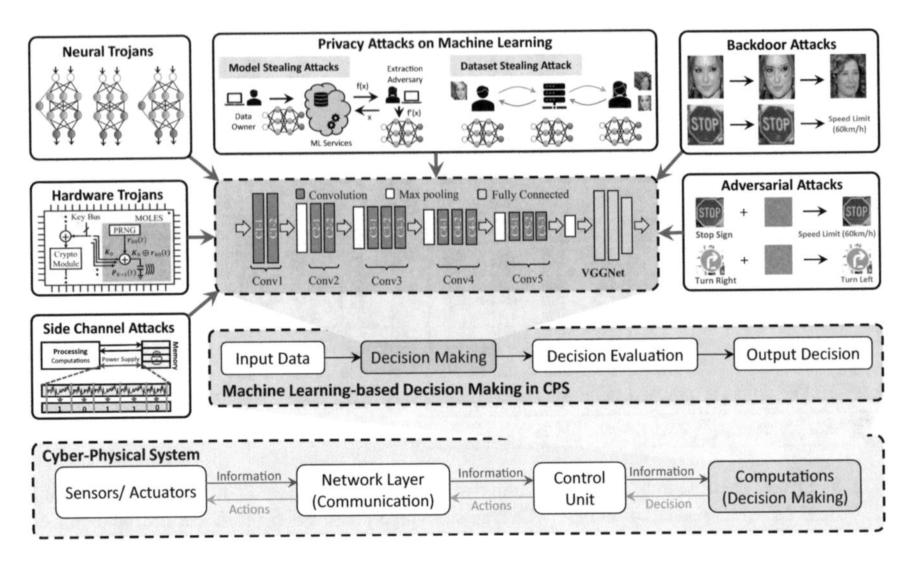

Fig. 10 Security attacks on ML systems in CPS

training dataset to introduce the backdoors. Moreover, it also includes the extent of the attacker's access to the targeted model (Black-box[1]and white-box[2]).
2. Payload: It is defined as the consequence of the performed attack, i.e., confidence reduction, random misclassification or targeted misclassification, model stealing, dataset stealing, etc.
3. Location: The design and development cycle phase of the ML system where the attacker performs an attack. For example, the attacker manipulates the data during training and uses this information to attack the ML during inference, or it directly manipulates the ML inference.

5.2 *Security Attacks*

Every phase of the design and development cycle of the ML is potentially vulnerable to security attacks. Depending upon the attack methodology and payload, these attacks are categorized as follows:

1. **Adversarial Attacks:** In these attacks, adversary either manipulates the training data to introduce the backdoors in the ML model (also known as poisoning

[1]Black-box model: When the attacker has only access to the inputs and outputs of the targeted model.

[2]White-box model: When the attacker knows the architecture, trained weights and all other parameters of the targeted model.

attacks or backdoors) [79–81] or manipulates the inference data by adding an imperceptible noise to perform a specific payload (also known as evasion attacks or adversarial examples) [82–85]. Moreover, the attacker can insert a parallel functionality by introducing an extra small model that performs a specific payload when triggered (also known as model-level Trojans) [86, 87].

2. **Hardware Attacks:** The hardware accelerators or hardware platforms for ML is also vulnerable to all hardware attacks, i.e., hardware Trojans [88], side-channel attacks (row-hammer attacks can be used to change the weights in the memory randomly).
3. **Model Stealing Attacks:** Typically, ML models leak the information about their different parameters and the training datasets. This vulnerability is exploited to steal the trained models [89] and the training datasets [90].

5.3 Defenses

To address the above-mentioned attacks, several defense strategies have been proposed. Depending upon the attack type, these defenses are categorized as follows:

1. **Defenses Against Adversarial Attacks:** To defend against poisoning attacks or backdoors, encrypting the training datasets, local training, redundant training, and the local training with pruning are being employed at different levels of the training process. The preprocessing of the inference data (i.e., quantization [91], filtering [92, 93], transformation [94]), gradient masking [95], and adversarial training [96] (training the model for known adversarial example) are used to defend against evasion attacks.
2. **Defenses Against Hardware Attacks:** All the defense strategies that can be applied to hardware attacks can also be applied to defend against the hardware attack on ML platform or accelerators [97, 98]. Some of the commonly used defenses are side-channel based online anomaly detection, hardware isolation, protected memories, etc.
3. **Defense Against Model/Dataset Stealing Attacks:** One of the most commonly used methods is encrypting the training dataset. However, the dataset encryption does not ensure the defense against model stealing. Therefore, other techniques like model obfuscation, rounding or runtime monitoring can be applied along with dataset encryption [99].

Note: though these defenses work under certain assumptions, they do not cover the entire attack surface. Therefore, to design a secure smart CPS, a comprehensive defense strategy is required to counter the attacks on ML along with the traditional security measures for CPS.

6 Conclusion

This chapter first presented a brief overview of the several security vulnerabilities and attacks at different CPS layers and the possible attack models with respect to attacker motive, attacker's access, and attacker's capabilities. To defend against these security attacks and attack models, we identify the associated research challenges to develop secure CPS. To address these challenges, we also provide a brief survey of the state-of-the-art static and adaptive security measures for CPS. Towards the end, we discussed and developed a preliminary analysis of the state-of-the-art online anomaly techniques that leverage the machine learning algorithms and property-specific language, respectively.

Acknowledgments This work is supported in parts by the Austrian Research Promotion Agency (FFG) and the Austrian Federal Ministry for Transport, Innovation, and Technology (BMVIT) under the "ICT of the Future" project, IoT4CPS: Trustworthy IoT for Cyber-Physical Systems.

References

1. R.F. Babiceanu et al., Big data and virtualization for manufacturing cyber-physical systems: a survey of the current status and future outlook. Comput. Ind. **81**, 128–137 (2016)
2. R. Rajkumar et al., Cyber-physical systems: the next computing revolution, in *IEEE DAC* (2010), pp. 731–736
3. M. Shafique et al., Intelligent security measures for smart cyber physical systems, in *Euromicro/IEEE DSD* (2018), pp. 280–287
4. D. Ratasich et al., A roadmap toward the resilient internet of things for cyber-physical systems. IEEE Access **7**, 13260–13283 (2019)
5. J. Giraldo et al., Security and privacy in cyber-physical systems: a survey of surveys. IEEE Des. Test **34**(4), 7–17 (2017)
6. A. Humayedet al., Cyber-physical systems security—a survey. IEEE Internet Things J. **4**(6), 1802–1831 (2017)
7. M. Pelino et al., *The Internet of Things Heat Map, 2017* (Forrester Research, Cambridge, 2017)
8. Statista. Internet of things (IoT) connected devices installed base worldwide from 2015 to 2025 (in billions) (2019). https://www.statista.com/statistics/471264/iot-number-of-connected-devices-worldwide/. Accessed 04 Nov 2019
9. F. Kriebel et al., Robustness for smart cyber physical systems and internet-of-things: from adaptive robustness methods to reliability and security for machine learning, in *IEEE ISVLSI* (2018), pp. 581–586
10. S. Rehman et al., Hardware and software techniques for heterogeneous fault-tolerance, in *IEEE IOLTS* (2018), pp. 115–118
11. R.B. Sowby, Hydroterrorism: a threat to water resources. Wasatch Water Rev. 1–4 (2016)
12. S.A. Timashev, Cyber reliability, resilience, and safety of physical infrastructures, in *IOP Conference Series: Materials Science and Engineering*, vol. 481 (2019), p. 012009
13. C. Cerrudo, An emerging us (and world) threat: cities wide open to cyber attacks. Secur. Smart Cities **17**, 137–151 (2015)
14. S.A. Groeneveld, N. Jongejan, A.T.L. Fiolet et al., Hacking into a pacemaker; risks of smart healthcare devices. Nederlands tijdschrift voor geneeskunde **163** (2019)

15. J.L. Beavers et al., Hacking NHS pacemakers: a feasibility study, in *IEEE ICGS3* (2019), pp. 206–212
16. Y. Shoukry et al., Non-invasive spoofing attacks for anti-lock braking systems, in *International Workshop on Cryptographic Hardware and Embedded Systems* (Springer, Berlin, 2013), pp. 55–72
17. S.N. Narayanan et al., Security in smart cyber-physical systems: a case study on smart grids and smart cars, in *Smart Cities Cybersecurity and Privacy* (Elsevier, Amsterdam, 2019), pp. 147–163
18. Y. Shoukry et al., Pycra: physical challenge-response authentication for active sensors under spoofing attacks, in *AM CCS* (2015), pp. 1004–1015
19. A. Francillon, B. Danev, S. Capkun, Relay attacks on passive keyless entry and start systems in modern cars, in *NDSS*, 2011
20. D. Preuveneers et al., The intelligent industry of the future: a survey on emerging trends, research challenges and opportunities in industry 4.0. J. Ambient Intell. Smart Environ. **9**(3), 287–298 (2017)
21. C.-T. Lin et al., Cyber attack and defense on industry control systems, in *IEEE Conference on Dependable and Secure Computing* (2017), pp. 524–526
22. D. Antonioli et al., Taking control: design and implementation of botnets for cyber-physical attacks with CPSBot (2018). Preprint. arXiv:1802.00152
23. P. Wang et al., Cyber-physical anomaly detection for power grid with machine learning, in *Industrial Control Systems Security and Resiliency* (Springer, Berlin, 2019), pp. 31–49
24. S. Jin et al., Changepoint-based anomaly detection for prognostic diagnosis in a core router system, in *IEEE TCAD*, 2018
25. T. Roth et al., Physical attestation of cyber processes in the smart grid, in *Springer ICIIS* (2013), pp. 96–107
26. H.R. Ghaeini et al., Patt: physics-based attestation of control systems, in *RAID* (2019), pp. 165–180
27. A. Essa et al., Cyber physical sensors system security: threats, vulnerabilities, and solutions, in *IEEE ICSGSC* (2018), pp. 62–67
28. H. Salmani et al., On design vulnerability analysis and trust benchmarks development, in *IEEE ICCD* (2013), pp. 471–474
29. B. Shakya et al., Benchmarking of hardware Trojans and maliciously affected circuits. J. Hardware Syst. Secur. **1**(1), 85–102 (2017)
30. J. Wurm et al., Introduction to cyber-physical system security: a cross-layer perspective. IEEE Trans. Multi-Scale Comput. Syst. **3**(3), 215–227 (2016)
31. S.R. Chhetri et al., Cross-domain security of cyber-physical systems, in *IEEE ASP-DAC* (2017), pp. 200–205
32. Y. Liu et al., False data injection attacks against state estimation in electric power grids. ACM Trans. Inf. Syst. Secur. **14**(1), 13 (2011)
33. E. Nakashima et al., Hackers have attacked foreign utilities, CIA analyst says. Washington Post, 19, 2008
34. H. Fawzi et al., Secure estimation and control for cyber-physical systems under adversarial attacks. IEEE Trans. Autom. Control **59**(6), 1454–1467 (2014)
35. F. Pasqualetti et al., Attack detection and identification in cyber-physical systems. IEEE Trans. Autom. Control **58**(11), 2715–2729 (2013)
36. Y. Mo et al., Detecting integrity attacks on SCADA systems. IEEE Trans. Control Syst. Technol. **22**(4), 1396–1407 (2013)
37. M. Zeller, Myth or reality—does the aurora vulnerability pose a risk to my generator?, in *IEEE Conference for Protective Relay Engineers* (2011), pp. 130–136
38. S. Islam et al., Physical layer security for the smart grid: vulnerabilities, threats and countermeasures. IEEE Trans. Ind. Inform. **15**, 6522–6530 (2019)
39. C. Li et al., Hijacking an insulin pump: security attacks and defenses for a diabetes therapy system, in *IEEE International Conference on e-Health Networking, Applications and Services* (2011), pp. 150–156

40. K. Koscher et al., Experimental security analysis of a modern automobile, in *IEEE Symposium on Security and Privacy* (2010), pp. 447–462
41. T. Hoppe et al., Security threats to automotive can networks—practical examples and selected short-term countermeasures. Reliab. Eng. Syst. Saf. **96**(1), 11–25 (2011)
42. S. Han et al., Intrusion detection in cyber-physical systems: techniques and challenges. IEEE Syst. J. **8**(4), 1052–1062 (2014)
43. C. Konstantinou et al., Cyber-physical systems: a security perspective, in *IEEE ETS* (2015), pp. 1–8
44. M. Conti, Leaky cps: inferring cyber information from physical properties (and the other way around), in *ACM Workshop on CPS* (2018), pp. 23–24
45. A. Chattopadhyay et al., Security of autonomous vehicle as a cyber-physical system, in *IEEE ISED* (2017), pp. 1–6
46. J.A. Stankovic, Research directions for the internet of things. IEEE Internet Things J. **1**(1), 3–9 (2014)
47. Q. Xu et al., Security-aware waveforms for enhancing wireless communications privacy in cyber-physical systems via multipath receptions. IEEE Internet Things J. **4**(6), 1924–1933 (2017)
48. Y. Mo et al., Secure control against replay attacks, in *IEEE Allerton* (2009), pp. 911–918
49. L. Peng et al., Energy efficient jamming attack schedule against remote state estimation in wireless cyber-physical systems. Neurocomputing **272**, 571–583 (2018)
50. Y. Li et al., Jamming attacks on remote state estimation in cyber-physical systems: a game-theoretic approach. IEEE Trans. Autom. Control **60**(10), 2831–2836 (2015)
51. Y. Won et al., An attack-resilient cps architecture for hierarchical control: a case study on train control systems. IEEE Comput. **51**(11), 46–55 (2018)
52. S. Ali et al., Wsn security mechanisms for cps, in *Cyber Security for Cyber Physical Systems* (Springer, Berlin, 2018), pp. 65–87
53. G. Hatzivasilis et al., SCOTRES: secure routing for IoT and CPS. IEEE Internet Things J. **4**(6), 2129–2141 (2017)
54. S.H. Bouk et al., Named data networking's intrinsic cyber-resilience for vehicular cps. IEEE Access **6**, 60570–60585 (2018)
55. Y. Zhou et al., A secure control learning framework for cyber-physical systems under sensor attacks, in *IEEE ACC* (2019), pp. 4280–4285
56. J. Shen et al., A game-theoretic method for cross-layer stochastic resilient control design in cps. Int. J. Syst. Sci. **49**(4), 677–691 (2018)
57. S.R. Chhetri et al., Fix the leak!: an information leakage aware secured cyber-physical manufacturing system, in *IEEE DATE* (2017), pp. 1412–1417
58. S.R. Chhetri et al., Information leakage-aware computer-aided cyber-physical manufacturing. IEEE Trans. Inf. Forensics Secur. **13**(9), 2333–2344 (2018)
59. J.-S. Wang et al., Data-driven methods for stealthy attacks on TCP/IP-based networked control systems equipped with attack detectors. IEEE Trans. Cybern. **49**(8), 3020–3031 (2018)
60. O. Al Ibrahim et al., Cyber-physical security using system-level PUFs, in *IEEE Wireless Communications and Mobile Computing Conference* (2011), pp. 1672–1676
61. C. Liu et al., Securing cyber-physical systems from hardware Trojan collusion. IEEE Trans. Emerg. Top. Comput. (2017)
62. J.S. Mertoguno et al., A physics-based strategy for cyber resilience of cps, in *Autonomous Systems: Sensors, Processing, and Security for Vehicles and Infrastructure*, vol. 11009 (2019), p. 110090E
63. A.A. Cardenas et al., Secure control: towards survivable cyber-physical systems, in *IEEE Conference on Distributed Computing Systems Workshops* (2008), pp. 495–500
64. B. Satchidanandan et al., On minimal tests of sensor veracity for dynamic watermarking-based defense of cyber-physical systems, in *IEEE COMSNETS* (2017), pp. 23–30
65. J. Siegel et al., A cognitive protection system for the internet of things. IEEE Secur. Priv. **17**(3), 40–48 (2019)

66. A. Petrovski et al., Designing a context-aware cyber physical system for detecting security threats in motor vehicles, in *ACM CSIN* (2015), pp. 267–270
67. F. Khalid et al., Simcom: statistical sniffing of inter-module communications for run-time hardware trojan detection (2018). Preprint. arXiv:1901.07299
68. F.K. Lodhi et al., Power profiling of microcontroller's instruction set for runtime hardware trojans detection without golden circuit models, in *IEEE DATE* (2017), pp. 294–297
69. F. Khalid et al., Behavior profiling of power distribution networks for runtime hardware trojan detection, in *IEEE MWSCAS* (2017), pp. 1316–1319
70. F.K. Lodhi et al., A self-learning framework to detect the intruded integrated circuits, in *2016 IEEE ISCAS* (2016), pp. 1702–1705
71. F. Khalid et al., Runtime hardware trojan monitors through modeling burst mode communication using formal verification. Integr. VLSI **61**, 62–76 (2018)
72. F. Khalid et al., Forasec: formal analysis of security vulnerabilities in sequential circuits (2018). Preprint. arXiv:1812.05446
73. F.K. Lodhi et al., Formal analysis of macro synchronous micro asychronous pipeline for hardware trojan detection, in *IEEE NORCAS: NORCHIP* (2015), pp. 1–4
74. I.H. Abbassi et al., Using gate-level side channel parameters for formally analyzing vulnerabilities in integrated circuits. Sci. Comput. Program. **171**, 42–66 (2019)
75. M. Shafique et al., An overview of next-generation architectures for machine learning: roadmap, opportunities and challenges in the iot era, in *IEEE DATE* (2018), pp. 827–832
76. M.A. Hanif et al., Robust machine learning systems: reliability and security for deep neural networks, in *2018 IEEE 24th International Symposium on On-Line Testing and Robust System Design (IOLTS)* (IEEE, Piscataway, 2018), pp. 257–260
77. J.J. Zhang et al., Building robust machine learning systems: current progress, research challenges, and opportunities, in *ACM/IEEE DAC* (2019), pp. 1–4
78. F. Khalid et al., Security for machine learning-based systems: attacks and challenges during training and inference, in *IEEE FIT* (2018), pp. 327–332
79. B. Chen et al., Detecting backdoor attacks on deep neural networks by activation clustering (2018). Preprint. arXiv:1811.03728
80. Y. Ji et al., Backdoor attacks against learning systems, in *IEEE CNS* (2017), pp. 1–9
81. T. Gu et al., BadNets: evaluating backdooring attacks on deep neural networks. IEEE Access **7**, 47230–47244 (2019)
82. A. Marchisio et al., Capsattacks: robust and imperceptible adversarial attacks on capsule networks (2019). Preprint. arXiv:1901.09878
83. F. Khalid et al., Red-attack: resource efficient decision based attack for machine learning (2019). Preprint. arXiv:1901.10258
84. A. Marchisio et al., SNN under attack: are spiking deep belief networks vulnerable to adversarial examples? (2019). Preprint. arXiv:1902.01147
85. F. Khalid et al., TrISec: training data-unaware imperceptible security attacks on deep neural networks, in *IEEE IOLTS* (2019), pp. 188–193
86. T. Liu, W. Wen, Y. Jin, SIN 2: stealth infection on neural network—low-cost agile neural trojan attack methodology, in *IEEE HOST* (2018), pp. 227–230
87. Y. Liu et al., Neural trojans, in *IEEE ICCD* (2017), pp. 45–48
88. Y. Zhao et al., Memory trojan attack on neural network accelerators, in *IEEE DATE* (2019), pp. 1415–1420
89. T. Orekondy et al., Knockoff nets: stealing functionality of black-box models, in *IEEE CVPR* (2019), pp. 4954–4963
90. A. Salem et al., Updates-leak: data set inference and reconstruction attacks in online learning (2019). Preprint. arXiv:1904.01067
91. F. Khalid et al., QuSecNets: quantization-based defense mechanism for securing deep neural network against adversarial attacks, in *IEEE IOLTS* (2019), pp. 182–187
92. F. Khalid et al., FAdeML: understanding the impact of pre-processing noise filtering on adversarial machine learning, in *IEEE DATE* (2019), pp. 902–907

93. H. Ali et al., SSCNets: Robustifying DNNs using Secure Selective Convolutional Filters. *IEEE Des. Test* **37**(2), (2020), pp. 58–65
94. E. Raff et al., Barrage of random transforms for adversarially robust defense, in *IEEE CVPR* (2019), pp. 6528–6537
95. I. Goodfellow, Gradient masking causes clever to overestimate adversarial perturbation size (2018). Preprint. arXiv:1804.07870
96. F. Tramèr et al., Ensemble adversarial training: attacks and defenses (2017). Preprint. arXiv:1705.07204
97. X. Xu et al., Detecting AI trojans using meta neural analysis (2019). Preprint. arXiv:1910.03137
98. Y. Gao et al., Strip: a defence against trojan attacks on deep neural networks (2019). Preprint. arXiv:1902.06531
99. M. Juuti et al., Prada: protecting against DNN model stealing attacks, in *IEEE EuroS&P* (2019), pp. 512–527

Design and Operation Framework for Industrial Control System Security Exercise

Haruna Asai, Tomomi Aoyama, Yuitaka Ota, Yoshihiro Hashimoto, and Ichiro Koshijima

1 Introduction

In recent years, a cyber-attack targeted the ICS (Industrial Control System) on operation sites becomes a threat that can not be ignored [1–4]. Authorities said that a malware named Stuxnet attacked a uranium enrichment facility of Iran in 2010 [5], unknown hackers severely damaged a German steel-mill in 2014 [6], and Russian hackers caused significant power outages in Ukraine in 2015 [7] and 2016 [8]. However, ICS emphasizes availability and its stable operation, security measures such as OS updates and application patches that affect availability are avoided. Therefore, it is essential for owners of the operation site to evaluate the installed security measures on their ICS [9] and prepare a company-wide response against cyber-atacks [10, 11].

Usually, in critical infrastructure companies, a production division prepares safety-BCP (Business continuity plan)s against physical troubles, such as fires, toxic spills, and natural disasters [12, 13]. An information system division also has pre-pared IT-BCPs to respond to cyber-incidents on the business network, such as information leaks and so on. Some cyber-attacks on the ICS cause hazardous situations in the plant, and as a result, it should invoke a particular BCP of the company [14–16]. There are, however, difficulties in integrating safety-BCPs and IT-BCPs because of a lack of experience of cyber-incidents that covers both BCPs.

For preparing the above situation, each company has to plan required corporate resources and educate their staffs and operators using a Safety-Security-Business Continuity (SSBC) exercise.

In this paper, the authors propose a framework based on the following points.

H. Asai (✉) · T. Aoyama · Y. Ota · Y. Hashimoto · I. Koshijima
Nagoya Institute of Technology, Nagoya, Aichi, Japan
e-mail: h.asai.852@nitech.jp; aoyama.tomomi@nitech.ac.jp; cjr17009@nitech.jp; hashimoto@nitech.ac.jp; koshijima.ichiro@nitech.ac.jp

H. Karimipour et al. (eds.), *Security of Cyber-Physical Systems*,
https://doi.org/10.1007/978-3-030-45541-5_3

1. Design methodology of a security training exercise for ICS
2. Evaluation methodology of ICS security exercise
3. Management methodology of ICS security exercise

2 Method for Developing Security Exercises Tailored to Individual Companies

Participants of the SSBC (Safety-Security-Business Continuity) exercise [7] need to learns not only safety methods but also security methods against cyber-attacks on a simulated plant with field control devices, ICS networks, and information networks that mimic corporate operation structure.

Critical infrastructure companies, therefore, need to prepare training facilities that include simulated plants with control systems. Using this facility, not only field operators but also IT staffs and managers learn the knowledge of process safety and practical procedures under cyber-attacks [17–19].

The SSBC exercise is conducted on a scenario that reflects the company's profile. Through this exercise, participants have to learn the knowledge of security measures and security-related operation processes on the simulated plant. The proposed design procedure of the SSBC exercise is shown in Fig. 1.

In the procedure, in the first step, a virtual company for the SSBC exercise is specified based on the actual company's profile, and possible attack scenarios to the company are selected. In the second step, process operations based on the company's standardized safety procedure are assigned to meet the simulated plant. In the third step, safety counteractions are taken into consideration, selected process signals affected by the result of the cyber-attack. In this step, additional conditions, resources to the existing safety-BCP will be clarified.

Then, security counteractions are specified to arrange the existing IT-BCPs where business impacts are considered based on the virtual company's profile. In this step, factors such as the company's security policies, and network and communication structures will be testified their readiness and resilience to bottom-up acransition from the field operation to the company's business operations.

2.1 Virtual Company Image

The company profile specifies the participant's roles and limitations while playing their roles. In setting up the company image, the following conditions are determined.

- Business contents
- Organizational structure
- Organization's role (Routine work, Skills)

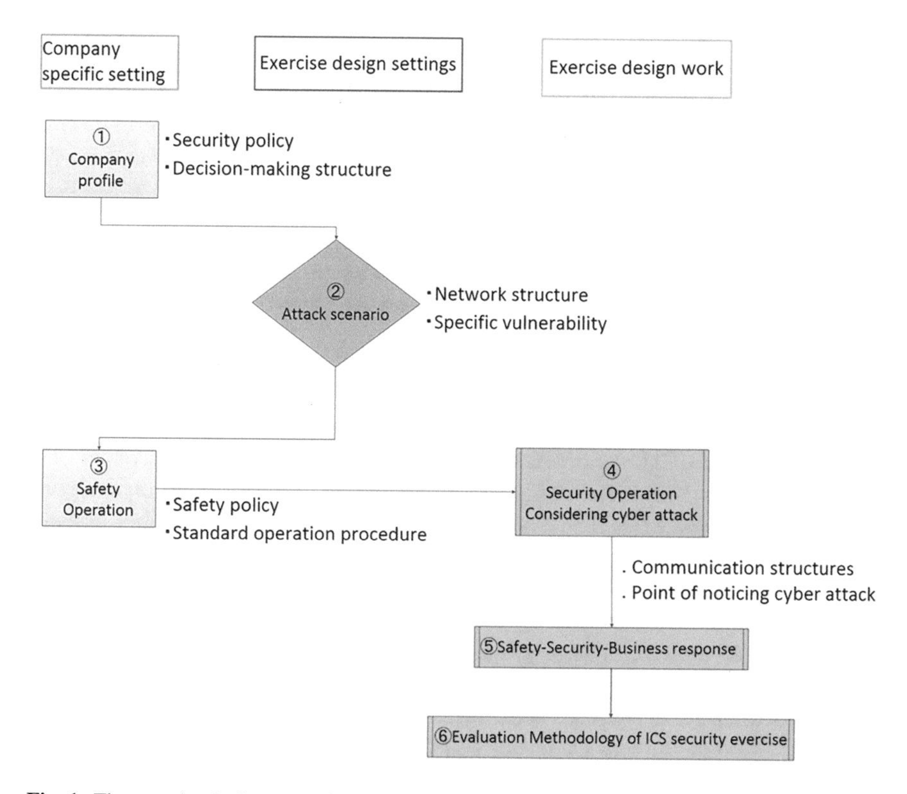

Fig. 1 The exercise design procedure and points of company uniqueness

- Communication role
- Plant structure
- Network structure

It is difficult to evaluate the impacts of the cyber-attacks on the company if the network structure does not match the actual structure characteristics. Accordingly, for example, it is also desirable to take into account the structure connecting between the local production sites and the headquarters. However, if the actual conditions are used for exercise, the SSBC exercise becomes complicated, so the selected conditions should be simplified. These selected conditions are temporary and evolutionary improved through the SSBC exercise-evaluation process (a kind of PDCA cycle).

2.2 *Attack Scenario*

After setting up the company image, the attack scenario is created. Currently, there is little recognition that cyber-attacks occur at control systems leading to many serious

Table 1 Procedure for the cyber-attack scenario

Cyber kill chain	Design cyber-attack scenario
1st-Reconnaissance	1. Maximum risk (objectives)
2nd-Delivery	2. Malicious operation (lateral movement)
3rd-Compromise/exploit	3. ICS hacking (C&C)
4th-infection/installation	4. Installation of ICS hacking (infection)
5th-Command and control	5. Prerequisite for attack (compromise)
6th-Lateral movement/pivoting	6. Recent situation scenario2 (delivery)
7th-Objectives/exfiltration	7. Recent situation scenario1 (reconnaissance)

accidents. Therefore, it is necessary for the cyber-attack to be recognized by the participants as a real problem with the importance of the security measures in the SSBC exercise [20, 21]. An SSBC exercise developer should create a scenario that enables the participants to notice the attacks through the security measures designed in the scenario. Likewise, if in the scenario, an intruder (external factor) intrudes from a place without security measures and causes the cyber-attack, the importance of security measures can be further recognized. The method for creating the cyber-attack scenario is shown below.

Typically, an attacker attacks based on Cyber Kill Chain [22, 23]. However, in the SSBC exercise, it is desirable to assume the worst possible scenario from the viewpoints of risk management and education thereof. In our created scenario, we have considered the flow of the Cyber Kill Chain in a reverse direction (Table 1) so that the maximum risk (maximum abnormality) is expected, and the attack targets are determined. It also provides an attack route through which intruders pass after intruding from areas with weak security measures.

First, the participants of the SSBC exercise are decided. In the SSBC exercise, in consideration of the safety measures, a discussion is made for mainly about changing the safety measures in a plant site. At the same time, a discussion is also made for the development of information. When the SSBC exercise is carried out to educate on-site operators, the scenario is created where measurements in the plant site will change drastically.

On the other hand, when the SSBC exercise is carried out to consider the security measurements for the company as a whole, the scenario is created. In the created scenario, not only the security countermeasure on the plant site but also the information network can be experienced by the participants. Also, in the scenario, it is preferable that target sites to be attacked should have a linked business structure (such as a supply-chain, a common market) so that business conflicts to be considered by the attacks being built into the SSBC exercise.

Second, abnormalities (risks) such as accidents and breakdowns not wanted to happen are identified. Regarding safety and security, abnormalities that can occur in the simulated plants are identified. In terms of businesses, possible management risks are identified. First of all, as a company, the maximum goal in the safety security business is raised. Next, risks that may hinder that goal are conceived. Finally, outliers that cause that risks are identified. In this way, specific plans can be listed in order so that various opinions are revealed easily. Then, the more plans are listed, the more the scenario options are obtained. It can be selected as an efficient method to brainstorm ideas asking for "Quantity over quality." For a similar purpose, in creating the scenario, it is desirable that persons belonging to various departments, such as site operators, IT engineers, and managers, involve creating the scenario.

Third, thus, identified abnormalities are summarized, and a trigger in the attack scenario is determined. The opinions are also set in the scenario to have branches based on the abnormalities incorporated. The possible abnormalities are roughly divided into those in safety, those in security, and those in business. After roughly dividing the abnormalities, the determined abnormalities are classified regarding the relationship between the result and the cause. By doing so, the abnormalities are further organized, and new ideas come out. In repeating this work, key events in the risks can be seen as the causes so that a choice of abnormalities to be considered in the scenario can be obtained.

After that, to experience conflict, that is a major object of the SSBC exercise, common abnormalities related to two or three of the safety, security, and business are selected from the determined abnormalities. Further, in the abnormalities in safety, critical (in importance) and trouble-some (on frequency) abnormalities are selected. Thus, the participants have increased some opportunities to consider his or her experiences, referring to the abnormalities in the SSBC exercise. In other words, safety measures considered in the SSBC exercise are likely to be reflected his or her business activities resulting in more practical exercise.

In the SSBC exercise, linking points between safety-security operation processes and business continuity operation processes are also implicated in recognizing the safety-security-business constraint of each linking point with the market impact. Therefore, it is necessary to select common anomalies related to safety, security, and business. The attack scenario can have more opportunities for participants to compare with their experiences.

Fourth, to cause the abnormalities, the attack route is selected from the viewpoint of the attacker. Depending on the network structure of the simulated plant, network elements on the attack route that have security holes and weak countermeasures are specified by the attacker's view. Along with the attack route, concealment of traces of intrusion should be considered to understand a delay in recognizing the cyber attack.

2.3 *Cyber Defense Scenario Based on Safety Operation*

The abnormality, which can occur at the site, does not change even in the case of cyber-attack nor equipment failure/malfunction, although the causes thereof are not identical. In other words, the on-site operators can put out regular safety measures for the abnormalities. Safety procedures are divided into several branches according to the situation. However, the defense scenario is designed based on one safety measure focused by incorporating the result of the safety measure (situation) into the defense scenario based on the attack scenario [24, 25].

By trying an attack similar to the attack in the attack scenario to the simulated plant, it is possible to create the defense scenario into which more accurate information is incorporated. Moreover, then, it is preferable to select a person who is involved in a security field or in on-site work as a designer of the defense scenario. It is also desirable to prepare a company outline, an organization structure, a plant outline, and a network diagram in advance to make it easier to reflect normal business activities to the defense scenario.

Once the safety measures are taken, the safety measures that take into consideration the cyber-attacks and the safety measures that take into consideration business impact are added. In consideration of the following matters, as many measures as possible should be added [26].

1. What is a new measure formed in considering the effect of the cyber-attack?
2. In what way is information shared (in the communication network)?
3. Who will decide the measures in the presence of the information?

In an existing safety measure, it is required to cope with actually occurred abnormalities. Also, under the influence of the cyber-attacks, since concealment and simultaneous occurrence of abnormalities may be performed, not only the abnormalities which may be caused at a place where the abnormality is at present not confirmed but also abnormalities caused on purpose should be watched out. Specifically, it should be considered to include whether abnormal signals detected on SCADA monitors reflect the actual plant process data. When an abnormal state is set in a control device, it is recognized that maintenance activities are necessary to confirm the status of the device by using the vendors provided engineering stations.

Besides, the degree of impacts from the cyber-attack changes communication among corporate departments. When an abnormality occurs, opportunities to cope with other departments (normally irrelevant departments) will increase. Also, the information sharing method should be considered so as not to become a bottleneck in the overall operation. In cooperation with departments in different technical fields, it is necessary to consider the information sharing protocol to reduce traffic volume and errors.

2.4 *Organizing Scenario (Safety-Security-Business)*

It is desirable to design the SSBC exercise so that the participants can focus on the safety measures newly added in consideration of the cyber-attacks. Therefore, only the place where the required safety measures are made is left separately from the existing safety measures to create the attack scenario and the defense scenario. However, when only the place to be added is left as an SSBC exercise, it does not add up. To cope with this, the necessary information in the front and back of the place is left.

Also, it is necessary to incorporate the conflicts that may occur in the actual measure into the SSBC exercise. Specifically, there is the conflict where the priority of measures cannot be determined easily in forming a workflow, the conflict where the measures cannot be concurrently performed but overlapped, and the conflict where it seems that a communication pass cannot be connected smoothly. Also, the afore-mentioned necessary information in the front, and the back of the place should be left. Each conflict installed may occur in the actual situation, and the participants should experience conflicts through the SSBC exercise.

2.5 SSBC Exercise Phase

When considering a company's response to cyber incidents, the company's behavioral indicators, such as response goals and priorities, vary with safety, security, and business work volume. The SSBC exercise separates phases in situations where work volume is a change. It is easy for participants to recognize changes in the purpose of the response. When there are many participants, the situation of the scenario varies from team to team according to the progress of the SSBC exercise. However, the same assumptions can be redone from the beginning of the new phase, making it easier to compare ideas between teams and lead to a deeper discussion. In addition, setting a timeline for each phase and add-ing time restrictions increases the reality, allowing participants to take action with priorities in mind. The SSBC exercise stage is divided as follows (Figs. 2 and 3) [27].

1. Until it is determined that the anomaly that occurred is a cyber attack (predictive phase)
2. After recognizing that it is a cyberattack, until the plant is safely continued or stopped (emergency phase)
3. After continuing or stopping the plant, consider preventing a recurrence (recovery phase)

First, if it cannot be determined a cyberattack, onsite operators deal with in the field as a normal, abnormal situation. Therefore, the predictive phase is centered on safety and security.

Next, once the incident is determined to be a cyberattack, the entire company, including the IT department, will respond. Since then, the number of responses from a business perspective has increased. This is because once the plant is temporarily stabilized, it is necessary to take measures to prevent recurrence and disclose information such as causes and effects. In the emergency response and recovery phase, security business response is more than safety.

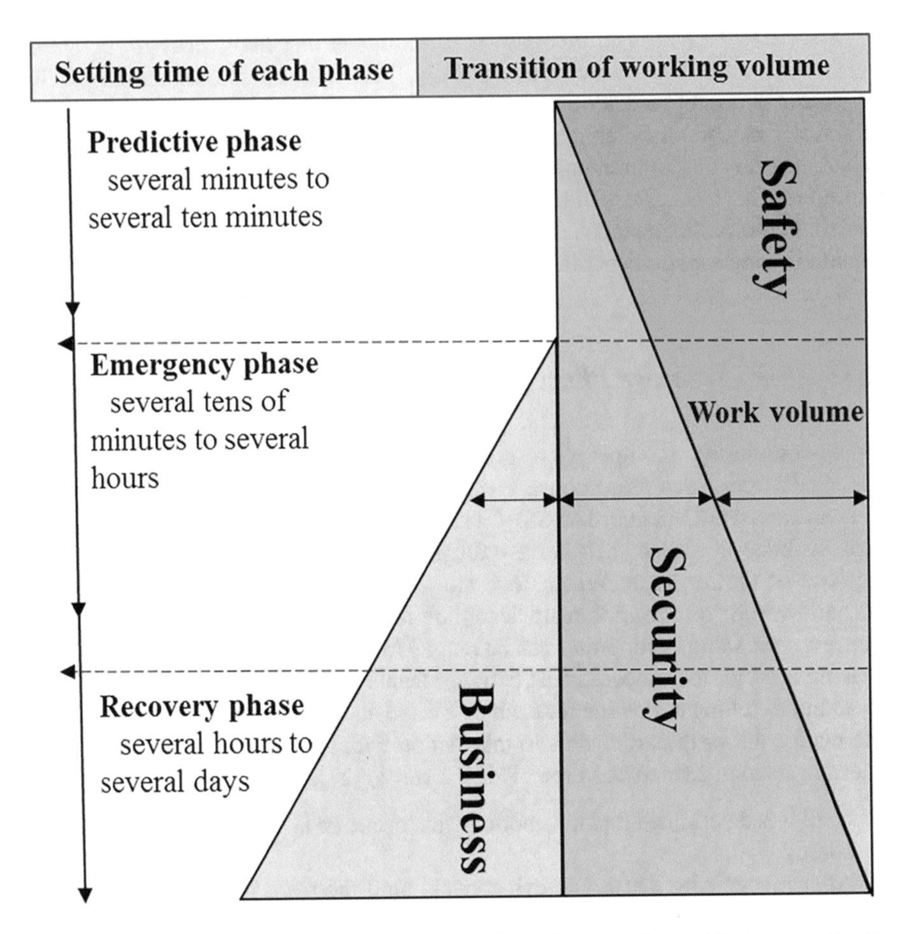

Fig. 2 Structure of exercise with workload image for safety, security, and business continuity activities

3 Evaluation Methodology of ICS Security Exercise

In the evaluation method, we will explain how to utilize the knowledge gained in the SSBC exercise and reflect it in the BCP of the company. The SSBC exercise de-signed by the methodology is immediately provided to the participants. The participants disclose important safety-security-business constraints, but will voluntarily reveal unknown and uncertain conditions, rules, and activities [28]. These published entities have been evaluated, some of which have been implemented. When the next SSBC exercise is designed, this design evaluation loop provides a PDCA cycle for less experienced cyber incidents concerning ICS. (Please refer to Fig. 4.)

Therefore, SSBC exercise workflow created by participants is important. No matter how active the discussion is, if you do not describe specific actions in the

Phase	Step	Security response	Safety response
Predictive phase	Detection of Events	Detection of activity on network different from usual	Detection of plant behavior different from normal operation
	Preliminary Analysis and Identification	Determine whether to treat it as cyber incident	Determine whether to treat it as normal abnormality or equipment failure
Emergency phase	Preliminary Response Action	Data collection for initial movement for defense, prevention of damage expansion and further cause analysis	Data collection for initial response for ensuring safety, propagation prevention of insecure state and further cause analysis
Recovery phase	Incident Analysis	Understand technical details, root cause and the potential impact of cyber incident	Understand technical details, root cause and the potential impact of plant unsafe conditions
	Response and Recovery	Recover the current situation of the affected part (soft, hard), prevent further damage, restore normal operation and prevent recurrence	Restore the current state of the affected equipment, prevent further damage and return to normal operation
	Post-Incident Analysis	Confirm the effectiveness and efficiency of incident handling	Confirm effectiveness and efficiency of safety response

Fig. 3 This shows the SSBC exercise design procedure and points of company uniqueness

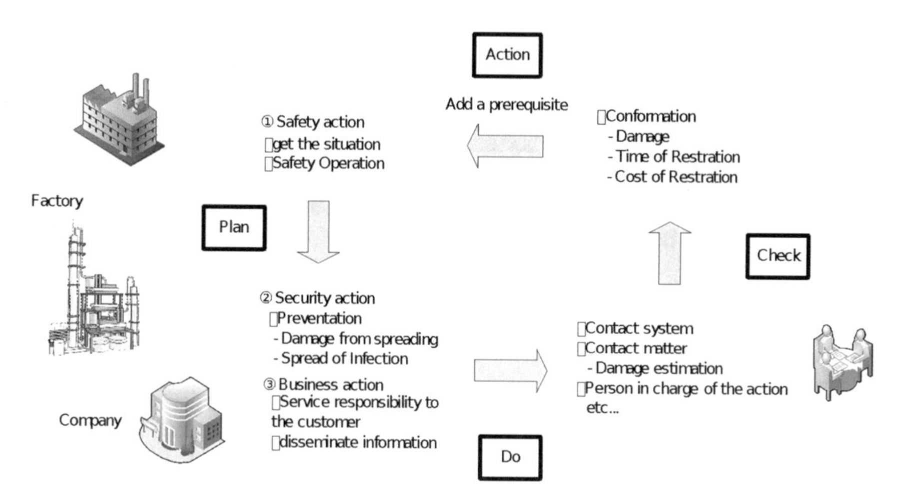

Fig. 4 PDCA circle of cybersecurity exercise for ICS

workflow, it will not be a consideration of corporate security measures. In order to work the evaluation method properly, participants need to think about the responses at the discussion points designed in the SSBC exercise and draw conclusions as a team.

After the SSBC exercise, the participants sort out "who", "what", and "at what timing" based on the workflow. As a result, the responses considered by the participants are concisely organized, and the necessary organizational structure, skills (including non-technical skills) and contact network for BCP can be determined.

The correspondence and ideas will make the amount of information easy to use in participant own work, and the versatility will increase further.

4 Exercise Management

Support from an admin side is indispensable to consider the concrete organizational response through the SSBC exercise. This chapter describes the items that administrators should be aware of.

4.1 White Teaming

The white team has a role managing the participant's activity and engagement to proceed with the SSBC exercises smoothly. In order to maximize the effect of exercise, it is necessary to have a coordinated team of trainers specialized in operations that can provide appropriate support to participants.

The exercise team consists of about five people, and one advisor per team is appropriate. The advisor does not have to be large. While teams should be aware that their role is to stimulate discussion rather than lead the discussion. Participants can learn most effectively when discussions are guided by their initiatives. In addition, since the variation of scenario changes during the exercise varies depending on the students, the exercise management side must have the ability to respond flexibly.

The white team has the roles of facilitator and advisor. More details are given below.

4.1.1 Facilitator

The facilitator guides participants through the SSBC exercise [29]. He/she explains the exercise at briefing, and describes the scene at each cycle. At the beginning of the SSBC exercise, the facilitator communicates the purpose, method, and scenario of the SSBC exercise to all participants. This allows participants to easily organize their work positions and tasks and focus on the points to discuss within a limited time. During the group work, the facilitator pays attention to each group's progress, while keeping track of time. He/she also supervises the discussion and debriefing. In debriefing, he/she helps participants to summarize results and lessons learned. When there are multiple teams, discussions deepen by pointing out differences between teams or asking questions from other teams, leading to the evaluation of security measures.

4.1.2 Advisor

During group work, the adviser walks among tables and gives suggestions to each group based on his/her expertise. He/she also asks questions that trigger more actions and discussion. Therefore, the role requires knowledge and experience in the field. During the discussion, the adviser provides positive feedback and comments for each group. We invited IT security specialists, ICS security researchers, and experts from ICS security agencies as advisers. These experts also helped during the process of scenario development.

For Advisor education, you should let him/her point conflicts. At this point, noticing is not just pointing out where to change. The purpose of the exercise SSBC exercise is to educate so that they can take advantage of the security knowledge they have learned in their normal work. To this end, participants need to learn the "thinking" of incident response specific to cyber-attack by experiencing an example of behavior that changes in response to normal equipment failures and cyber-attacks through this exercise. When advisors join a team through the SSBC exercises, they are not required to participate actively in group work, but to take a step back and support according to the participant's conversation and work situation.

4.2 *Participant*

The impact of cyber incidents may not only spread throughout the company, but it may become necessary to work with other companies and vendors in the same industry. Therefore, the team should be composed of people from different departments and positions.

4.2.1 In Case of Implementation in a Single Company

When an SSBC exercise is conducted for one organization, the organization may implement for two purposes; (A) BCP development and (B) in-house security training. Here, we introduce the ideal participants for each case.

It is desirable to form a team of employees who are in charge of the business, such as working at the headquarters, where the participants are on-site, IT security, and corporate employee. Participants can examine security measures for the company's own system in the SSBC exercise. Therefore, SSBC exercise is the best activity for the purpose of (A) BCP development. In this case, the participants are preferably the employee who is in a position to make a decision in the department mentioned above.

The SSBC exercises are also a useful tool for learning the concept of cyber incident response. Therefore, it can be used for (B) in-house training purposes. In this case, some may participate from the same department. When teams are formed with people from the same department, participants tend to fail to keep the broader view, because they are busy dealing with their own departments' detailed task. For instance, when multiple people from the IT department participate, participants tend to focus on time-consuming root cause analysis like a log analysis, often underestimating the importance of prompt field response and communication. For the education purposes, if the student notices a failure, a good enlightenment effect can be obtained. Therefore, in order to intentionally narrow the student's field of view, the student may form a team with only a single department.

4.2.2 Teaming of Multiple Organizations

When implemented in multiple companies, the team must be composed of people from different backgrounds. Typically, participants in exercises con-ducted for multiple companies are often leading security awareness program as a leader in their organizations. In this case, the participants need to reflect their learning to their skills but also reflect the ideas needed to create a customized exercise and BCP for own company. In addition, the mixed group al-lows the participants from different departments and industries that are not normally involved in daily communication. This environment stimulates the learning of the participants to gain a new perspective. However, to complete the exercise workflow, it is essential that the team has at least one individual from a control system engineering background and from the IT department.

Furthermore, not only ICS asset owners but also vendor companies that handle customers with ICS should participate. By joining the exercise together with asset owners, vendor companies can learn the communication and action that customer expects from them.

Another benefit of conducting this exercise to the several organizations is the possibility to discuss the industry-wide problems. For example, the past case shows that participants actively discuss how to utilize advisory and intelligence provided by the CERT community.

4.3 Exercise Procedure

This chapter explains the specific procedure for the SSBC exercise. The SSBC exercise requires participants to think about systematic countermeasures and collaborative communication. Therefore, no specific role is given to group members.

The exercise consists of five steps (Fig. 5): description of prerequisites, Safety response and security perspective discussion, group work, organizing issues, and presentation and discussion. As mentioned in the previous section, the exercise is divided into three phases. These steps occur in each phase as one cycle.

First, before the SSBC exercise, the facilitator explains the purpose of the SSBC exercise, differences of cyber incident response, and the flow of the SSBC exercise. This ensures that participants confirm the purpose of this exercise itself and that the

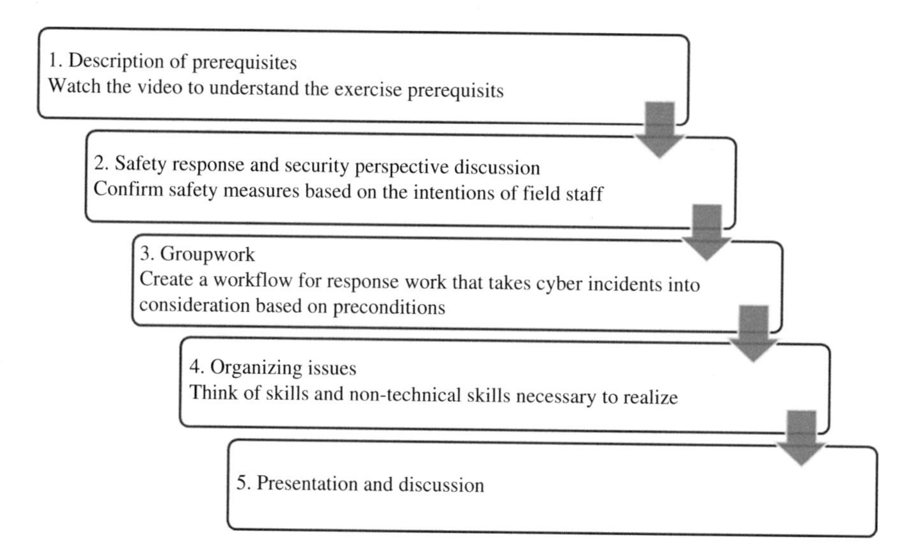

Fig. 5 SSBC exercise procedure

items to be discussed do not deviate. In addition, it also helps them to understand the entire, so it becomes easier to concentrate on what they should during the SSBC exercise. It is important for the facilitator to recognize that the cyber attack is a familiar incident and to make the participants aware that it is a problem that must be addressed. It is good to explain using real incident examples. Since the difference between the SSBC exercise objectives and the cyber attack is considered above, the exercise flow is described below.

First, ask participants to understand the prerequisites and to review existing safety measures that are prepared to deal with equipment failures. By creating a workflow, participants can focus on the different responses to cyber-attacks that need to be learned in the SSBC exercises. After creating the workflow, organize your thoughts, discuss between teams, deepen your understanding, and complete the SSBC exercises so that they can leverage that knowledge in the future. The purpose of each item is shown below.

4.3.1 Description of Prerequisites

Participants are given the preconditions prepared for each phase. This precondition is used as a hint to understand the attack scenario prepared in the SSBC exercise. Specifically, Prerequisites include the state of the plant, the state of the IT network, and the company situation. At this time, t is good to prepare a video, etc. so that the participants can easily imagine the situation other than their own department. In fact, people in IT departments sometimes have difficulty understanding how to operate field instruments. On the other hand, it is difficult for field operators to understand how to use the control network configuration and security tools. Responses and ideas derived from the experience of each department may be explained by the participants in the discussion of the SSBC exercise, but the operations and technical terms that appear in the SSBC exercise set-tings are within the scope supported by the white team.

4.3.2 Safety Response and Security Perspective Discussion

What you need to prepare here is an A0-size worksheet and supplementary material for the SSBC exercise prerequisites. A precondition and safe response are printed on each worksheet. First, participants understand the safety measures described on the A0 form while confirming the scenario and phase prerequisites of the SSBC exercise. Not only deeply understand the prerequisites, but also check ideas and points of discussion. If they have a thorough understanding of the scenario, they will spend more time learning in the next workflow creation, so it is desirable to make sure they understand the SSBC exercise premise within this time.

4.3.3 Group Work

In group work, prepare markers and two-color sticky notes. In the SSBC exercise, participants add actions with actions and reports/instructions and add them with different colored sticky notes. Participants write their thoughts on a sticky note and affix them to the corresponding points on workflow (Fig. 6). After that, consider the connection between the added response and the safety response, and connect the

Fig. 6 An example of team discussion in group work

response with arrows. Through this exercise, participants can visualize the response to cyber-attacks and the structure of the organization's collaboration.

We discuss two viewpoints. The first is a discussion that takes advantage of my expertise. Participants share what should be handled, and the degree of influence that should be considered in the positions of the SSBC exercises from a standpoint that is close to the usual work contents. Even if you have little knowledge of security, not only can you think about the response, it also leads to an opportunity for teams to express their opinions and an environment where all teams can discuss.

The second is a discussion by the team. This is an important discussion for developing optimal measures for the entire company, taking into account the overall balance and relationships when incorporating proposals that make the most of the company's expertise into the response of the entire company.

No matter how much security you know, if they don't understand the entire incident, they create a workflow with a unified response. In addition, if the bias is too large for the overall grasp, the parts of the safety response that are individually incorporated into the SSBC exercise design process should be overlooked. There is also a method that the white team does not dare to recognize the bias of thinking of participants. However, the original purpose is to create a BCP plan that takes into account the professional and business and management perspectives. Therefore, advisors need to point out it.

4.3.4 Organizing Issues

The time to organize the responses is very important in order to make use of the responses they have made in group work in the BCP of the company. During this time, the correspondence added in the group work is organized, and the organization cooperation as a whole company, task cooperation, and the organization that leads the task are confirmed. As a result, it is possible to sort out the ideal organizational structure and necessary skills/non-technical skills for the organization to respond promptly in the event of a cyber attack.

4.3.5 Presentation and Discussion

Each group presents their work flow created in the group work and skills needed to achieve them while displaying their worksheets to all participants. In this presentation, each group explains not only the added responses but also why they were added.

By exchanging opinions with each other, it is possible to share the knowledge examined during the SSBC exercise and deepen the understanding of the security response to the cyber attack. The facilitator then asks questions and answers and supports the participants so that they can easily exchange opinions. Participants can gain new knowledge by discussing the parts of the team that has different responses and problems.

5 Illustrative Exercise

The SSBC exercise is established for a simulated plant of our study room through the methodology mentioned in Chap. 2 in this paper.

5.1 *NIT-Exercise Workshop*

The target company in the SSBC exercise prepared in this paper is a plant having the structure shown in the figure and a company holding the internal network. The service that this company does is a service that generates energy to move air conditioning and supplies energy to the area. The tank 1 is a tank possessed by a supplier, and the tank 2 is a tank holding a supply destination. The plant has the following functions.

1. The heater warms the water in the lower tank
2. Hot water is supplied to the upper tank using a pump

Zoning and firewalls are also introduced as network security measures. Control of Valve 2 and heater by Single Loop Controller (SLC) in the different zone makes it possible to detect empty firing events caused by lowering of the liquid level of Tank 1 and continuation of heater operation. Also, by looking at the level of each tank on the SCADA screen in each Zone, you can notice abnormality even if one screen is concealed by a cyber-attack. Moreover, by installing a firewall, it is possible to detect and block suspicious communication from outside.

Participants understand the impact of concurrency and concealment of abnormalities by cyber-attacks on correspondence through SSBC exercise. It is used to learn the skills and elements necessary to prepare the organization and communication system required to deal with cyber-attacks.

5.2 *Virtual Company Image*

The virtual company's profile is as follows;

1. Project outline: District heating and cooling service business
2. Organizational structure: Outline of the simulated plant (Fig. 7), network structure (Fig. 8)
3. Roles with a communication network (Fig. 9)

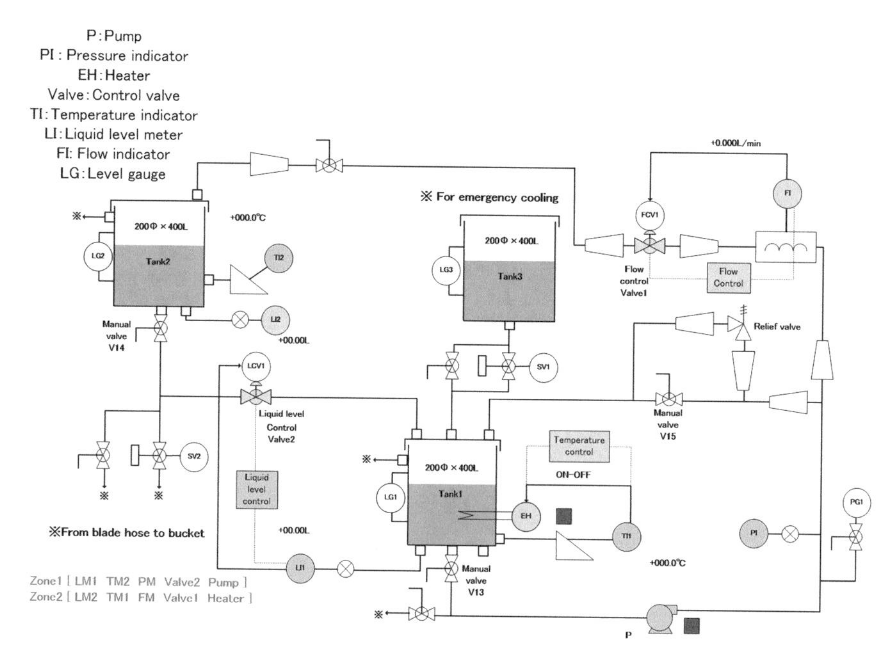

Fig. 7 A simulated plant heats water with a heater in the lower tank and circulates the heated water through a simple pipeline to the upper tank (using a pump)

5.3 Attack Scenario

The object of the SSBC exercise to be created this time is a virtual company. Therefore, education for improving the security of the company is the SSBC exercise purpose. Specifically, the purpose is to allow the participants to consider the difficulty of early warning of the cyber-attacks and the measures in the company as a whole. Therefore, the participants of the SSBC exercise are all the members belonging to the virtual company. The designer creates exercises con-figured so that the participants think about three aspects of safety, security, and business.

Next, the designer identifies possible abnormalities as many as possible. The participants should be aware that the abnormalities due to the cyber-attacks may lead to serious accidents. Also, the participants should realize in the SSBC exercise that such abnormalities are events related to the life of persons at the supply destination and employees. For that reason, we will aim for safe operation at safety and maximum continuous operation for business as the maximum targets. Likewise, companies that do not take security measures and education are more likely to deal with cyber-attacks late. Therefore, the maximum targets are pre-venting the damage and the spread of infection by cyber-attacks. By determining the maximum goal, it becomes possible to discover the risks of impeding the achievement of the goal. Therefore, the following are listed as risks.

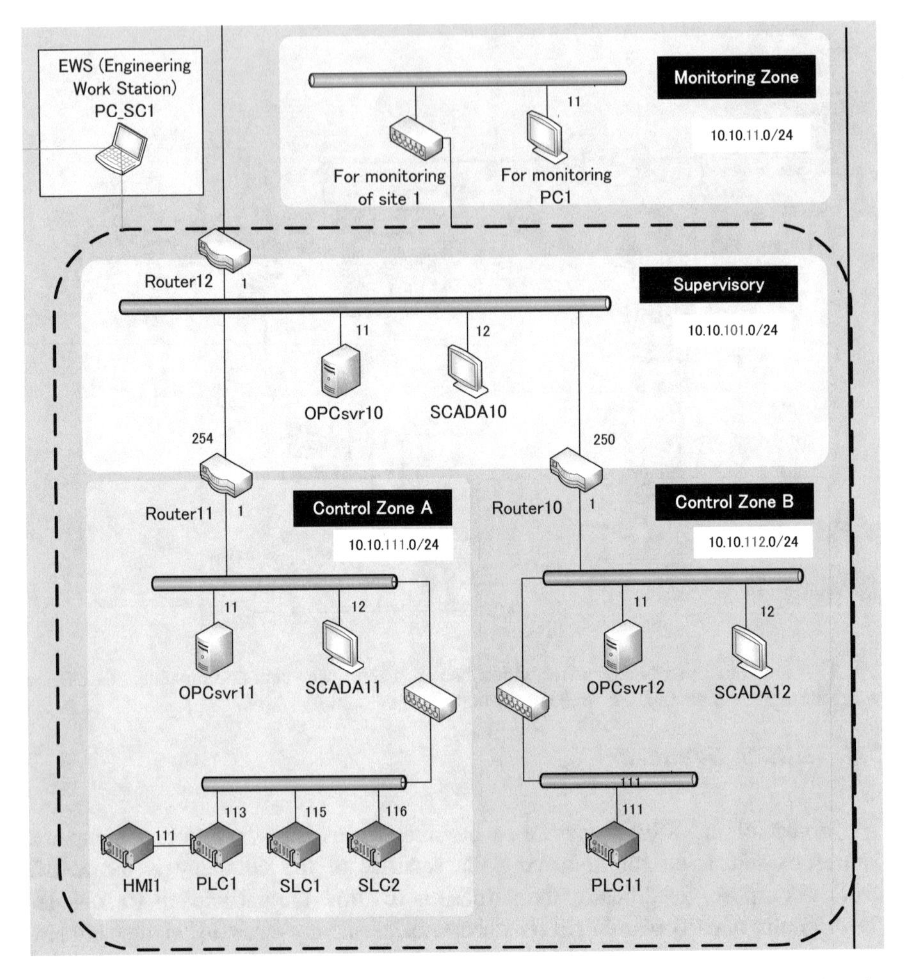

Fig. 8 In this ICS network, OPC servers collect and exchange process data, and monitor them by using SCADA function included in the OPC Servers

1. Safety: An abnormality occurs in the plant
2. Security: Damage caused by the cyber-attacks, infection of terminals
3. Business: Shut down of the plant

The participant uses the brainstorming method to clarify the events that cause these risks. Table 2 shows the revealed events. In Table 2, in the safety viewpoint, the first line indicates the results caused by the risk, and the second and subsequent lines indicate the causes thereof.

The participant selects, from the revealed events, an event to be generated by the scenario of the cyber-attack. The selected abnormality must be a common abnormality related to multiple risks to the safety, security, and business. The business risks are associated with the safety risks and the security risks if the cause

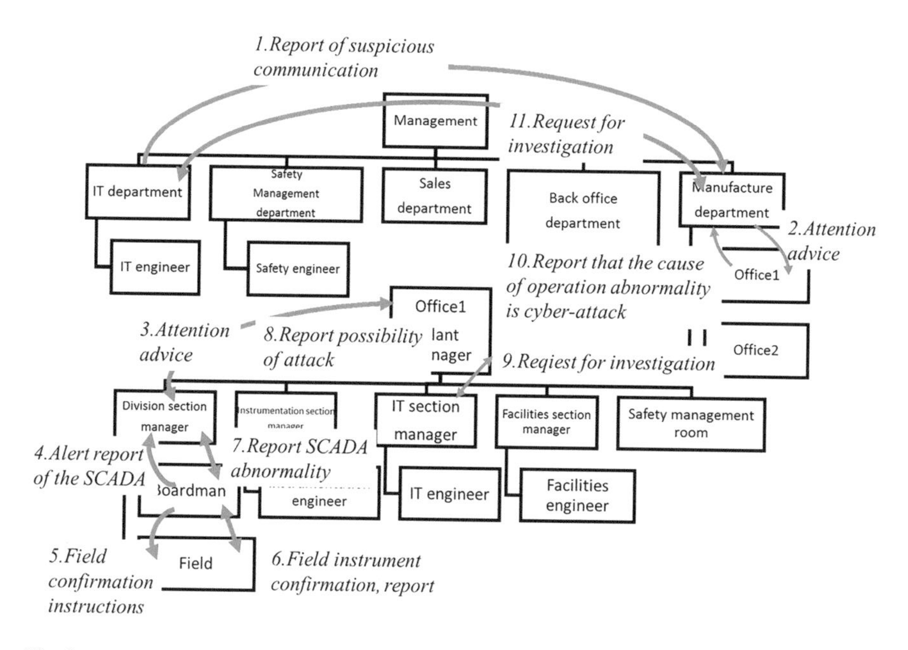

Fig. 9 Emergency communication management

Table 2 The maximum goal and risk of the company for cyber-attack

	Goal	Risk
Safety	Safe operation of the plant	An abnormality occurs in the plant
Security	Prevention of damage and spread of infection	Damage and infection of devices
Business	Continuing plant operation	Shut down plant

of the business risk as "the event where conveyance to the customers is failed" in Table 2 is abnormal at the plant. That is, the risk is a common abnormality in all aspects of safety, security, and business. Therefore, the risk is selected as the event generated by the attack scenario.

Next, the participant selects another event where the services cannot be supplied to the customer in the abnormality of the safety/security viewpoint. From the viewpoint of safety risk where an abnormality occurs in the plant site, if one of the events occurs when Valve 2 is not fully closed, or Pump is stopped, the water does not circulate to the Tank 2 as the supply destination. As a result, the liquid level of Tank 2 drops, and the hot water supply service becomes impossible.

Even if the manual valve of Tank 2 is closed to prevent the liquid level of Tank 2 from lowering, the hot water does not circulate, and the service quality gradually deteriorates, as shown in Fig. 10. The flow of water is indicated by an arrow, and the part where the flow is stopped (Valve 2 and Heater) is designated by x. Also, the valves and pumps are likely to be failed. Therefore, the on-site operator firstly suspects equipment failure and responds accordingly. Also, since the same

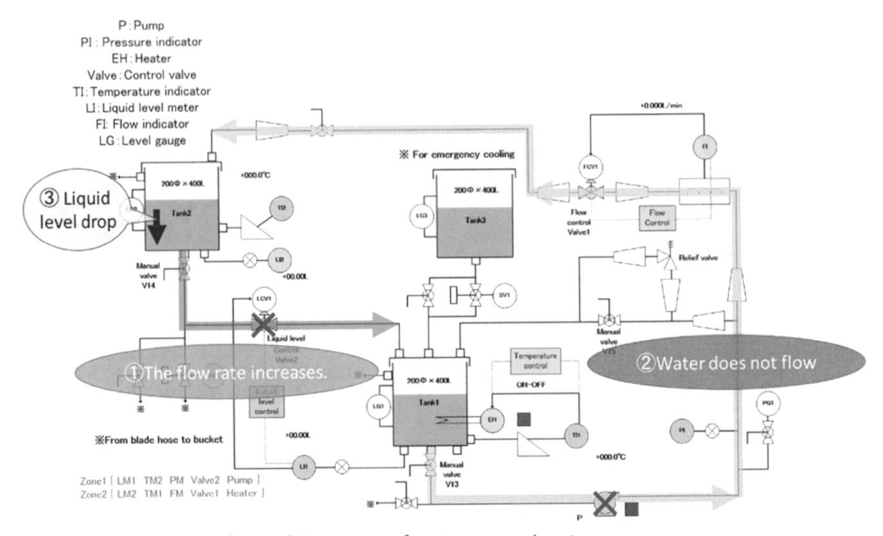

Fig. 10 Abnormality caused in the plant

SLC controls the Valve 2 and Pump in the network diagram, the network is easily attacked. From the above, it is difficult to conclude that the event where the Valve 2 does not close or the pump stops is recognized as a cyber-attack. Therefore, this event is considered to be optimal for an attack scenario, as it not only causes an influential incident but also causes an attacker to create a structure that is easy to attack.

Concealment of cyber-attack is also important. The attacker simultaneously causes a plurality of malicious abnormalities. In doing so, the attacker operates (concealment) that delays the detection of abnormality to prolong the time where the attacker freely attacks. Specifically, in this scenario, the attacker conceals the monitoring screen (SCADA screen) to delay the detection of the abnormality. Therefore, in the attack scenario, the event "the instruction is not reflected on the SCADA screen" in Table 2 is selected. Based on the above, the events, which will be incorporated in the attack scenario, are colored in Table 3.

In the attack scenario, it is important that the participant recognizes the necessity of security measures. In a company network system, the firewall installed between the headquarters and business sites can block the cyber-attacks. Therefore, in the attack scenario, the intrusion is performed at the place (within the plant site) where the firewall is not installed. Although a serious accident cannot be caused only by Zone splitting, a scenario is created where the attacks are repeated within the same zone, and the events selected from Table 3 are generated. The attack scenario corresponding to the procedure of the Cyber Kill Chain is organized, as shown in Table 4.

Table 3 Selected events—in safety, the first line caused the risk, and as a result, the second and subsequent lines indicate the cause

Safety			
Power outage	Empty accident		
Cannot recover power	Water in tank 1 runs out	Heater can not stop	Heater stop
	Water leakage in the plant	Sensor breakdown	Pump stop
	Leaking water at supply destination	No signal is output	Pump trips
	Overflow		
	The supplied flow rate can not keep up with the demand	The control valve breaks down	
	Water supply problem	The controller breaks down	
	Reduction in supply pressure	Instructions on the SCADA screen are not reflected	The monitoring screen can not be seen
Security			
Communication line slows down			
Communication expires			
OPC server data bug			
Business			
Loss of customer information			
Loss of attendance information			
Manufacturing orders do not come			
Manufacturer does not come up			
Supply temperature out of range			
Leak in the drainage line			
It will not flow to customers			

Table 4 Design procedure of the cyber-attack scenario (NIT exercise)

Cyber-attack scenario template	NIT exercise—cyber-attack Scenario
1. Maximum risk (objectives)	1. Stopping the plant
2. Malicious operation (lateral movement)	2. Instructions for stopping the pump full indication of valve 2
3. ICS hacking (C&C)	3. Program change of SLC
4. Installation of ICS hacking (infection)	4. Malware infection due to execution of the attached file
5. Prerequisite for attack (compromise)	5. Open the attached file on user (supervisory zone employee PC)
6. Recent situation scenario2 (delivery)	6. Send mail with malware (enterprise/supervisory zone employee PC)
7. Recent situation scenario (reconnaissance)	7. External vulnerability scanning send phishing email (to the company)

In this scenario, the information system department belonging to the headquarters warns that "Recognizing that suspicious e-mails are increasing in the company recently." The attackers attack with the above procedure. They send e-mails containing the virus inside the headquarters and office. Since companies do not have security education, they both open e-mails.

A firewall that can't intrude by the attacker is set up at the headquarters. On the other side, the office does not have a firewall, so the attackers can intrude. After that, they take over the SLC through OPC Server 1 of Plant 1. They rewrite the program so that operation on the SCADA screen and the on-site panel is not reflected, and open Valve 2 and stop Pump. The operators can be turned on them manually. However, an incorrect command is continuously sent from the rewritten

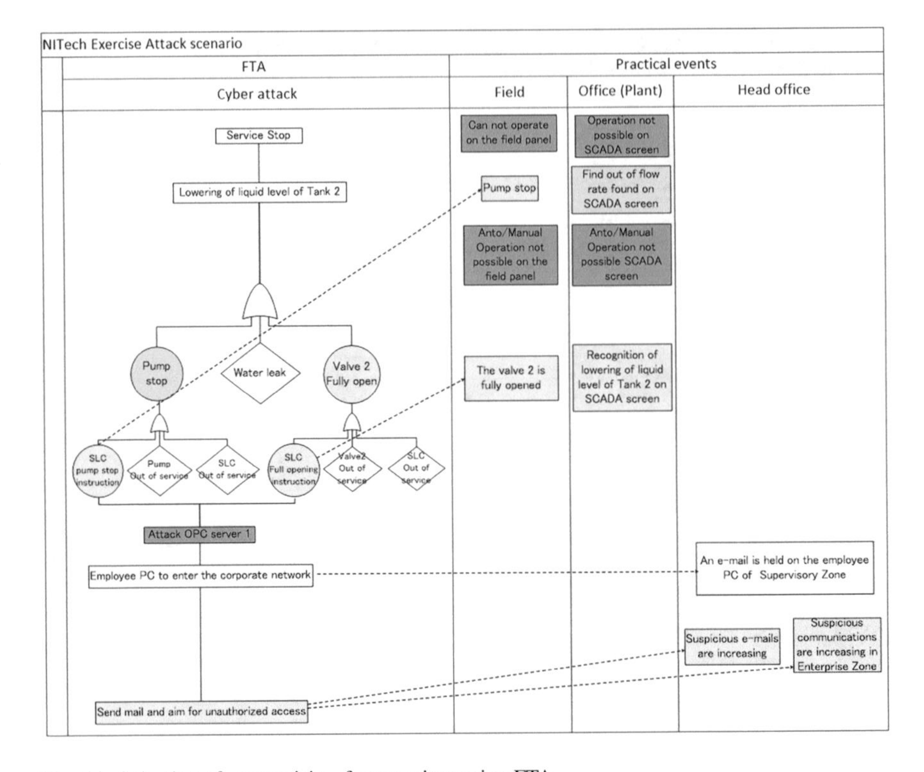

Fig. 11 Selection of prerequisites for exercises using FTA

program; the command turns off Pump immediately and does not start it. As a result, unknown abnormalities occur frequently and simultaneously, and the plant is forced to shutdown.

The designed attack procedure can be indicated by FTA (Fault Tree Analysis) [30]. By issuing an event corresponding to the attack procedure issued by FTA, situations on the site that can occur in the SSBC exercise scenario can be seen and can be reflected as a premise (Fig. 11). It is also good to create an illustration of the network that added what kind of route to attack like Fig. 12. It helps designers to consider the correspondence and assume the influence range of cyber-attack at the same time.

5.4 Defense Scenario

A safety response is created in the attack scenario. The designer creates an at-tack scenario based on the abnormality occurring at the site according to the ordinary work procedure. After that, the designer adds security counteractions and business counteractions in consideration of the cyber-attack. Specifically, the response from

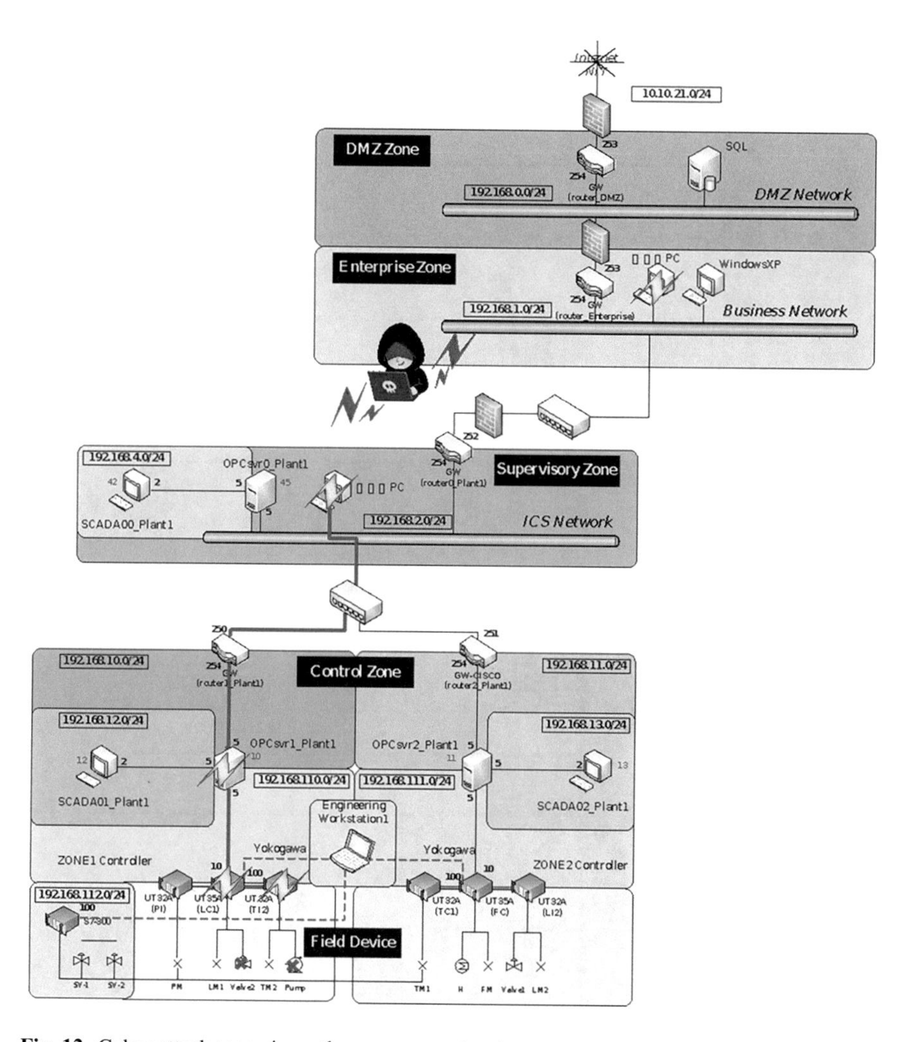

Fig. 12 Cyber-attack scenario on the company network

the site such as on-site confirmation and inspection spots are increased due to multiple simultaneous abnormalities. Moreover, when considering the business impact, new information developed from the worksite to the head office and reflection of decisions based thereon are considered. When creating the defense scenario, the measures are changed as shown in Figs. 13 and 14.

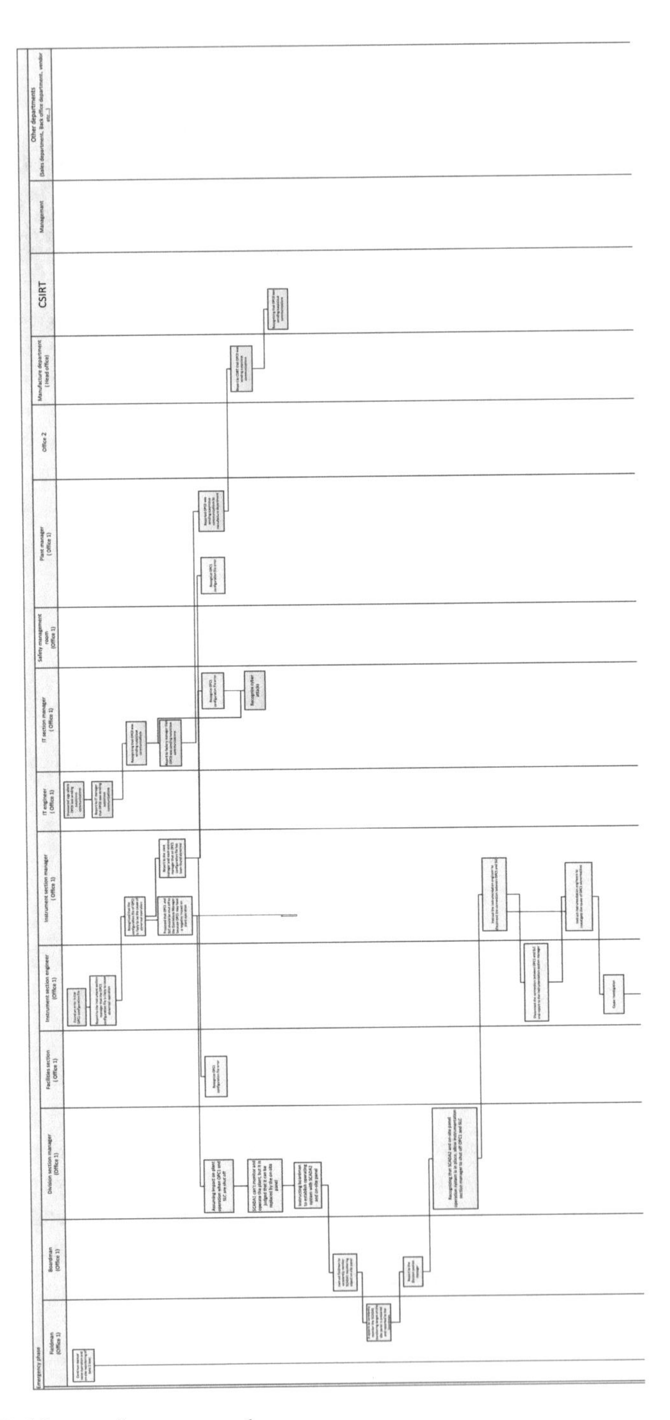

Fig. 13 Workflow—safety correspondence

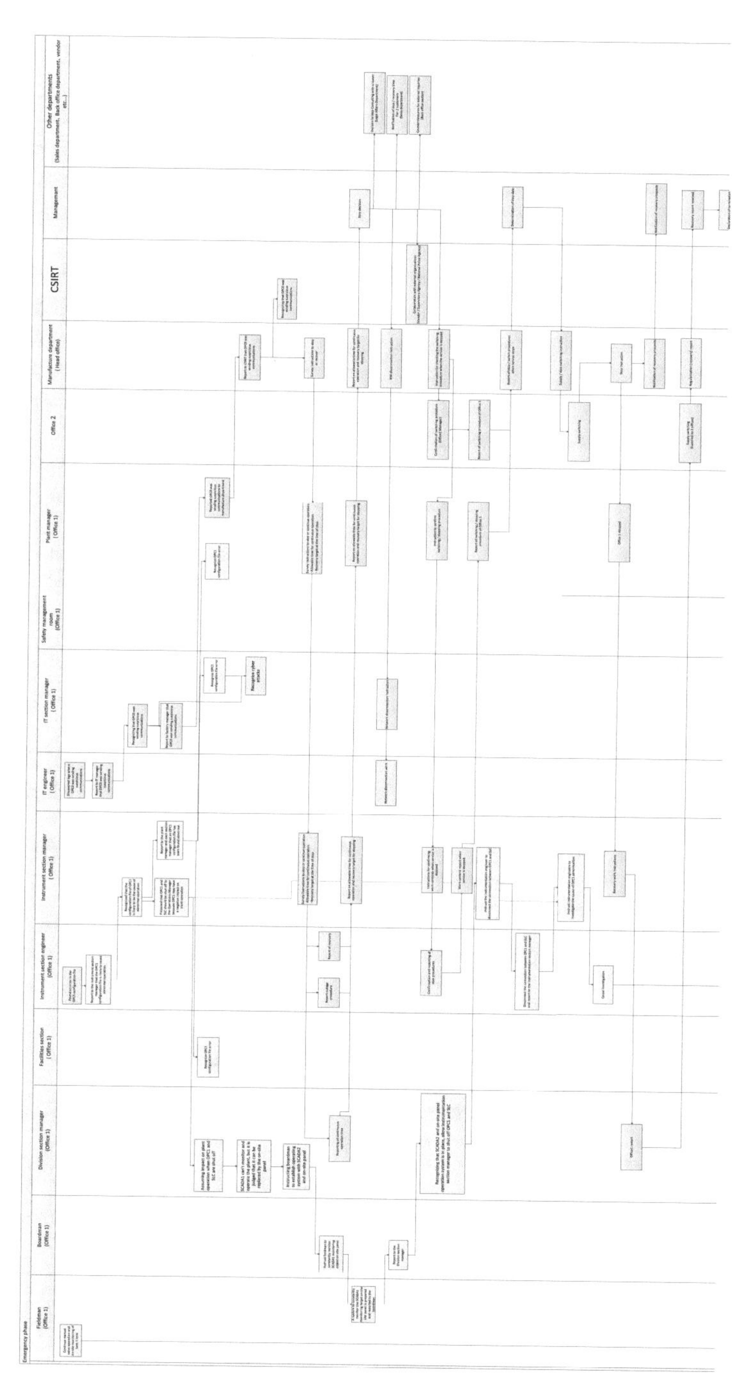

Fig. 14 Workflow—safety correspondence considering cyber attack

6 Concluding Remarks

In this paper, the authors proposed a framework for designing company-specific SSBC exercises to create BCPs, considering cyber-attacks. NIT exercise could be constructed according to the proposed procedure. In creating the SSBC exercise, it is possible to identify the weak points in the current security measure and review the applicability of the current organization.

As for the SSBC exercise development framework, it is necessary to verify even the details of the method by proceeding with the experiment in collaboration with actual companies. Through practical trials, several security exercises were designed based on the framework and improved by operation technology and information technology experts in "Industrial Cyber Security. Center of Excellence" organized by the Japanese Ministry of Economy, Trade, and Industry.

Acknowledgments This research is partially supported by the Ministry of Education, Science, Sports and Culture, Grant-in-Aid for Scientific Research (A), No.16H01837 (2016) and Council for Science, Technology and Innovation (CSTI), Cross-ministerial Strategic Innovation Promotion Program (SIP), "Cyber-Security for Critical Infra-structure" (Funding Agency: NEDO), however, all remaining errors are attributable to the authors.

References

1. IPA, Security of industrial control system (2018), https://www.ipa.go.jp/security/controlsystem/
2. H. Karimipour, S. Geris, A. Dehghantanha, H. Leung, Intelligent anomaly detection for large-scale smart grids, in *2019 IEEE Canadian Conference of Electrical and Computer Engineering (CCECE)* (IEEE, 2019), pp. 1–4. https://doi.org/10.1109/CCECE.2019.8861995
3. H. Karimipour, A. Dehghantanha, R.M. Parizi, K.K.R. Choo, H. Leung, A deep and scalable unsupervised machine learning system for cyber-attack detection in large-scale smart grids. IEEE Access **7**, 80778–80788 (2019)
4. Kaspersky, Stuxnet: zero victims (Securelist), https://securelist.com/stuxnet-zero-victims/67483/
5. Bundesamt für Sicherheit in der Informationstechnik, Informationstechnik, in *Die Lage der IT-Sicherheit in Deutschland 2014* (BSI, Bonn, 2014), p. 44
6. SANS Industrial Control Systems Security Blog: (2017), https://ics.sans.org/blog/2016/01/09/confirmation-of-a-coordinated-attack-onthe-ukrainian-power-grid
7. The Center for Strategic Cyberspace + Security Science: (2017) http://cscss.org/CS/2017/01/19/ukraine-confirms-december-kiev-blackout-was-cyber-sabotage/
8. F. Yoshihiro, Countermeasures and challenges to cyber-attacks at factories and plants. A well-understood book (2015)
9. S. Mohammadi, H. Mirvaziri, M. Ghazizadeh-Ahsaee, H. Karimipour, Cyber intrusion detection by combined feature selection algorithm. J. Inf. Secur. Appl. **44**, 80–88 (2019)
10. H. Karimipour, V. Dinavahi, Parallel relaxation-based joint dynamic state estimation of large-scale power systems. IET Gener. Transm. Distrib. **10**, 452–459 (2016)
11. H. Karimipour, V. Dinavahi, Parallel domain-decomposition-based distributed state estimation for large-scale power systems. IEEE Trans. Ind. Appl. **52**(2), 1265–1269 (2016)

12. H. Karimipour, V. Dinavahi, Extended Kalman filter-based parallel dynamic state estimation. IEEE Trans. Smart Grid **6**, 1539–1549 (2015)
13. A.N. Jahromi, J. Sakhnini, A. Dehghantanha, A deep unsupervised representation learning approach for effective cyber-physical attack detection and identification on highly imbalanced data, in *Proceedings of the 29th Annual International Conference on Computer Science and Software Engineering* (ACM, 2019), pp. 14–23
14. J. Sakhnini, H. Karimipour, A. Dehghantanha, Smart grid cyber attacks detection using supervised learning and heuristic feature selection, in *2019 IEEE 7th International Conference on Smart Energy Grid Engineering (SEGE)* (IEEE, 2019), pp. 108–112. https://doi.org/10.1109/sege.2019.8859946
15. S. Mohammadi, V. Desai, H. Karimipour, Multivariate mutual information-based feature selection for cyber intrusion detection, in *2018 IEEE Electr. Power Energy Conf. EPEC 2018* (IEEE, 2018), pp. 1–6. https://doi.org/10.1109/EPEC.2018.8598326
16. H. Asai, T. Aoyama, I. Koshijima, Design and operation framework for industrial control system security exercise. Adv. Intell. Syst. Comput. **782**, 171–183 (2019)
17. E.M. Dovom, A. Azmoodeh, A. Dehghantanha, D.E. Newton, R.M. Parizi, H. Karimipour, Fuzzy pattern tree for edge malware detection and categorization in IoT. J. Syst. Archit. **97**, 1–7 (2019). https://doi.org/10.1016/j.sysarc.2019.01.017
18. F. Ghalavand, B.A.M. Alizade, H. Gaber, H. Karimipour, Microgrid islanding detection based on mathematical morphology. Energies **11**(10), 2696 (2018). https://doi.org/10.3390/en11102696
19. H. Karimipour, Robust massively parallel dynamic state estimation of power systems against cyber-attack. IEEE Access **6**, 2984–2995 (2018)
20. H. HaddadPajouh, R. Parizi, A survey on internet of things security: requirements, challenges, and solutions. Internet Things (2019). https://doi.org/10.1016/j.iot.2019.100129
21. A.N. Jahromi, S. Hashemi, A. Dehghantanha, K.K.R. Choo, H. Karimipour, D.E. Newton, R.M. Parizi, An improved two-hidden-layer extreme learning machine for malware hunting. Comput. Secur. **89**, 101655 (2020). https://doi.org/10.1016/j.cose.2019.101655
22. J. Sakhnini, H. Karimipour, A. Dehghantanha, R.M. Parizi, G. Srivastava, Security aspects of internet of things aided smart grids: a bibliometric survey. Internet Things (2019). https://doi.org/10.1016/j.iot.2019.100111
23. A. Shostack, *Threat Modeling: Designing for Security* (Wiley, Indianapolis, In, 2014)
24. M. Begli, A layered intrusion detection system for critical infrastructure using machine learning, in *2019 IEEE 7th Int. Conf. Smart Energy Grid Eng* (IEEE, 2019), pp. 120–124
25. S. Geris, Joint state estimation and cyber-attack detection based on feature grouping, in *2019 IEEE 7th Int. Conf. Smart Energy Grid Eng* (IEEE, 2019), pp. 26–30
26. H. Hirai, Y. Takayama, T. Aoyama, Y. Hashimoto, I. Koshijima, Development of the cyber exercise for critical infrastructures focusing on inter-organization communication. Comput. Aided Chem. Eng. **44**, 1669–1674 (2018)
27. Y. Ota, T. Aoyama, D. Nyambayar, I. Koshijima, Cyber incident exercise for safety protection in critical infrastructure. Int. J. Saf. Secur. Eng. **8**, 246–257 (2018)
28. Kochi Prefectural Office, Business continuity training manual, pp. 2–5
29. T. Aoyama, K. Watanabe, I. Koshijima, Y. Hashimoto, Developing a cyber incident communication management exercise for CI stakeholders. Lect. Notes Comput. Sci. (including Subser. Lect. Notes Artif. Intell. Lect. Notes Bioinform.) **10242 LNCS**, 13–24 (2017)
30. The Nikkan Kogyo Shimbun, Mech. Des. **62**, 41–44 (2018)

Cascading Failure Attacks in the Power System

Weixian Liao and Pan Li

1 Introduction

Electric power systems are critical infrastructure and the failure of these systems can lead to severe economic, social, and security consequences. Thus, the security of these systems is crucial. The recent development of the Internet of Things (IoT) technologies helps traditional electric power systems to be transformed into smart grids, and offers tremendous promise of future smart grids [1]. In particular, IoT technologies enable power systems to support various network functions throughout the generation, transmission, distribution, and consumption of energy by incorporating IoT devices (such as smart sensors, actuators and smart meters), as well as by providing the connectivity, automation, etc. [2]. However, the use of such IoT devices also brings new security challenges. The security of power systems has now been further aggravated by various malicious cyber attacks that can be launched on the IoT devices such as Denial-of-Service (DoS) attacks [3], false data injection attacks [4], unobservable cyber attacks through topology errors [5], etc. Due to the expansive geographical coverage and complex interdependencies among system components, protecting the power system is data and computing intensive and hence extremely challenging [6].

Cascading failures are a very concerning security problem in the power system. They are system failures where the failure of a system component can trigger the successive components and a series of unpredictable chain events in the system that can possibly result in a large-scale collapse of the system. Taking the cascading

W. Liao
Towson University, Towson, MD, USA
e-mail: wliao@towson.edu

P. Li (✉)
Case Western Reserve University, Cleveland, OH, USA
e-mail: lipan@case.edu

H. Karimipour et al. (eds.), *Security of Cyber-Physical Systems*,
https://doi.org/10.1007/978-3-030-45541-5_4

failure in transmission networks [7] as an example, when a transmission line fails, it will shift its load it has been supplying to the other lines that share the same bus with it. Those connected lines may be pushed beyond their line capacities, become overloaded, and further shift their loads to other lines. Such sudden load spikes could induce overloaded lines into loss of service due to the operation of the protection system or failure, which quickly spreads to other lines before the system operator can conduct any countermeasures, hence finally taking down the entire system in a very short period of time [8]. This is exactly what happened in the 2003 Northeastern blackout, where the failure of a critical transmission line triggered a cascade of failures, resulting in shutting down a portion of the power system that affected more than 55 million people in the Eastern U.S. and Canada [9]. Cascading failures have attracted intensive attention because of their criticality in the power system operations. Chen et al. [7] propose a hidden failure model to assess the cascading dynamics in power systems. In [10], Rahnamay-Naeini et al. construct a probabilistic model for cascading failures while retaining key physical attributes and operating characteristics of bulk power systems.

As cascading failures can lead to catastrophic damages in the power system and possibly take down the entire system, there is strong motivation for attackers to launch deliberate attacks by taking advantage of it, which we call "*cascading failure attacks (CFAs)*". For example, a malicious attacker can launch CFAs to trip the critical transmission lines and in turn induce a massive cascading failure [11]. In the literature, Motter et al. [12] study cascading-based attacks on complex networks. Zhu et al. [13] assess the line vulnerability and attack strategies from an attacker's perspective in the smart grid. Yan et al. [14] also investigate the topology and cascading attacks in the smart grid. These previous works mainly focus on the impact of cascading attacks but do not consider the defense strategies in such systems. In fact, analyzing CFAs in the power system is a very challenging problem because of the unpredictable cascading effect, the complex interactions between the attacker and the system operator, the extremely high problem dimensionality in a large-scale system, and so on [15].

In this chapter, we explore CFAs in the power system, from a game theoretic perspective. Specifically, defending critical infrastructures against malicious attacks requires system operators to make optimal decisions about where to deploy limited resources to improve system resilience against adversaries. Game theory can naturally be used to provide the system operators with guidance on strategies for infrastructure protection [16–19]. For instance, Salmeron et al. [17] formulate the competition between a defender and an attacker as a leader-follower game. Chen et al. [18] propose a static game framework for defending the power system against deliberate attacks. Rao et al. [19] study a Stackelberg game while taking both the infrastructure survival probability and costs into account. These works consider the competition between the attacker and defender as a one-time event. However, power system protection can be a continuous process where an attacker and a defender interact with each other many times across different dynamic states [20]. For example, the nationwide power system in Yemen suffered from repeated attacks on transmission lines in 2014, which very soon left Yemen in total darkness [21].

Therefore, an attack-defense interaction model that explicitly considers the temporal aspects of the dynamic system states and the long-term effects is indispensable.

To this end, we formulate a zero-sum stochastic game to characterize the long-term interactions between an attacker and a system operator in CFAs. Specifically, we consider that an attacker deploys limited resources to disrupt the components in the power system, such as transmission lines and substations, through either physical attacks or cyber attacks on the IoT devices. Maximizing the amount of load shedding due to disruption is usually adopted as the objective of the attacker in previous studies. However, loads on different transmission lines are of different importance to the system, and each transmission line contributes differently to the overall system reliability and security [7]. Therefore, we consider that the attacker's objective is to maximize the total cost of the load shedding that is defined as a non-decreasing function of the total amount of shedding load, making the problem more challenging. On the other hand, a system operator deploys limited resources to minimize the total cost of load shedding by taking actions such as reinforcing a vulnerable transmission line or repairing a damaged line. Because the objectives of the attacker and the system operator are conflicting, we model the interactions in dynamic environments between two players as a zero-sum stochastic game.

Stochastic games are difficult to solve due to the possible large problem dimensionality and their stochastic nature. Value iteration and policy iteration [22], such as iteratively improving the value functions or policies, respectively, have been developed to solve this problem. Unfortunately, such dynamic programming based algorithms need to enumerate all the system states, the number of which is obviously too large in a large-scale power system for the solution to be tractable. Thus, these algorithms suffer from the well known "curse of dimensionality" problem [23]. Furthermore, although such approaches are proven to converge to the optimum, they are under the assumption that all the dynamic system parameters, for example, reward functions and transition probabilities, are always available for the players, which may not always hold in practice, especially to the attacker in the power system. Some previous works on stochastic game analysis also assume complete *a priori* system information. Instead of having such strong assumptions, we develop a Q-CFA learning algorithm to solve our stochastic game which can address the dimensionality problem and does not need any *a priori* system information. The intuition behind the learning process is that learning through past experience facilitates more intelligent decision making and performance optimization.

The rest of this chapter is organized as follows. Section 2 introduces our system models in detail, including DC power network model, cascading hidden-failure model, as well as the threat and defense models. We formulate the zero-sum stochastic game in the dynamic environment in Sect. 3, which is solved by the proposed Q-CFA learning algorithm in Sect. 4. In Sect. 5, we conduct extensive simulations to validate the convergence and efficiency of our scheme, followed by the conclusion drawn in Sect. 6.

2 System Models

In this section, we introduce DC power network model, cascading hidden failure model, as well as the threat and defense models used in our paper, respectively.

2.1 DC Power Network Model

We consider a power network consisting of $\mathcal{N} = \mathcal{G} \cup \mathcal{D}$ buses and $\mathcal{L} = \{1, \cdots, l, \cdots, L\}$ transmission lines. We assume that each bus is either a generation bus, denoted by $g \in \mathcal{G}$, or a load bus, denoted by $d \in \mathcal{D}$. Bus n_1 is identified as the reference bus. Similar to [24, 25], we use DC power flow approximation of the AC system. Denote by $\mathbf{\Theta} = [\theta_1, \cdots, \theta_n, \cdots, \theta_N]^T$, $\mathbf{P^G} = [p_1^G, \cdots, p_g^G, \cdots, p_G^G]^T$ and $\mathbf{D} = [d_1, \cdots, d_d, \cdots, d_D]$ as the bus voltage angle vector, the real power generation vector and the load demand vector, respectively (note that $N = |\mathcal{N}|$, $G = |\mathcal{G}|$, and $D = |\mathcal{D}|$). Then, the DC power flow equations are formulated as:

$$\mathbf{P^{inj}} = \mathbf{K_g P^G} - \mathbf{K_d D}, \tag{1}$$

$$\mathbf{\Theta} = \mathbf{B P^{inj}}, \tag{2}$$

$$f(l) = b_{ij}(\theta_i - \theta_j), \tag{3}$$

where $\mathbf{P^{inj}} = [p_2^{inj}, \cdots, p_n^{inj}, \cdots, p_N^{inj}]^T$ is a vector of nodal injection power for buses $2, \ldots, N$, $\mathbf{K_g}$ is the bus-generation incidence matrix, and $\mathbf{K_d}$ is the bus-load incidence matrix. θ_i and θ_j are the phase angles of bus i and bus j, respectively, that are connected by transmission line l. $f(l)$ is the real power flow on line l. $\mathbf{B}$ is the $N \times N$ system susceptance matrix, in which $b_{ii} = -\sum_{j \in S_i, j \neq i} \frac{1}{x_{ij}}$ and $b_{ij} = \frac{1}{x_{ij}}$, where x_{ij} is the reactance between bus i and bus j. Notice that in this DC power network model, (1) is the power balance constraint. Equation (2) calculates the phase angles for all the buses, which is used for the power flow calculation on each line in the network as shown in (3).

2.2 Cascading Hidden Failure Model

Cascading failures are system failures where the failure of a system component triggers the successive components and possibly spread among the entire system. Hidden failures are among the top reasons for cascading failures in the power system [7, 10]. Particularly, hidden failure remains undetected until it is triggered by another system failure [26]. In this paper, we study the line protection hidden failure by considering how protective relays work. Protective relays are designed to trip the circuit breakers on the transmission lines when any fault is detected. They may incorrectly trip a transmission line with a load-dependent probability [7], which may in turn lead to more and more lines tripped due to the increased load, i.e., the cascading failure. Hidden failures are undetectable during normal operation but will be exposed as a direct consequence of other system disturbances such as a sudden attack or natural disasters. For example, a malicious attacker can launch false data injection attacks on selected IoT devices, or physically sever critical transmission lines, and in turn induce a massive cascading hidden failure [11]. Such sudden disturbances may cause the protective relay systems to inappropriately and incorrectly disconnect circuit elements. In particular, when transmission line l trips, hidden failures on all the lines connected with it will be exposed such that those lines are then exposed to incorrect tripping probabilistically because of the redistribution of the loads from the tripped line [27]. Furthermore, if an exposed line trips, then the lines that are connected to this tripped line will be further exposed and subject to tripping probabilistically as well, which could eventually cause a cascade of failures and in the worst case, may spread across the entire power system and result in a blackout.

In order to quantify the effects by the cascading failure, we follow a general cascading hidden failure model in [7]. Specifically, we consider line protection hidden failures in the power system. Assuming that an attacker launches a successful attack and takes down a transmission line l, it will trigger the cascading effects in the power system. That is, lines that are connected to this tripped line will be exposed because of the load redistribution from the tripped line, which may result in the total flow through the remaining lines to be larger than the nominal capacities. Based on the observations in NERC events [28], the probability for an exposed line to be tripped incorrectly is very low and considered as a constant p, when the load on this line is below its rated capacity, denoted by $F^{max}(l)$, and increases linearly to 1 as the load approaches $1.4F^{max}(l)$. Furthermore, when the load on the line is or above $1.4F^{max}(l)$, this line will be tripped immediately for security purposes. Thus, the probability of an exposed line tripping incorrectly, also known as the load-dependent probability, defined as $P_t(l)$, is

$$P_t(l) = \begin{cases} p, \text{ if } 0 \leq f(l) \leq F^{max}(l); \\ \frac{5(1-p)f(l)+7pF^{max}(l)-5F^{max}(l)}{2F^{max}(l)}, \\ \text{if } F^{max}(l) \leq f(l) \leq 1.4F^{max}(l); \\ 1, \text{ if } 1.4F^{max}(l) \leq f(l). \end{cases} \tag{4}$$

Based on (4), we are able to determine if exposed lines will be further tripped after the initial line tripping as a chain of cascading effects. If the exposed line trips, then the lines that are connected to this new tripped line will be further exposed and tripped based on (4). Therefore, we can model the potential spread of cascading hidden failure in the power system by the protective relays in all the transmission lines. Notice that our framework accounts for the possibility that lines that are not connected to failed lines may also fail. For example, let us assume that line l_1 is connected to l_2, l_2 is connected to l_3, but l_1 is not connected to l_3. When line l_1 is tripped, l_2 may be tripped, and l_3 could further be tripped based on its load-dependent probability. Therefore, in each round the number of tripped lines is a variable and can be greater than 1. This procedure will go on until there are no further line trippings in the system, then the system will conduct the optimal power flow for the current system configuration, which will be clear in Sect. 3 after we formulate the zero-sum stochastic game.

2.3 *Threat Model*

In the power system, an attacker aims to disrupt the system by either physical attacks such as severing transmission lines, damaging critical infrastructure like transmission towers, or cyber attacks on IoT devices, e.g., false data injection attacks and DoS attacks on sensors [4]. We also assume that the attacker has the knowledge of the system topology and that is able to launch a combination of cyber and physical attacks that can affect many more components across geographical locations. The attacks can be launched on any components of the power system. Without loss of generality, in this paper we use attacks on transmission lines as an example, which are one of the most common and far-ranging targets in the power system [29].

We first define two binary variables as follows:

$$\epsilon(l) = \begin{cases} 1, & \text{if line } l \text{ is attacked;} \\ 0, & \text{otherwise.} \end{cases} \tag{5}$$

$$\delta(l) = \begin{cases} 1, & \text{if line } l \text{ is exposed;} \\ 0, & \text{otherwise.} \end{cases} \tag{6}$$

where $\epsilon(l)$ is equal to 1 if the transmission line l is attacked by the attacker, and $\delta(l)$ is equal to 1 if line l is exposed according to the cascading hidden failure model in Sect. 2.2.

For practicality, we assume that the malicious attacker has limited resources to launch attacks. Specifically, it can only attack a limited number of transmission lines in one action. Therefore, the attacker's action is constrained by

$$\sum_{l \in \mathcal{L}} \mathbf{1}_{\epsilon(l)=1} = b_a, \tag{7}$$

where b_a denotes the attacker's limited resources, i.e., the maximum number of transmission lines that can be attacked in one action, and $\mathbf{1}_A$ is an indicator function that is equal to 1 when the event A is true and zero otherwise.

Subject to resource constraints, the objective of the attacker is to cause the most damage to the power system. In the past, damage is simply measured as the total amount of loads that have to be shed due to line failures [8]. However, because different loads may have different adverse impacts on the power system, it is more appropriate if we use the *costs of load shedding* as the objective of the attacker instead of the *amount of load that is shed*. To this end, we denote the cost function for the transmission line l as $u_l(\cdot)$, which is a nondecreasing function with regard to the shed load on the transmission line l, i.e., $\hat{d}(l)$. Consequently, the objective of the attacker is to maximize the total cost of the loads that are shed in the power system, i.e., to maximize $\mathcal{U} = \sum_{l \in \mathcal{L}} u_l(\hat{d}(l))$.

2.4 Defense Model

Similarly, a system operator, who could be the power system operator or a third-party system protector, aims to protect the power system from the attack. For illustrative purposes, we define the available actions by the system operator as repairing a damaged line or comprised IoT devices, or reinforcing an important line such that:

$$\beta(l) = \begin{cases} 1, & \text{if line } l \text{ is repaired or reinforced.} \\ 0, & \text{otherwise.} \end{cases} \tag{8}$$

where $\beta(l)$ indicates if the system operator chooses to repair the transmission line l or reinforce it. We note that by reinforcement, we mean that the system operator can reinforce the protection on a specific line by adding physical barriers or deploying additional security personnel. A system operator can also adopt new malicious data analysis schemes that can detect compromised PMUs and perform healing actions such as firmware updates to defend cyber attacks. Since some particular lines are more likely to start a cascading failure, system operators are willing to allocate more resources to these lines to enhance the security of their system. We also assume that the system operator has limited resources to protect the power system, i.e.,

$$\sum_{l \in \mathcal{L}} \mathbf{1}_{\beta(l)=1} = b_o, \tag{9}$$

where b_o denotes the system operator's limited resources, i.e., the maximum number of transmission lines that it can repair or reinforce in one action. Besides, the objective of the system operator of the power system is to find the best strategy that minimizes the total cost of load shedding in the power system, i.e., to minimize $\mathcal{U} = \sum_{l \in \mathcal{L}} u_l(\hat{d}(l))$.

Therefore, as the objectives of the system operator and the attacker are conflicting and two players compete with each other through dynamic system states, we formulate a zero-sum stochastic game that will be introduced in the next section.

3 A Zero-Sum Stochastic Game for CFAs

As presented above, the objectives of the attacker and that of the system operator in CFAs are opposite to each other. Therefore, in this section, we formulate a zero-sum stochastic game for the attacker and the system operator in the power system.

Before delving into details of the formulation for the zero-sum stochastic game, we first briefly introduce stochastic games. In game theory, a stochastic game is a dynamic game with probabilistic transitions played by several players [30], which can be considered as an extension of Markov Decision Processes [31]. The game is played in a sequence of stages. Specifically, at the beginning of each stage, the game is in a given state and players select actions independently and simultaneously based on their own resources and constraints at the current state, and each player will then receive an *immediate reward* that results from the chosen actions and the current state. Thereafter, the game moves to a new random stage, the transition probability of which is determined by both the actions from the players and the previous state. This procedure repeats continuously for a number of stages and each player endeavors to maximize their *long-term reward*, that is defined as the discounted sum of the *immediate rewards* at all stages.

3.1 States, Actions, and State Transitions

By considering the interactive competition between the attacker and the system operator, we now formulate the CFAs as a stochastic game G. In this game G, there are a set of system states, denoted by $\mathcal{S}$, in which each state $s \in \mathcal{S}$ is a vector that denotes the current status of all the transmission lines. we use time-slot based system as the temporal resolution in our model [20]. Without loss of generality, we define the status of each transmission line as "up", denoted by u, or "down", denoted by w, when the line is functioning well or malfunctioning after being attacked, respectively. The stochastic game proceeds in a time-slotted fashion. Specifically, in each time slot, each player will choose an action based on the current system state so as to optimize its own objective. We denote by $\mathcal{M}_A(s)$ and $\mathcal{M}_O(s)$ the set of all the possible actions that the attacker and the system operator can take at state s, respectively. As discussed in Sects. 2.3 and 2.4, for the attacker, each $a \in \mathcal{M}_A(s)$ indicates the set of transmission lines to be attacked. On the other hand, for the system operator, each $o \in \mathcal{M}_O(s)$ refers to a set of transmission lines to be repaired (if not working) or reinforced (if still working but vulnerable to attacks). Each action $a \in \mathcal{M}_A(s)$ and $o \in \mathcal{M}_O(s)$ will be selected by the attacker and the system operator, respectively, in each state s, with a certain probability denoted by $\pi_a(s)$ and $\pi_o(s)$.

Recall that each player selects their actions independently and simultaneously in each stage. We denote p_{uwr} and p_{uw} as the probabilities that a functioning transmission line fails upon attack with and without reinforcement by the system operator in the same time slot, respectively. Similarly, we denote p_{wua} and p_{wu} as the probabilities that a non-functioning line recovers after repair with and without being attacked in the same time slot, respectively. The following constraints must be satisfied, $0 \le p_{uwr} < p_{uw} \le 1$ and $0 \le p_{wua} < p_{wu} \le 1$ and in practice, these probabilities can be obtained by either conducting simulations or observing historical records. We can see that these probabilities determine the transition probability $T(a, o, s, s')$ from state s to state s' under the actions a and o by the attacker and the system operator, respectively. For example, suppose at the beginning all lines in the system are up and there are no actions from the attacker or the system operator. Then, when the attacker and the system operator choose the same line to attack and reinforce, respectively, the probability for the power system to remain in the same state is $1 - p_{uwr}$. Similarly, when the attacker attacks a line l and the system operator chooses to reinforce another line l', the probability for the system to move to another state where only line l is down is p_{uw}.

3.2 Immediate Rewards

As mentioned before, the objectives of the attacker and the system operator are opposite; maximizing/minimizing the total cost of the load shedding in the power system. At each stage of the game, both the attacker and the system operator will receive an *immediate reward* defined by the actions taken by them (the attacker a, the system operator o) at state s. For example, the immediate reward for the attacker, denoted by $U_A(a, o, s)$, is the total cost for load shedding.

We show in Fig. 1 what happens sequentially in one stage of the game where the attacker and the system operator take actions a and o, respectively, at state s. Particularly, after both players take actions, some transmission lines might be tripped, and hence the system immediately adjusts according to the power equations (1)–(3) [32]. Then the system checks whether there are any lines overloaded. If so, the protective relays trip the overloaded lines and the system re-adjusts accordingly until there are no overloaded lines. Otherwise, the exposed lines, which share the same bus with the tripped lines, are tripped with probability $P_t(l)$, based on the cascading model in Sect. 2.2. The cascading effect continues until there are no line outages. Finally, the power system performs security constrained optimal power flow, which is formulated as an optimization problem to minimize the total cost of load shedding, i.e., $\mathcal{U}(a, o, s)$, in the current configuration of power system:

$$\textbf{Minimize} \quad \mathcal{U}(a, o, s) = \sum_{l \in \mathcal{L}} u_l(\hat{d}_l),$$

$$\textbf{s.t.} \quad \sum_{g \in \mathcal{G}} P_g + \sum_{l \in \mathcal{L}} \hat{d}_l - \sum_{l \in \mathcal{L}} d_l = 0 \tag{10}$$

$$P_g^{min} \le P_g \le P_g^{max}, \quad \forall g \in \mathcal{G} \tag{11}$$

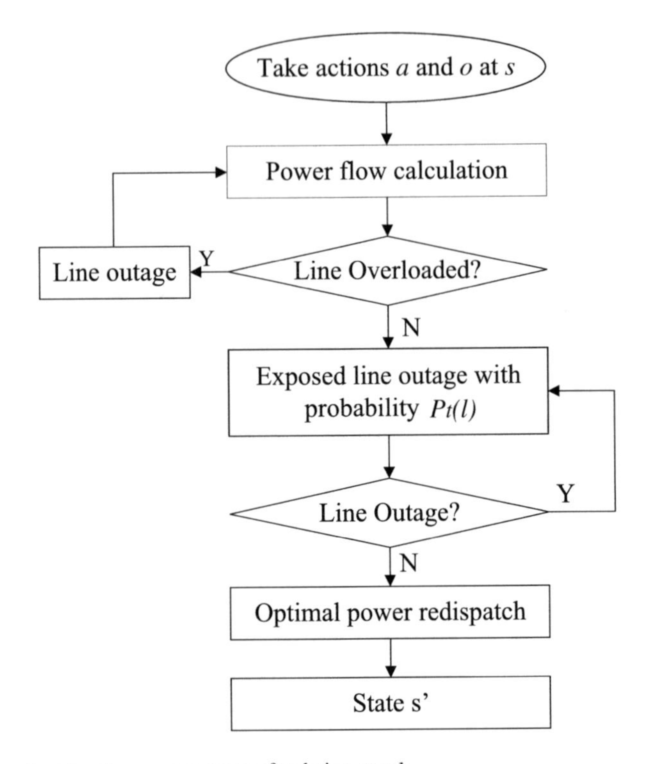

Fig. 1 Flow chart for the power system after being attack

$$-F^{min}(l) \leq f(l) \leq F^{max}(l), \quad \forall l \in \mathcal{L} \tag{12}$$

$$0 \leq \hat{d}_l \leq d_l, \quad \forall l \in \mathcal{L} \tag{13}$$

where (10) is the power balance constraint, (11) is the generation capacity constraint for each generation unit, (12) limits the maximum power flow on each transmission line, and (13) indicates that the shed load cannot exceed the original load on the load bus. After solving the above minimization problem, we can shed loads when necessary. In practice, utility companies have several tools to reduce users' load demands. For example, industrial users, which account for 60% of total energy consumption [33], often have contracts with utility companies where they commit to reduce their load after a request from the utility companies in exchange for reduced energy prices; under real-time energy pricing, all users can be incentivized to reduce their energy consumption by significantly increasing prices; and system operators can disconnect complete sections of the power system by opening switches.

Therefore, we have the immediate rewards for the attacker and the system operator, known as the payoff of the game at state s given by $\mathcal{U}(a, o, s)$ for all $a \in \mathcal{M}_A(s)$ and $o \in \mathcal{M}_O(s)$. Notice that this framework can also account for the case where the system is disconnected into non-connected islands. Specifically, when the system is disconnected to several islands, both players still take actions in the whole system subject to the limited resources b_a and b_o. After there are no more line trippings in the system, we conduct the OPF for every island and then the system state transits to the next state.

Because the objective function is convex and all the constraints are linear, this problem can be easily solved and we can obtain the *immediate rewards* for each player at any system status. Actions a and o executed at state s will bring the system state to the next state, resulting in further immediate rewards, i.e., $\mathcal{U}(a', o', s')$, at the next state s'. Thus, actions taken at dynamic states will finally accrue a long-term reward as the game continues. The objective of both players is to obtain the optimal expected long-term reward, which will be discussed next.

4 Optimal Strategies of the Stochastic Game

In this section, we first present the definition of optimal strategies. Then, we develop a Q-CFA learning algorithm to find the optimal strategies for the zero-sum stochastic game.

4.1 Optimal Strategies

We refer to the *optimal strategies* as the mixed strategies of all actions chosen by the players that maximize their expected long-term rewards [34]. In this paper, we consider the case of stationary policies where the action selection probabilities, i.e., $\pi_A(s)$ and $\pi_O(s)$, do not change over time. In other words, we are interested in finding the stationary policies for each player at each state s.

From the attacker's point of view, we let $V_A(s)$ denote the attacker's expected long-term reward under the optimal strategies when the game starts at state s, and $Q_A(a, o, s)$ as the expected long-term reward for taking action a while the system operator selects the action o when the game starts at state s. Specifically, we have

$$V_A(s) = \max_{\pi_A(s)} \min_{\pi_O(s)} \sum_{a \in \mathcal{M}_A(s)} \sum_{o \in \mathcal{M}_O(s)} \pi_a(s) Q_A(a, o, s) \pi_o(s), \tag{14}$$

where $\pi_A(s) = \{\pi_a(s) | a \in \mathcal{M}_A(s)\}$, $\pi_O(s) = \{\pi_o(s) | d \in \mathcal{M}_O(s)\}$, and

$$Q_A(a, o, s) = \mathcal{U}(a, o, s) + \gamma \cdot \sum_{s' \in \mathcal{S}} V_A(s') \cdot T(a, o, s, s'). \tag{15}$$

$V_A(s)$ and $Q_A(a, o, s)$ are also called the *value* of the state $s \in \mathcal{S}$ and the *quality* of the state s given actions a and o, respectively, for the attacker. $T(a, o, s, s')$ is the state transition probability from state s to state s' after taking actions a and o. Here the *maxmin* function can be interpreted as follows. Because our game is a fully competitive stochastic game where each player selects an action independently and simultaneously at each system state, we need opponent-independent algorithms to solve this problem [35]. The *maxmin* function makes (14) opponent-independent in which the attacker attempts to maximize its own expected long-term reward under the worst case assumption that the system operator will always endeavor to minimize the payoff. Besides, note that (15) states that $Q_A(a, o, s)$ is equal to the immediate reward plus the discounted expected optimal value attainable from the next state s'. In (15), $\gamma \in [0, 1)$ is a discount factor that represents how much impact the current decisions can have on the long-term reward. Particularly, when γ equals 0, the game becomes a one-time-event game [17–19]. When γ is larger than 0, a smaller value of γ emphasizes more the immediate rewards and a larger γ gives higher weight to the future rewards.

Similarly, the system operator's expected long-term reward under the optimal strategies when the game starts at state s, denoted by $V_O(s)$, is

$$V_O(s) = \min_{\pi_O(s)} \max_{\pi_A(s)} \sum_{a \in \mathcal{M}_A(s)} \sum_{o \in \mathcal{M}_O(s)} \pi_a(s) Q_O(a, o, s) \pi_o(s), \tag{16}$$

where $Q_O(a, o, s)$ is the expected long-term reward for taking action o while the attacker selects the action a, known as the *quality* of the state s for the system operator, and is formulated as

$$Q_O(a, o, s) = \mathcal{U}(a, o, s) + \gamma \cdot \sum_{s' \in \mathcal{S}} V_O(s') \cdot T(a, o, s, s'). \tag{17}$$

We note that generally $V_A(s) \leq V_O(s)$ due to weak duality, where $V_A(s)$ and $V_O(s)$ correspond to the primal problem and the dual problem, respectively. However, in a zero-sum stochastic game, strong duality holds and we have $V_A(s) = V_O(s) = V(s)$ (Section 5.4.5 in [36]). Consequently, the optimal solutions computed individually by the two players, i.e., $\boldsymbol{\pi}_{\mathbf{A}}^*(s)$ and $\boldsymbol{\pi}_{\mathbf{O}}^*(s)$, are the best responses to each other. We denote by $\pi^*(s) = \{\boldsymbol{\pi}_{\mathbf{A}}^*(s), \boldsymbol{\pi}_{\mathbf{O}}^*(s)\}$ the *optimal strategy pair* [37], which is known as the *Nash equilibrium point* in a stochastic game and defined as follows.

Definition 1 (Nash Equilibrium) In a zero-sum stochastic game $\mathbf{G}$, the Nash equilibrium for any state $s \in \mathcal{S}$ is an optimal strategy pair $\pi^*(s) = \{\boldsymbol{\pi}_{\mathbf{A}}^*(s), \boldsymbol{\pi}_{\mathbf{O}}^*(s)\}$ satisfying

$$V^{\pi^*(s)}(s) \geq V^{\{\boldsymbol{\pi}_{\mathbf{A}}(s), \boldsymbol{\pi}_{\mathbf{O}}^*(s)\}}(s),$$

$$V^{\pi^*(s)}(s) \leq V^{\{\pi_{\mathbf{A}}^*(s), \pi_{\mathbf{O}}(s)\}}(s).$$

Therefore, by finding the Nash equilibrium for each state s, we can obtain the attacker's and the system operator's optimal strategies, specifically, the probability mass distributions on their action sets $\mathcal{M}_A(s)$ and $\mathcal{M}_O(s)$, which results in the optimal expected long-term reward for the attacker and the system operator, respectively.

From the attacker's perspective, the optimal strategies $\pi_A^*(s)$ $(s \in \mathcal{S})$ can be obtained by solving (14) using algorithms like "value iteration" [22]. Particularly, at the kth iteration, for each $s \in \mathcal{S}$, the attacker needs to solve the following problem:

$$V_A^k(s) = \max_{\{\pi_a(s)\}} \min_{o \in \mathcal{M}_O(s)} \sum_{a \in \mathcal{M}_A(s)} Q_A^k(a, o, s) \cdot \pi_a(s)$$

$$\textbf{s.t. } Q_A^k(a, o, s) = \mathcal{U}(a, o, s) + \gamma \cdot \sum_{s' \in \mathcal{S}} V_A^{k-1}(s') \cdot T(a, o, s, s')$$

$$\sum_{a \in \mathcal{M}_A(s)} Q_A^k(a, o, s) \geq V_A^{k-1}(s)$$

$$\sum_{a \in \mathcal{M}_A(s)} \pi_a(s) = 1$$

$$\pi_a(s) \geq 0, \forall a \in \mathcal{M}_A(s)$$

where $V_A^k(s)$ is the value of the state s in the kth iteration. The basic idea of value iteration is that it iteratively estimates the value of $Q_A(a, o, s)$ and $V_A(s)$ using (14) and (15) for each $s \in \mathcal{S}$ in each iteration until convergence. The optimal strategies can then be obtained after scanning all the available states and action spaces. The system operator can find its optimal strategies $\pi_{\mathbf{O}}^*(s)$ $(s \in \mathcal{S})$ by following a similar approach, which is omitted here due to space limit.

Value iteration has been proved to converge to the optimal results in stochastic games [38]. However, it assumes that the system information, such as the state transition probabilities $T(a, o, s, s')$'s, is *a priori* knowledge for both players, which may not be the case in most practical applications. Moreover, this algorithm needs to enumerate all the system states and available actions in each iteration in order to obtain the optimal strategies. Nevertheless, the number of states and actions grows exponentially with the number of transmission lines, which obviously makes such algorithms infeasible for large-scale system applications.

4.2 A Q-CFA Learning Algorithm

In order to account for the drawbacks of previous algorithms, we develop a machine learning based method named Q-CFA learning algorithm that is based on the minimax-Q learning framework [34]. The proposed algorithm can gradually learn the optimal strategies without having any *a priori* knowledge of system information such as the state transition probabilities, i.e., $T(a, o, s, s')$'s. Besides, unlike value iteration and other previous algorithms, it does not need to scan all the states and actions in each iteration, and hence is very efficient for power system applications.

The main idea of the proposed algorithm is as follows. Different from that in (15), we rewrite the quality of state s for the attacker under actions a and o by the attacker and the system operator, respectively, i.e., $Q_A(a, o, s)$, at the kth iteration into:

$$\begin{aligned} Q_A^k(a, o, s) = (1 - \alpha(k)) \cdot Q_A^{k-1}(a, o, s) \\ +\alpha(k) \cdot [\mathcal{U}(a, o, s) + \gamma V_A^{k-1}(s')] \end{aligned} \tag{18}$$

where $\alpha(k) = \frac{1}{k+1}$ is the learning rate that decays over time, and s' is the next state after actions are executed in the current state s. In other words, $Q_A^k(a, o, s)$ is updated by mixing the previous Q-value with a correction from the new estimate at a learning rate $\alpha(k)$. Then, the value of state s at the kth iteration, i.e., $V_a^k(s)$, can be updated accordingly by (14). Note that the quality and the value of state s for the system operator can be updated in the same fashion.

Specifically, because of their limited resources, both the attacker and the system operator only have a limited number of actions at each stage of the game, which could be very diverse at different states. At the beginning of each state s_k, the algorithm firstly checks whether the current state has been observed in previous stages. If so, then both players use the previous profiles at state s_k to initialize parameters such as the action sets, along with Q and V values. Otherwise, the algorithm initializes all the variables, and then adds the current state s_k into the *observation history set* denoted by H_s that contains profiles at all the past states. Subsequently, each player chooses an action. In particular, with a probability of p_{exp}, the attacker and the system operator choose to explore their available action spaces, i.e., $\mathcal{M}_A(s)$ and $\mathcal{M}_O(s)$, respectively, and uniformly and randomly select actions. This process is called *exploration*. On the other hand, with a probability of $1 - p_{exp}$, they choose to take the same actions selected in the previous initialization step, that is called *exploitation*. The intuition here is that the players in Q-learning can either randomly try out one of the available action profiles to possibly achieve higher reward in the long run, namely exploration, or attempt to maximize the reward by choosing the best known action, namely exploitation [39]. Looking into (18), the Q-CFA learning algorithm only uses the previous predicted state value, i.e., $V_a^{k-1}(s)$, which avoids enumerating all the possible future states for current state s. After both players take actions, they obtain their *immediate rewards*, update their Q and V function values, policies $\pi_A^*(s_k)$ and $\pi_O^*(s_k)$, and learning rates $\alpha(k)$, respectively, and then update the profiles for state s_k in the observation history set H_s. Thereafter, the game transits to the next state s_{k+1}. This procedure goes on until the policies in all states have converged. The details of the proposed Q-CFA learning algorithm are described in Algorithm 1.

Notice that in order to update the profiles for each state, i.e., $(\pi_A^*(s_k), \pi_O^*(s_k))$, $V_A(s_k)$ and $V_O(s_k)$, we need to solve the subproblem of $\max_{\pi_A(s_k)} \min_{\pi_O(s_k)} \sum_{a\in\mathcal{M}_A(s_k)} \sum_{o\in\mathcal{M}_O(s_k)} \pi_a(s) Q_A^k(a, o, s_k)\pi_o(s_k)$ in the learning process, which turns out to be a matrix game where the strategies of the attacker and system operator form the rows and columns of the matrix, respectively, with payoffs $Q_A^k(a, o, s)$ and $Q_O^k(a, o, s)$ and we have that $Q_A^k(a, o, s) = Q_O^k(a, o, s_k) = Q^k(a, o, s_k)$. Therefore, we formulate the matrix game as:

$$\max_{\pi_A(s_k)} \min_{\pi_O(s_k)} \sum_{a\in\mathcal{M}_A(s_k)} \sum_{o\in\mathcal{M}_O(s_k)} \pi_a(s_k) Q^k(a, o, s_k)\pi_o(s_k) \tag{19}$$

However, the above optimization problem cannot be solved directly. In order to achieve the optimal strategies, i.e., $(\pi_A^*(s_k), \pi_O^*(s_k))$, we begin by assuming that the attacker's strategies are fixed. Then the problem is reduced to:

$$\min_{\pi_O(s_k)} \sum_{a\in\mathcal{M}_A(s_k)} \pi_a(s_k) Q^k(a, o, s_k) \sum_{o\in\mathcal{M}_O(s_k)} \pi_o(s_k) \tag{20}$$

As $\sum_{a\in\mathcal{M}_A(s_k)} \pi_a(s_k) Q^k(a, o, s_k)$ is a vector, the solution to problem (20) is equivalent to searching for the smallest element in the vector, i.e., $\min_i [\sum_{a\in\mathcal{M}_A(s_k)} \pi_a(s_k) Q^k(a, o, s_k)]_i$. Thereafter, the matrix game (19) can be reformulated as:

$$\max_{\pi_A(s_k)} \min_i [\sum_{a\in\mathcal{M}_A(s_k)} \pi_a(s_k) Q^k(a, o, s_k)]_i \tag{21}$$

Next, we define $x = \min_i [\sum_{a\in\mathcal{M}_A(s_k)} \pi_a(s_k) Q^k(a, o, s_k)]_i$ and we have that $[\pi_a(s_k) Q^k(a, o, s_k)]_i \geq x$. Therefore, problem (19) can be further rewritten as:

$$\max_{\pi_A(s_k)} \quad x$$

$$s.t. \quad [\sum_{a\in\mathcal{M}_A(s_k)} \pi_a(s_k) Q^k(a, o, s_k)]_i \geq x \tag{22}$$

$$\sum_{a\in\mathcal{M}_A(s_k)} \pi_a(s_k) = 1 \tag{23}$$

$$\pi_a(s_k) \geq 0, \forall a \in \mathcal{M}_A(s_k) \tag{24}$$

Finally, we can transform this to a linear programming problem by viewing x as another variable:

$$\max_{\pi'} \quad \mathbf{0}_{aug}^T \pi'$$

$$s.t. \quad Q'\pi' \leq \mathbf{0} \tag{25}$$

$$\sum_{a \in \mathcal{M}_A(s_k)} \pi_a(s_k) = 1 \tag{26}$$

$$\pi_a(s_k) \geq 0, \forall a \in \mathcal{M}_A(s_k) \tag{27}$$

where $\pi' = [\boldsymbol{\pi}_{\mathbf{a}(\mathbf{s_k})}, x]^T$, $Q' = ([\mathbf{0} \quad \mathbf{1}] - [\mathbf{Q^k}(\mathbf{a}, \mathbf{o}, \mathbf{s_k}) \quad \mathbf{0}])$. $\mathbf{0}_{aug}^T = [\mathbf{0}^T \quad 1]$ is used to augment the original variable vector $\boldsymbol{\pi}_{\mathbf{a}(\mathbf{s_k})}$ by viewing x as another variable so that we can transform the problem into the standard form of a linear program. Because (25) is a linear program, we can find the optimal solution of the matrix game. Furthermore, as we optimally solve the subproblem, our algorithm converges to the Nash equilibrium of the game, which is proved in the next section.

4.3 *Proof of the Nash Equilibrium*

In what follows, we prove that our proposed algorithm converges to the Nash equilibrium in the formulated zero-sum stochastic game. The general idea is that, we firstly prove the convergence of our algorithm, then prove that the obtained result is the Nash equilibrium of the game as defined in Sect. 4.1.

Before we prove the convergence of the proposed algorithm, we have the following assumptions and lemma [40]:

Assumption 1 Every state and action have been visited infinitely often. □

Assumption 2 The learning rate, $\alpha(k)$, satisfies the following conditions:

1. $1 < \alpha(k) < 1$;
2. $\sum_{k=0}^{\infty} (\alpha(k))^2 < \infty$. □

Lemma 1 (Conditional Averaging Lemma) *Under Assumptions 1 and 2, the process* $V(k+1) = (1 - \alpha(k))V(k) + \alpha(k)\omega(k)$ *converge to* $\mathbb{E}(\omega|h(k), \alpha(k))$, *where* $h(k)$ *is the history at time stamp* k.

Then, we arrive at a theorem for the convergence of our algorithm.

Theorem 1 *In the proposed Algorithm 1, for any state* $s \in \mathcal{S}$, *the attacker's and the system operator's policies, i.e.,* $\pi_A(s)$ *and* $\pi_O(s)$, *converge to the Nash equilibrium point.*

Algorithm 1 Q-CFA learning algorithm

1: **At State** $s_k, k = 0, 1, \ldots$
If state s_t has been observed in any previous iteration, i.e., $s_t \in H_s$
initialize π_a, π_o, Q, V with the recorded profiles in H_s
Otherwise,
generate action sets $\mathcal{M}_A(s_k)$ and $\mathcal{M}_O(s_k)$,
initialize $Q(a, o, s_k) \leftarrow 1$, for all $a \in \mathcal{M}_A(s_k)$ and $o \in \mathcal{M}_O(s_k)$,
initialize $\pi_A(s_k) \leftarrow \frac{1}{|\mathcal{M}_A(s_k)|}$ and $\pi_O(s) \leftarrow \frac{1}{|\mathcal{M}_O(s_k)|}$,
2: **Choose an action pair $\{\pi_a, \pi_o\}$ at state s_k:**
With probability p_{exp}, uniformly and randomly select an action in the action sets;
Otherwise, return the action pair $\{\pi_a, \pi_o\}$ obtained in the initialization;
3: **Learn and Update:**
Update $Q_A^k(a, o, s_k)$ according to (18), and $Q_O^k(a, o, s_k)$ similarly
Update the optimal strategies $\pi_A^*(s_k)$ and $\pi_O^*(s_k)$ by

$$\pi_A^*(s_k) \leftarrow \arg \max_{\pi_A(s)} \min_{\pi_O(s)} \sum_{a \in \mathcal{M}_A(s_k)} \sum_{o \in \mathcal{M}_O(s_k)} \pi_a(s_k) Q_A^k(a, o, s_k) \pi_o(s_k),$$

$$\pi_O^*(s_k) \leftarrow \arg \min_{\pi_O(s_k)} \max_{\pi_A(s_k)} \sum_{a \in \mathcal{M}_A(s_k)} \sum_{o \in \mathcal{M}_O(s_k)} \pi_a(s_k) Q_O^k(a, o, s_k) \pi_o(s_k)$$

Update $V_A(s_k)$ and $V_O(s_k)$ according to (14) and (16),
Update $\alpha(k+1) \leftarrow \frac{1}{k+1}$;
4: **The system transits to the next state s_{k+1};**
5: **If all states' policies have converged, stop; otherwise, go to step 1.**

Proof In Algorithm 1, we have that the decaying learning rate $\alpha(k)$ is equal to $\frac{1}{k+1}$. Therefore, we can see that $0 < \alpha(k) < 1$, and $\sum_{k=1}^{\infty} (\alpha(k))^2 = \sum_{k=1}^{\infty} (\frac{1}{k+1})^2 < \sum_{k=1}^{\infty} (\frac{1}{k+1} \frac{1}{k}) = \sum_{k=1}^{\infty} (\frac{1}{k} - \frac{1}{k+1}) < \infty$.

For the attacker, by substituting (18) into (14), we get that for any $s \in \mathcal{S}$,

$$\begin{aligned} & V_A^k(s) \\ = & \max_{\pi_A(s)} \min_{\pi_O(s)} \sum_{a \in \mathcal{M}_A(s)} \sum_{o \in \mathcal{M}_O(s)} \pi_a(s) \cdot \big[(1 - \alpha(k)) \cdot \\ & Q_A^{k-1}(a, o, s) + \alpha(k) \cdot (\mathcal{U}(a, o, s) + \gamma V_A^{k-1}(s'))\big] \cdot \pi_o(s) \\ = & (1 - \alpha(k)) V_A^{k-1}(s) + \alpha(k) \max_{\pi_A(s)} \min_{\pi_O(s)} \sum_{a \in \mathcal{M}_A(s)} \sum_{o \in \mathcal{M}_O(s)} \\ & \pi_a(s) \big(\mathcal{U}(a, o, s) + \gamma V_A^{k-1}(s')\big) \pi_o(s). \end{aligned}$$

Define a mapping function T^k as

$$T^k V_A^k(s) = \mathbb{E}_{s'}\Bigg[\max_{\pi_A(s)} \min_{\pi_O(s)} \sum_{a \in \mathcal{M}_A(s)} \sum_{o \in \mathcal{M}_O(s)} \pi_a(s) \cdot \\ \big(\mathcal{U}(a, o, s) + \gamma V_A^{k-1}(s')\big)\pi_o(s)\Bigg].$$

According to the Conditional Averaging Lemma, we can know that as the iterations in Algorithm 1 continue, $V_A^k(s)$ converges to $T^k V_A^k(s)$.

Next, we show that $T^k V_A^k(s)$ converges to the optimal value. Specifically, we can rewrite $T^k V_A^k(s)$ into:

$$\begin{aligned} T^k V_A^k(s) &= \max_{\pi_A(s)} \min_{\pi_O(s)} \sum_{a \in \mathcal{M}_A(s)} \sum_{o \in \mathcal{M}_O(s)} \pi_a(s) \cdot \\ &\quad \sum_{s' \in \mathcal{S}} T(a, o, s, s')\big(\mathcal{U}(a, o, s) + \gamma V_A^{k-1}(s')\big)\pi_o(s) \\ &= \max_{\pi_A(s)} \min_{\pi_O(s)} \sum_{a \in \mathcal{M}_A(s)} \sum_{o \in \mathcal{M}_O(s)} \pi_a(s) \cdot \\ &\quad \big(\mathcal{U}(a, o, s) + \gamma \sum_{s' \in \mathcal{S}} V_A^{k-1}(s') T(a, o, s, s')\big)\pi_o(s). \end{aligned}$$

We define another mapping function Z^{k-1} as

$$Z^{k-1} V_A^{k-1}(s) = \pi_a(s)\big(\mathcal{U}(a, o, s) + \\ \gamma \sum_{s' \in \mathcal{S}} V_A^{k-1}(s') T(a, o, s, s')\big)\pi_o(s).$$

Z^{k-1} has been proved to be a contraction mapping in [41]. Therefore, $T^k V_A^k(s)$ is a contraction mapping as well.

Thus, we have

$$\begin{aligned} T^k (V_A^k)^*(s) &= \sum_{a \in \mathcal{M}_A(s)} \sum_{o \in \mathcal{M}_O(s)} \pi_a^*(s) \cdot \\ &\quad \big(\mathcal{U}(a, o, s) + \gamma \sum_{s' \in \mathcal{S}} V_A^{k-1}(s') T(a, o, s, s')\big)\pi_o^*(s) \\ &= (V_A^k)^*(s), \end{aligned}$$

which means that $(V_A^k)^*(s)$ is the fixed point of T^k. According to Theorem 1 in [40], $V_A^k(s)$ converges to $(V_A^k)^*(s)$, i.e., $V^*(s)$, with probability 1.

Similarly, we can prove that $V_O^k(s)$ converges to $V^*(s)$ with probability 1 as well. Thus, this theorem directly follows. □

5 Simulation Results

In this section, we conduct extensive simulations to demonstrate the efficacy and efficiency of the proposed scheme. We first demonstrate the convergence of our proposed Q-CFA algorithm in different systems. Then, we analyze the system operator's optimal strategies in different scenarios. Finally, we compare the system operator's expected long-term cost in our scheme with that in other existing schemes.

5.1 Convergence of Q-CFA

We first study the convergence of the proposed Q-CFA algorithm using the IEEE standard 9-bus, 30-bus and 118-bus systems, respectively, and the MATPOWER toolbox [42]. As IEEE 118 bus test system does not include flow limits, we employ the flow limits in Table 3 (the transmission line data) in [43]. In Fig. 2 we show the configuration of standard IEEE bus systems used in our experiments. To initialize the simulation, we set the transition probabilities $p_{uw} = 0.5$, $p_{uwr} = 0.3$, $p_{wu} = 0.5$, $p_{uwa} = 0.3$, the discounting factor $\gamma = 0.3$ and the exploration probability $p_{exp} = 0.6$. For illustrative purposes, we consider that the resources of each player are normalized to one, particularly, each player can affect one transmission line in one time slot. Because each transmission line is of different importance to the entire system, we set different load shedding cost for each line. Specifically, we define the load shedding cost as a linear function of the amount of shed loads on line l and is given by

$$u_l(\hat{d}_l) = c_l\hat{d}_l, \tag{28}$$

where c_l is a given positive constant for line l. We conduct experiments on a desktop with a 3.41 GHz i7-6700 CPU, 16 GB RAM and a 1 TB hard disk drive. To demonstrate the convergence of our proposed Q-CFA, we show in Figs. 3, 4, and 5 the learning curves of the system operator's and the attacker's strategies at certain states in the IEEE 9-bus, 30-bus, and 118-bus systems, respectively. For instance, line 3 and line 7 are the most important lines in the IEEE 9-bus system, which become the main targets in the players' optimal strategies as shown in Fig. 3. In particular, the attacker and the system operator tend to attack and

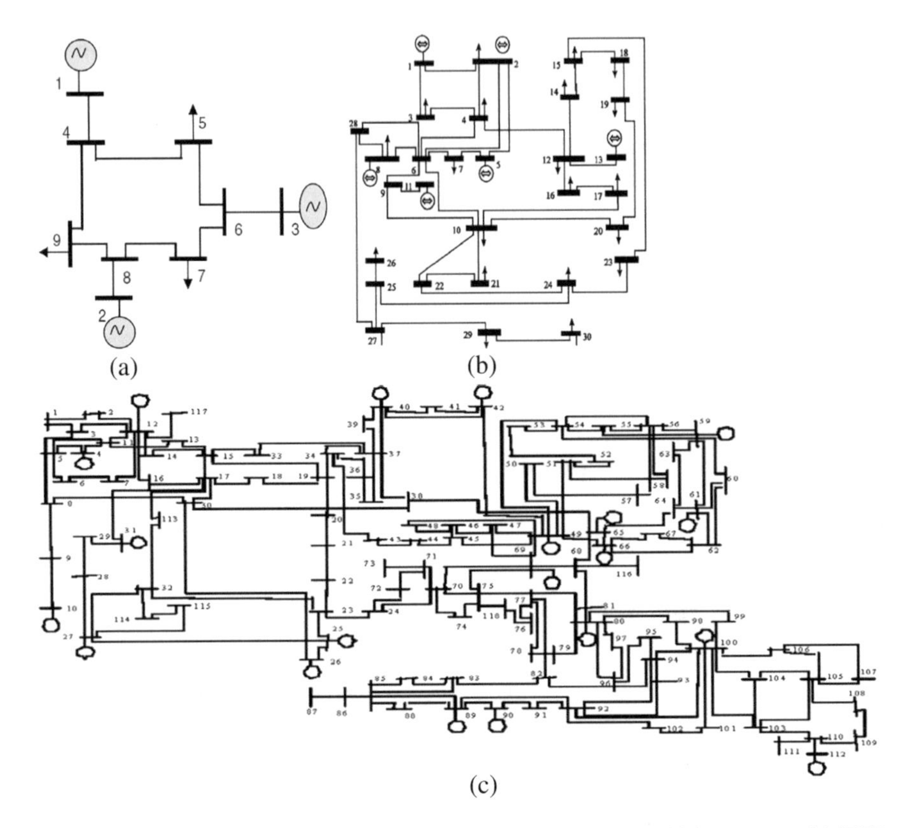

Fig. 2 IEEE standard bus systems. (**a**) IEEE 9-bus system. (**b**) IEEE 30-bus system. (**c**) IEEE 118-bus system

defend, respectively, the transmission line 7 when the game starts. It indicates that when all the transmission lines are well functioning, the most critical line in the IEEE 9-bus system is the line 7. As the iteration goes by, both the attacker and defender's strategies converge and the obtained strategies are stationary, which means the mixed strategies do not change over time. When the state of the game transits to state 7 where line 7 is malfunctioning, as shown in Fig. 3c, d, we can see that the system operator is more likely to repair line 7 but the attacker more likely turns to attack line 3. We can also observe similar results in the IEEE 30-bus and 118-bus systems. Noticeably, from Figs. 3, 4, and 5 we can find that both players' strategies converge within 200, 250, 400 iterations in the IEEE 9-bus, 30-bus and 118-bus systems, respectively. Since we have proved that the converged strategies are the Nash equilibrium points, the results in the simulation are optimal under dynamic environments. Moreover, from a game-theoretic perspective, the strategies obtained by our proposed algorithm will serve as guidance for the system operator to deploy either reinforcement or repair on system components in different system configuration under the condition that the attacker targets the most critical

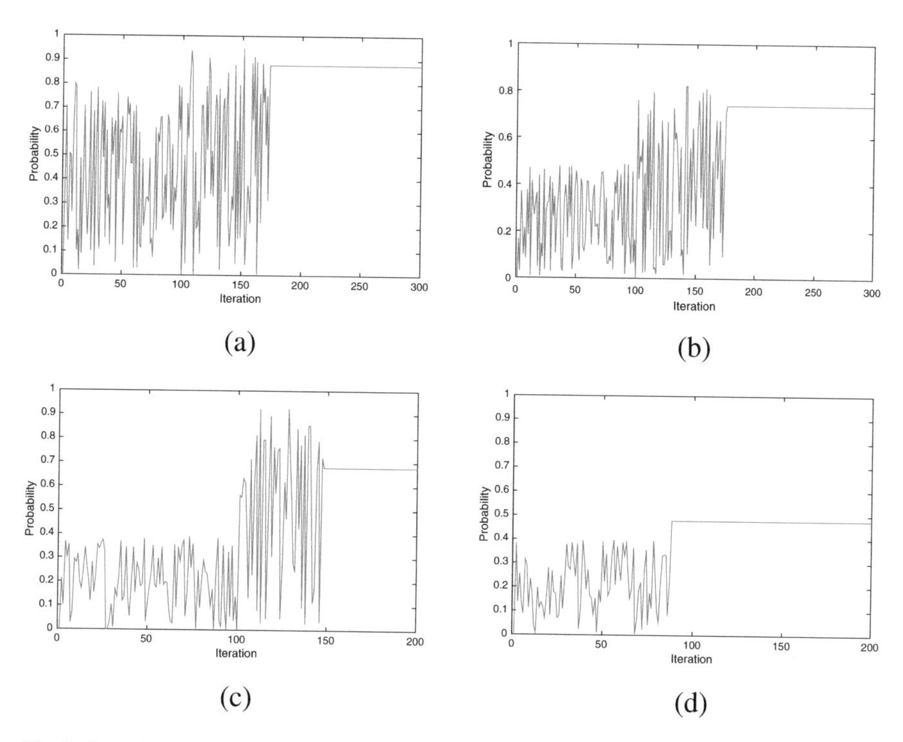

Fig. 3 Learning curves of the attacker and the system operator in the IEEE 9-bus system. (**a**) Attacker's strategy on line 7 at state 0. (**b**) System operator's strategy on line 7 at state 0. (**c**) Attacker's strategy on line 3 at state 7. (**d**) System operator's strategy on line 7 at state 7

system components. By doing so, the system operator can reduce the risk of having cascading failures, and hence the expected long-term costs.

Besides, the computing time of our proposed algorithm is dominated by the solution of a linear program (LP), i.e., (25)–(27), at every iteration. For example, one iteration of our algorithm for the 118-bus system takes 2.72 s of which 2.57 s are due to the solution of the LP. Therefore, the computational complexity of our algorithm depends on the size of the LP and the number of iterations. Specifically, according to (25)–(27), the number of variables and constraints in the LP, depends on the number of actions, which grows linearly with the number of lines in the system. By employing the Simplex algorithm, the complexity of solving one LP is a polynomial function of the number of lines in the system. Moreover, according to our simulation results, the number of iterations of our algorithm also grows linearly with the number of lines in the system. Therefore, the overall computational complexity of our algorithm is a polynomial function of the number of lines in the system. Moreover, in Figs. 3, 4, and 5, the number of iterations needed for convergence does not linearly increase as the number of system buses increases, which makes our algorithm scalable even for large systems.

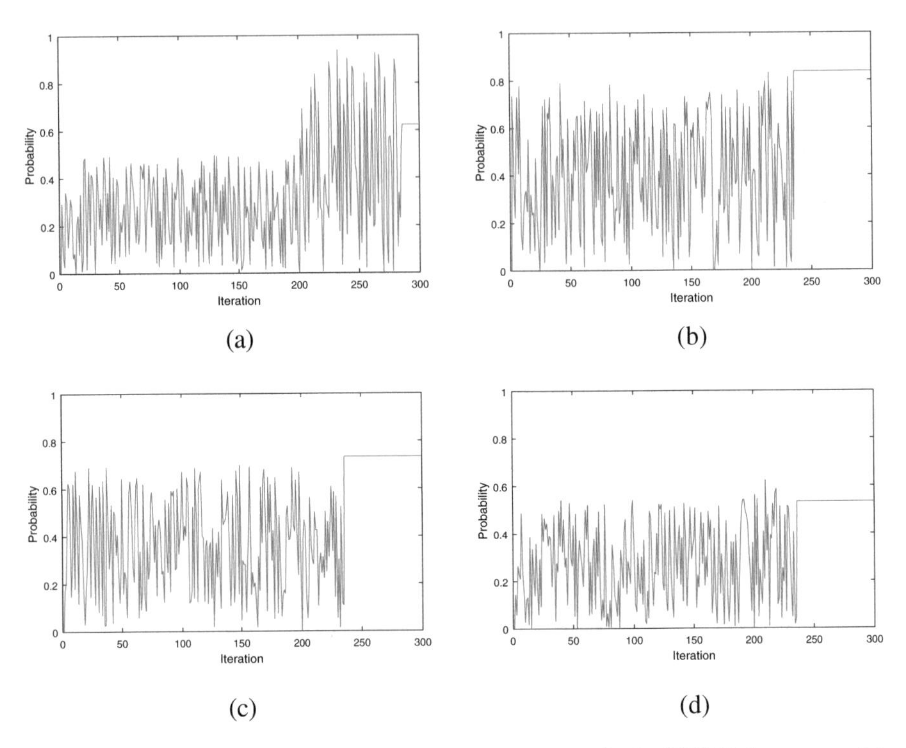

Fig. 4 Learning curves of the attacker and the system operator in the IEEE 30-bus system. (**a**) Attacker's strategy on line 27 at state 0. (**b**) System operator's strategy on line 29 at state 0. (**c**) Attacker's strategy on line 16 at state 27. (**d**) System operator's strategy on line 27 at state 27

5.2 *Strategy Analysis*

Next, we analyze the system operator's optimal strategies in the stochastic game when the discount factor γ varies, with γ being equal to 0, 0.3, 0.8. Recall that $\gamma \in [0, 1)$ represents the impact that current decisions can have on the long-term reward. Particularly, when γ equals 0, the game becomes a static game. When γ is larger than 0, a smaller value of γ emphasizes more on the immediate rewards and a larger γ gives a higher weight to the future rewards. In Fig. 6a, compared with the results in the static game where $\gamma = 0$, the performance in the stochastic games where $\gamma > 0$ is much better. This is because in the stochastic games, players not only care about current rewards, but also take the future states into consideration. By considering both the current and future rewards, players are able to obtain optimal expected long-term rewards. In addition, we can see that the higher γ is, the lower expected long-term load shedding cost the system operator can achieve. This is because when γ increases, the system operator emphasizes more on the future states and can better react to the dynamic environments, which result in more savings in the long-term cost. On the other hand, Fig. 6a, b together demonstrate the tradeoff

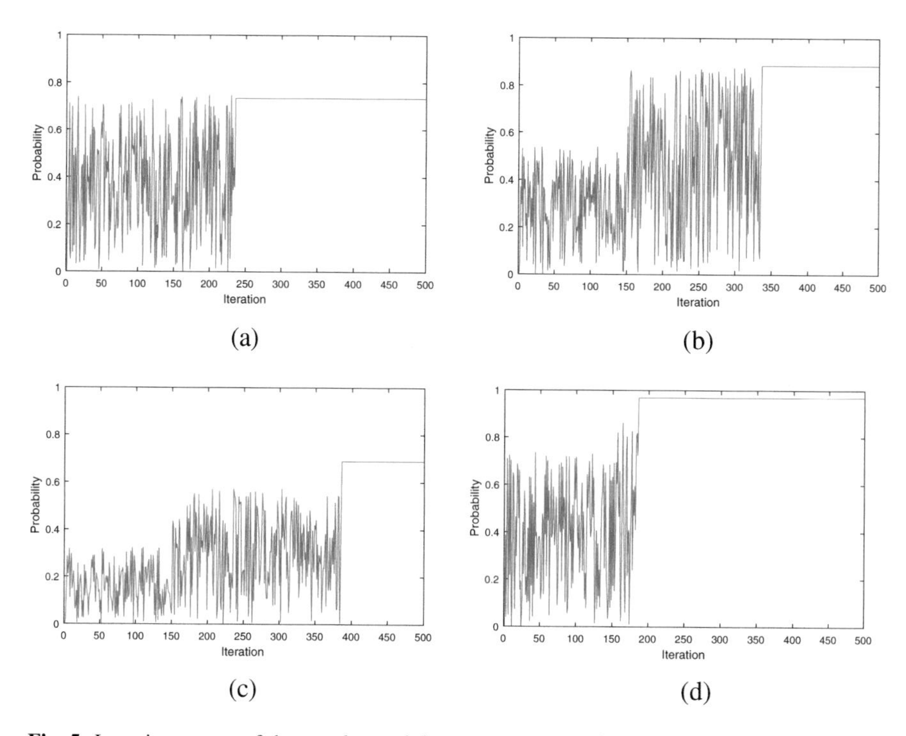

Fig. 5 Learning curves of the attacker and the system operator in the IEEE 118-bus system. (**a**) Attacker's strategy on line 9 at state 0. (**b**) System operator's strategy on line 7 at state 0. (**c**) Attacker's strategy on line 8 at state 9. (**d**) System operator's strategy on line 9 at state 9

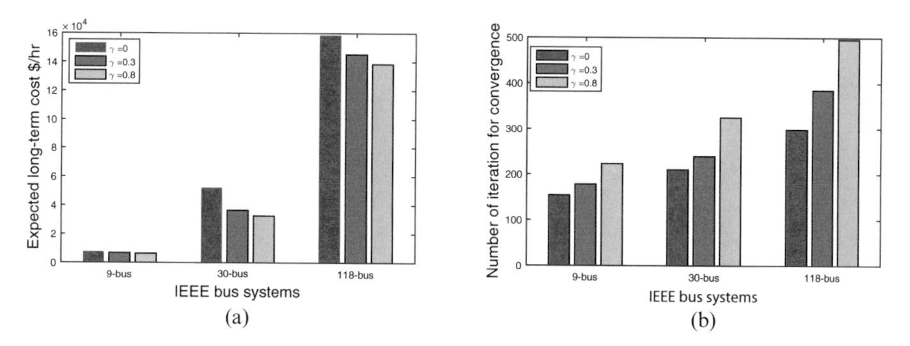

Fig. 6 Strategy analysis for stochastic game. (**a**) Performance analysis with regard to different γ. (**b**) Convergence analysis with regard to different γ

between performance and computational cost. As shown in Fig. 6b, the number of iterations needed for convergence increases as γ increases. This is because when we emphasize more on the future rewards, it takes more iterations to search for the optimal solution.

5.3 Performance Comparison

Finally, from the system operator's perspective, we compare the performance of the optimal strategies obtained by our Q-CFA algorithm with that of two other strategies, i.e., the fixed strategy and the myopic learning strategy. In particular, in the fixed strategy, the system operator will draw an action o uniformly from the available action space, i.e., $\mathcal{M}_O(s)$, for each state s. In the myopic learning strategy where the game is a static game ($\gamma = 0$), the system operator only considers immediate rewards and ignores the impact of the current action on future rewards. Note that it is of paramount importance to select initiating events in each algorithm because it allows the attacker to determine if the initial event can cause a cascading failure. In the three benchmark algorithms, the selections of "important line" are different. In particular, our proposed scheme optimizes the expected long-term rewards, so the selection of initiating events takes the opponent's strategy and the dynamic environments into consideration. However, as the myopic strategy is a static-game strategy, selection of initiating event only considers the opponent's strategy in current state and the strategy can be explained as trying to launch a one-time attack to cause cascading failure and achieve the maximum immediate reward. On the other hand, the fixed strategy is a uniform strategy for comparison. So the selection of initial event is uniformly distributed. We compare the optimal expected long-term cost in these three strategies in Fig. 7.

We can find that the optimal costs obtained by our proposed Q-CFA and the myopic learning strategy are much lower than that obtained by the fixed strategy. This is because both of our proposed Q-CFA and the myopic learning strategy try to minimize the attacker's maximal reward, while the fixed strategy only uniformly

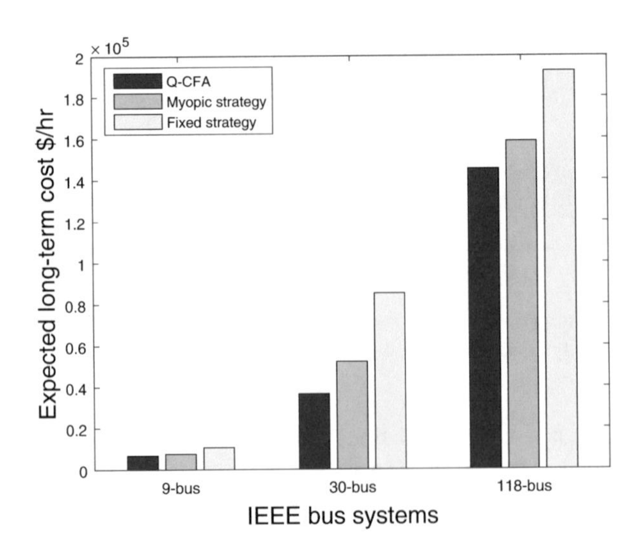

Fig. 7 Performance comparison among three strategies

chooses actions from the available action set without taking the opponent's possible strategies into consideration. In addition, because our Q-CFA algorithm optimizes the expected long-term reward while the myopic learning strategy only focuses on optimizing the strategies at the current state, our scheme outperforms the myopic learning strategy in the long run. Therefore, as a power system operator, adopting our proposed Q-CFA algorithm to defend the power system can both adapt to the dynamic state changes and attacker's intelligent strategies, which results in the best performance in the long run.

6 Conclusions

In this chapter, we have investigated cascading failure attacks (CFAs) in power systems. Specifically, we have formulated a zero-sum stochastic game to analyze the interactions between an attacker and a system operator in dynamic environments for power systems. This problem is very complex and computationally intensive. Different from previous work where complete enumeration of the system states is required, making the algorithms computationally intractable for large-scale power system applications, we propose an efficient Q-CFA learning algorithm that only searches certain related possible actions for each player in the game, making the scheme scalable with fast convergence. We have also theoretically proven that the proposed algorithm achieves the Nash equilibrium. Moreover, considering that real-time statistics and sensitive data like system transition probabilities may not be accessible in practice, which unfortunately is an indispensable assumption in previous algorithms, our scheme works efficiently without requiring *a priori* knowledge of the system transition states. Simulation results show that by considering the system dynamics and the opponent's possible strategies, the optimal policy obtained by our proposed Q-CFA algorithm can achieve much better performance compared to other algorithms on several benchmark schemes.

References

1. Internet of things and the myth of the killer app, Jan 2017. Available: https://www.metering.com/magazine_articles/smart-grid-and-iot/
2. Y. Saleem, N. Crespi, M.H. Rehmani, R. Copeland, Internet of things-aided smart grid: technologies, architectures, applications, prototypes, and future research directions (2017). Preprint. arXiv:1704.08977
3. S. Liu, X.P. Liu, A. El Saddik, Denial-of-service (dos) attacks on load frequency control in smart grids, in *IEEE PES Innovative Smart Grid Technologies (ISGT)*, Feb 2013
4. Y. Liu, P. Ning, M.K. Reiter, False data injection attacks against state estimation in electric power grids. ACM Trans. Inf. Syst. Secur. **14**(1), 1–13 (2011)
5. A. Ashok, M. Govindarasu, Cyber attacks on power system state estimation through topology errors, in *IEEE Power and Energy Society General Meeting*, Jul 2012 (IEEE, Piscataway, 2012)

6. I. Group et al., Managing big data for smart grids and smart meters. IBM Corporation, whitepaper, May 2012
7. J. Chen, J.S. Thorp, I. Dobson, Cascading dynamics and mitigation assessment in power system disturbances via a hidden failure model. Int. J. Electr. Power Energy Syst. **27**(4), 318–326 (2005)
8. I. Dobson, B.A. Carreras, V.E. Lynch, D.E. Newman, Complex systems analysis of series of blackouts: cascading failure, critical points, and self-organization. Chaos: Interdiscip. J. Nonlinear Sci. **17**, 026103 (2007)
9. B. Liscouski, W. Elliot, Final report on the Aug. 14, 2003 blackout in the United States and Canada: causes and recommendations. A report to US Department of Energy, vol. 40, no. 4, Dec 2004
10. M. Rahnamay-Naeini, Z. Wang, N. Ghani, A. Mammoli, M. Hayat, Stochastic analysis of cascading-failure dynamics in power grids. IEEE Trans. Power Syst. **29**(4), 1767–1779 (2014)
11. L. Liu, M. Esmalifalak, Q. Ding, V. Emesih, Z. Han, Detecting false data injection attacks on power grid by sparse optimization. IEEE Trans. Smart Grid **5**(2), 612–621 (2014)
12. A.E. Motter, Y.-C. Lai, Cascade-based attacks on complex networks. Phys. Rev. E **66**(6), 1–4 (2002)
13. Y. Zhu, J. Yan, Y. Tang, Y.L. Sun, H. He, Joint substation-transmission line vulnerability assessment against the smart grid. IEEE Trans. Inf. Forensics Secur. **10**(5), 1010–1024 (2015)
14. J. Yan, H. He, X. Zhong, Y. Tang, Q-learning based vulnerability analysis of smart grid against sequential topology attacks. IEEE Trans. Inf. Forensics Secur. **12**(1), 200–210 (2017)
15. Z. Bao, Y. Cao, G. Wang, L. Ding, Analysis of cascading failure in electric grid based on power flow entropy. Phys. Lett. A **373**(34), 3032–3040 (2009)
16. A.J. Holmgren, E. Jenelius, J. Westin, Evaluating strategies for defending electric power networks against antagonistic attacks. IEEE Trans. Power Syst. **22**(1), 76–84 (2007)
17. J. Salmeron, K. Wood, R. Baldick, Analysis of electric grid security under terrorist threat. IEEE Trans. Power Syst. **19**, 905–912 (2004)
18. G. Chen, Z.Y. Dong, D.J. Hill, Y.S. Xue, Exploring reliable strategies for defending power systems against targeted attacks. IEEE Trans. Power Syst. **26**(3), 1000–1009 (2011)
19. N.S. Rao, S.W. Poole, C.Y. Ma, F. He, J. Zhuang, D.K. Yau, Cyber and physical information fusion for infrastructure protection: a game-theoretic approach. Oak Ridge National Laboratory (ORNL), Tech. Rep., Jan 2013
20. C.Y. Ma, D.K. Yau, X. Lou, N.S. Rao, Markov game analysis for attack-defense of power networks under possible misinformation. IEEE Trans. Power Syst. **28**(2), 1676–1686 (2013)
21. O. Adaki, Attack on power lines leaves Yemen in total darkness, Jun 2014. Available: http://www.longwarjournal.org/threat-matrix/archives/2014/06/aqap_attack_on_power_lines_lea.php
22. T. Raghavan, J. Filar, Algorithms for stochastic games, a survey. Z. Oper. Res. **35**(6), 437–472 (1991)
23. D.P. Bertsekas, *Dynamic Programming and Optimal Control*, vols. 1 and 2, 2nd edn. (Athena Scientific, Belmont, 2007)
24. J. Salmeron, K. Wood, R. Baldick, Worst-case interdiction analysis of large-scale electric power grids. IEEE Trans. Power Syst. **24**(1), 96–104 (2009)
25. Y. Wang, R. Baldick, Interdiction analysis of electric grids combining cascading outage and medium-term impacts. IEEE Trans. Power Syst. **PP**(99), 1–9 (2014)
26. D.C. Elizondo, J. de la Ree, A.G. Phadke, S. Horowitz, Hidden failures in protection systems and its impact on power system wide-area disturbances, in *Proceeding of IEEE Power Engineering Society Winter Meeting*, Jan 2001
27. J. Thorp, A. Phadke, S. Horowitz, S. Tamronglak, Anatomy of power system disturbances: importance sampling. Int. J. Electr. Power Energy Syst. **20**(2), 147–152 (1998)
28. J. Chen, J. Thorp, M. Parashar, Analysis of electric power system disturbance data, in *Proceedings of the 34th Annual Hawaii International Conference on System Sciences*, Washington, Jan 2001

29. X. Liu, K. Ren, Y. Yuan, Z. Li, Q. Wang, Optimal budget deployment strategy against power grid interdiction, in *Proceedings of IEEE INFOCOM*, Turin, Apr 2013
30. A. Neyman, S. Sorin, *Stochastic Games and Applications*, vol. 570 (Springer, Berlin, 2003)
31. J. Filar, K. Vrieze, *Competitive Markov Decision Processes* (Springer Science & Business Media, Berlin, 1996)
32. S. Tamronglak, S. Horowitz, A. Phadke, J. Thorp, Anatomy of power system blackouts: preventive relaying strategies. IEEE Trans. Power Deliv. **11**(2), 708–715 (1996)
33. U.S. Energy Information Administration, "International energy outlook," May 2016. Available: https://www.eia.gov/outlooks/ieo/
34. M.L. Littman, Markov games as a framework for multi-agent reinforcement learning, in *Proceedings of the Eleventh International Conference on Machine Learning*, vol. 157, Jul 1994, pp. 157–163
35. L. Busoniu, R. Babuska, B. De Schutter, A comprehensive survey of multiagent reinforcement learning. IEEE Trans. Syst. Man Cybern. C **38**(2), 156–172 (2008)
36. S. Boyd, L. Vandenberghe, *Convex Optimization* (Cambridge University Press, Cambridge, 2004)
37. J.C. Harsanyi, R. Selten, *A General Theory of Equilibrium Selection in Games*, vol. 1 (MIT Press Books, Cambridge, 1988)
38. G. Owen, *Game Theory* (Academic, Cambridge, 1982)
39. Y. Gwon, S. Dastangoo, C. Fossa, H. Kung, Competing mobile network game: embracing antijamming and jamming strategies with reinforcement learning, in *IEEE Conference on Communications and Network Security (CNS)* (IEEE, Piscataway, 2013)
40. C. Szepesvári, M.L. Littman, A unified analysis of value-function-based reinforcement-learning algorithms. Neural Comput. **11**(8), 2017–2060 (1999)
41. J. Hu, M.P. Wellman et al., Multiagent reinforcement learning: theoretical framework and an algorithm, in *International Conference on Machine Learning* (ACM, New York, 1998)
42. R.D. Zimmerman, C.E. Murillo-Sánchez, R.J. Thomas, Matpower: steady-state operations, planning, and analysis tools for power systems research and education. IEEE Trans. Power Syst. **26**(1), 12–19 (2011)
43. IEEE 118 bus case flow limits, Illinois Institute of Technology. Available: http://motor.ece.iit.edu/data/JEAS_IEEE118.doc

The Risk of Botnets in Cyber Physical Systems

Farnaz Derakhshan and Mohammad Ashrafnejad

Abbreviations

ANN	Artificial neuron networks
API	Application programmable interface
C&C	Command and control
DDNS	Dynamic domain name system
DDoS	Distributed denial of service
DGA	Domain generation algorithm
DoS	Denial of service
GAN	Generative adversarial networks
HIDS	Host-based intrusion detection system
HTTP	Hyper text transfer protocol
ICMP	Internet control message protocol
IDS	Intrusion detection system
IoT	Internet of things
IRC	Internet relay chat
NIDS	Network-based intrusion detection system
P2P	Peer-to-peer
Pentester	Penetration tester
SMTP	Simple mail transfer protocol
SVM	Support vector machine
TCP	Transmission control protocol
UDP	User datagram protocol

F. Derakhshan (✉) · M. Ashrafnejad
University of Tabriz, Tabriz, Iran
e-mail: derakhshan@tabrizu.ac.ir

H. Karimipour et al. (eds.), *Security of Cyber-Physical Systems*,
https://doi.org/10.1007/978-3-030-45541-5_5

1 Introduction

In recent years, the growth in the use of computers has raised threats in this field, and the need for research on cybersecurity is more than ever for defending against crackers hacking different computer systems all over the world [1, 2].

Statistics [3] indicate that more than 350,000 new malwares are recognized annually, which show 18% increase in 2019 in comparison with 2018. In Fig. 1, AV-TEST Institute shows number of discovered malwares with annually for past 10 years.

In this chapter, we specifically focus on the detection of botnet malwares. Botnet is one of the newest online malwares. According to industry estimates, botnets have caused over 9 billion dollars in losses to U.S. victims and over 110 billion dollars in losses globally. Botnets infect almost 500 computers annually [5]. Based on these statistics, researchers feel more motivated to work in this field than in other areas. In order to examine methods for detecting and confronting botnets, it is necessary to get acquainted with their characteristics and structure; afterwards it is possible to analyze methods and algorithms of botnet detection.

The word "botnet" is a blend of two other words "robot" and "network" [6], which stands for the machines working on network platform pursuing a specific objective. Basically, bot refers to an infected device that is compromised by malware

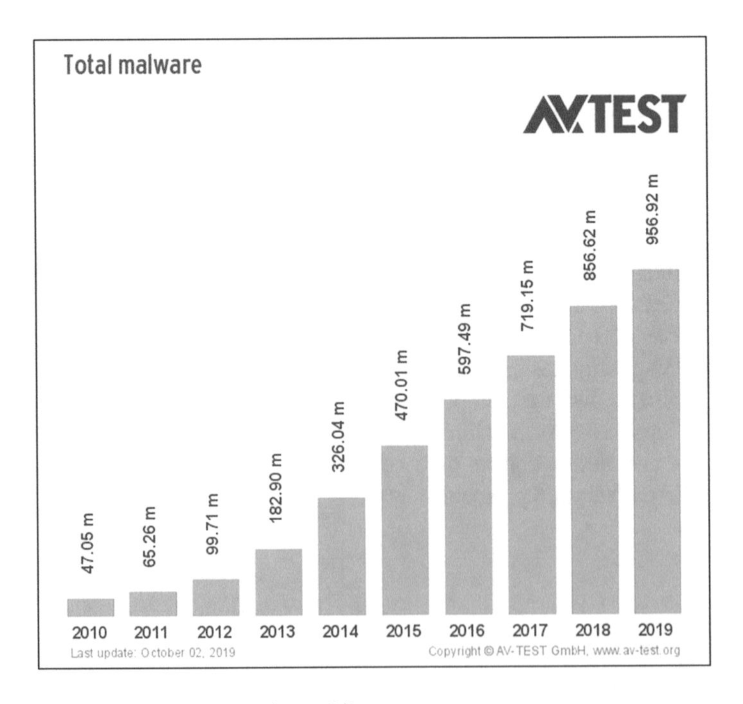

Fig. 1 No. of detected malwares each year [4]

through one of various ways (such as a vulnerable operating system or software). This malware, which is generally a script file, takes control of the machine without the user's awareness. Botnets are mostly applied in DDoS attacks, phishing, click fraud, and bitcoin mining.

Intrusion detection systems (IDS) encounter difficulties detecting botnets since they exchange a normal amount of traffic. Besides, botnets use different ways such as cryptography to hide their content connections; hence, the detection process becomes more complicated [7–9].

We outlined the structure of this chapter as follows:

In Sect. 2, we get acquainted with features and structure of botnets to analyze methods and algorithms to detect and confront them more efficiently. In Sect. 3, we discuss related works in botnet detection field. In Sect. 4, we will study some of the critical malware detection challenges that can be used in future works.

2 What Are Botnets?

In this section, we analyze the theoretical concept of botnets. Primarily, we get familiar with definitions, history, and life cycle of botnets; after that, we will generally explain topology and communication protocols of botnets.

2.1 An Introduction to Botnets

Since botnets are subcategories of malwares; we need to define malwares prior to analyzing botnets.

2.1.1 Malware Definition

Malware is a malicious software which damages computer sources through stealing their vital data and gaining access to the victim system [10, 11].

You can find a short description of some malwares in the following:

- *Worm*: It is a malware which infects the operating system, then procreates itself and send the copies through network platform [12].
- *Virus*: It is a type of malware that normally spreads to systems through an infected tool or a file, e.g. an infected removable disk [13].
- *Trojan*: It acts as a regular application and attracts the user to run it; after the user runs the Trojan application, Trojan takes control of the system resources and disrupts the operating system services [12].
- *Rootkit*: It is a malware, without the user's knowledge, takes over the control of the operating system, and permits hackers to access to all the settings [14].

2.1.2 Botnet Definition

In this section, we present the definition of botnets. Researchers proposed many definitions, which some features are shared. So the definition we are going to provide covers all common features.

In the following, we provide some definitions of botnets:

- A botnet is an infected network of computers by malware that is used for malicious activities [15].
- A botnet includes a large number of compromised computers with powerful computing capacity that enables hackers to operate major attacks using its high bandwidth. Nowadays, botnets are significant threats to cyber security of governments, industries, universities, and basic infrastructures [16].
- A network of computers controlled by hackers used for malicious purposes is called a botnet [17].

Different references proposed numerous definitions for botnets, distinguished botnets from other malwares such as Trojan in terms of connection-making capability. But these days, most of the advanced malwares have mentioned feature; thus, in this study, we will call a malware a botnet if it covers below three features:

- Connection-making in the network platform
- Purposeful attacks
- Consistency
- Hiding capability

2.2 *The History of Botnets*

According to [18], Greg Lindahl provided first botnet, GM, in 1989. It was a benevolent botnet that would play a game of "Hunt the Wumpus" with IRC users. In 1993, a botnet named *eggdrop* was developed, which used the IRC protocol.

In 1998, the first malicious botnet by the name GTBot was developed, which counted as one of the first botnets used solely for DDoS attacks [19].

2.2.1 Well-Known Botnets

In this section, the characteristics of some famous botnets will be explained. All features have been extracted from various surveys and trusted sources. As a matter of fact, in different sources, diverse features have been mentioned for botnets, but we have done our best to use the newest and the most trusted ones (Table 1). Also, a timeline of famous botnets is provided (Fig. 2).

Table 1 Historical list of botnets

No.	Name	Year	No. of bots	Topology	Architecture
1	Eggdrop	1993		Central	IRC
2	GTBot	1998	140,000	Central	IRC
3	SDBot	2002	7000	Central	IRC
4	AgoBot	2003		Central	IRC
5	SpyBot	2003		Central	IRC
6	Sinit	2003		P2P	ICMP
7	Slapper	2003		P2P	
8	RBot	2003		Central	IRC
9	Bagle	2004	230,000	Central	SMTP
10	Phatbot	2004		P2P	HTTP/IRC
11	Bobax	2004	100,000	Central	HTTP
12	Torpig	2004	180,000	Central	HTTP/IRC
13	SpamThru	2006	12,000	Hybrid	IRC
14	Rustock	2006	150,000	Central	IRC/HTTP
15	Nugache	2006	160,000	P2P	IRC
16	Mega-D	2006	250,000	P2P	HTTP
17	Donbot	2006	125,000	Central	HTTP/TCP
18	Jrbot	2006	–	Central	IRC
19	Rxbot	2006	–	Central	IRC
20	Storm	2007	160,000	Hybrid	
21	Zeus	2007	3,600,000	Central	UDP, TCP, HTTP
22	Cutwail	2007	1,500,000	–	SMTP/encrypted HTTP
23	Akbot	2007	1,300,000		IRC
24	Srizbi	2007	450,000	Central	HTTP/IRC
25	Lethic	2008	260,000	Central	IRC
26	MayDay	2008		P2P	HTTP
27	Waledac	2008	80,000	Hybrid	HTTP
28	Mariposa/buttery	2008	13,000,000	P2P	IRC/HTTP
29	Conficker	2008	+10,500,000	P2P	HTTP
30	Sality	2008	1,000,000	P2P	Email, HTTP, UDP, TCP
31	BredoLab	2009	30,000,000		HTTP/SMTP
32	Grum	2009	560,000	Central	HTTP
33	Festi	2009	25,000	Central	HTTP
34	Stuxnet	2010		P2P	HTTP
35	Kelihos	2010	+300,000	Hybrid	
36	Ramnit	2011			HTTP
37	ZeroAccess	2011	9,000,000	P2P	TCP/UDP
38	TDL-4	2011	4,500,000	P2P	
39	GameOver Zeus	2011	3,600,000	P2P	HTTP
40	Wordpress/QBot	2013	500,000		
41	newGOZ	2014		P2P	
42	Mirai	2016			

Fig. 2 Timeline of botnets

- *Eggdrop*: It revolutionized the concept of the botnet, as first bot using IRC protocol [19, 20].
- *GTBot*: It has been known as the first malicious botnet [19, 21].
- *SDBot*: It uses advanced key-logging software to collect personal information [6].
- *AgoBot*: It is a persistent and modular malware which is particularly flexible [15]
- *SpyBot*: It is a malware driven from AgoBot [15].
- *Sinit*: It is the first P2P botnet using random protocol for communication, due to weak connection and high probing traffic, its detection is easy [15, 22].
- *Slapper*: It lets the nodes to use encrypted algorithms for routing. In fact, the botmaster encrypts a command with a key and sends it to the network, the only node having the key can run the command [22, 23].
- *RBot*: It is an IRC-based botnet written in Ruby.
- *Bagle*: It spams and affects windows operating systems [24].

- *Phatbot*: It is an AgoBot type of malware using Gnutella cache server. It deactivates the antivirus and is used to send spam emails [17, 22, 23].
- *Bobax*: It runs on particular versions of Windows which attaches itself to emails collected from victim systems [19, 25].
- *Torpig*: It is used to collect personal and organizational data [26, 27].
- *SpamThru*: It uses a customized p2p protocol to share data [17].
- *Rustock*: It was responsible for sending 30 billion messages per day in 2010 [19, 24].
- *Nugache*: It is hard to detect and very flexible due to the absence of C&C server which is used to infect emails. The shortcoming of this malware is its dependency on the botmaster to launch.
- *Mega-D*: It was responsible for sending 33% of all spam emails at 2008 based on [15], and according to [17], responsible for sending 10 billion spam emails every day. As in [19], it was responsible for sending 58.3% of all spam emails.
- *Donbot*: This botnet is especially used to send pharmaceutical and stock-based e-mail spam [15].
- *Jrbot*: It enables unauthorized log-in and hands control of the system to attackers [15].
- *Rxbot*: It works on windows platform used for DDoS attacks [15].
- *Storm*: It is one of the strongest botnets capable of DDoS attacks via fast-flux [17, 23].
- *Zeus*: It used many C&C servers in early versions, but its structure changed from central to a hybrid in later versions [23].
- *Cutwail*: It is used to send spam emails, and also responsible for sending 17% of spam emails until the end of 2009 [19, 24].
- *Akbot*: It was written by an 18-year-old hacker infecting more than one million systems [24].
- *Srizbi*: It is able to secretly run in kernels of compromised systems and bypass the firewall using rootkit techniques [19].
- *Lethic*: It was mostly used for sending pharmaceutical spam emails [15, 24].
- *MayDay*: It uses encrypted ICMP traffic and steals browser proxy [17].
- *Waledac*: It is one of the most advanced p2p botnets which spreads via social engineering and client vulnerability. It is used for sending spam emails, DDoS and other attacks [15, 17, 19, 28].
- *Mariposa/Buttery*: It is a self-spreading, code injecting, and anti-debugging malware as its main features [17].
- *Conficker*: It has infected systems in more than 200 countries [15, 24].
- *Sality*: It was used to steal data especially passwords in early versions, and it is hard to detect due to the use of adjacent C&C codes that update in every 40 min [23, 24].
- *BredoLab*: It mainly downloads other malwares in infected computers [1, 24].
- *Grum*: It is a two-variable malware, one for configuration updates and the other for botnet formation, used to send spam emails [24].
- *Festi*: It is used to send spam emails and carry out DDoS attacks [24].

- *Stuxnet*: It was responsible for causing substantial damage to Iran's nuclear program which spread using accessory storage devices [20, 23, 29].
- *Kelihos*: It is majorly used for carrying out DDoS attacks and sending spam emails, also used for stealing bitcoin wallets [24, 30].
- *Ramnit*: It is a spying malware used to steal data from financial and social media accounts; it can disable windows firewall as well [31].
- *ZeroAccess*: It is a malware composed of the restricted number of C&C codes only used for updating bots and loading Trojan [17, 23].
- *TDL-4*: It is one of the newest and most complex botnets using DGA algorithm and cryptography to be stable. It infected about 4.5 million systems within the first 3 months of 2011 [15, 17].
- *GameOver Zeus*: It is one of the largest botnets on Zeus platform that is used for stealing emails and social media accounts log-in data [3, 17].
- *Wordpress/QBot*: It is a malware for cracking windows passwords and sniffing transactions [17].
- *newGOZ*: It is the malware using DGA algorithms to produce domain names. It can create 1000 domain names every day through Cutwail templates in sending spam emails [17].
- *Mirai*: It is a This malware is mostly used to infect IoT systems, which is estimated to affect more than 20 billion devices until 2020 [32].

2.3 The Structure of Botnets

Every botnet is composed of three main elements [23]:

- A number of bots
- Command & Control Server (C&C)
- BotMaster

A botmaster uses a particular communication platform called C&C, through which commands are sent to other bots. So the detection process of botmaster becomes a very complicated procedure.

To evade detection, the BotMaster can optionally employ a number of proxy machines, called stepping-stones, between the C&C server and itself (Fig. 3). Devices are infected by means of a malicious executable program referred to as bot binary. Bots belonging to the same botnet form the bot family. The ultimate goal of a botnet is to carry out malicious activities or attacks on behalf of its controller [33].

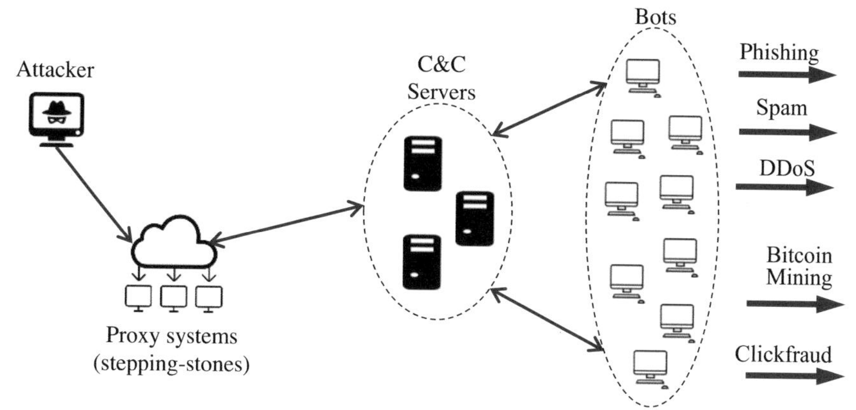

Fig. 3 A graphical picture of a botnet structure

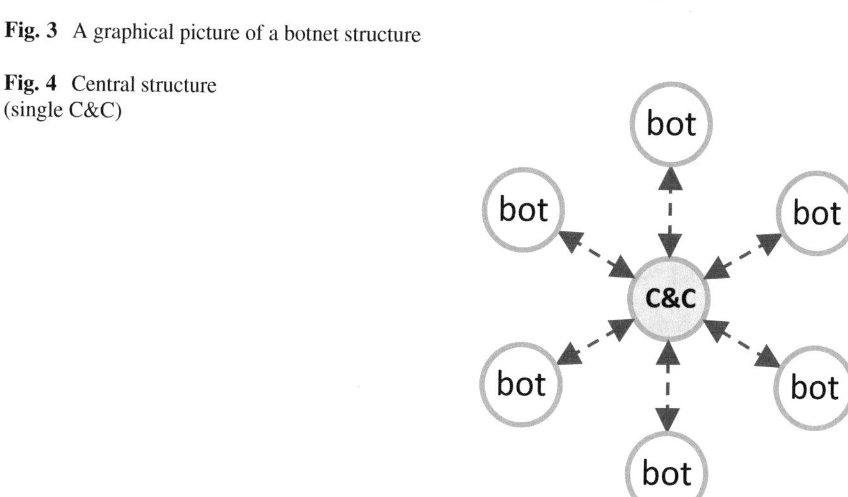

Fig. 4 Central structure (single C&C)

2.4 Botnets Topology

Botnet topology is an organized network structure, through which bots communicate together.

2.4.1 Centralized Topology

The most straightforward communication pattern of botnets is called central topology. In this topology, one or more central C&C servers are used that bots connect to them directly (Figs. 4 and 5). Which, there are a low delay and high scalability [15].

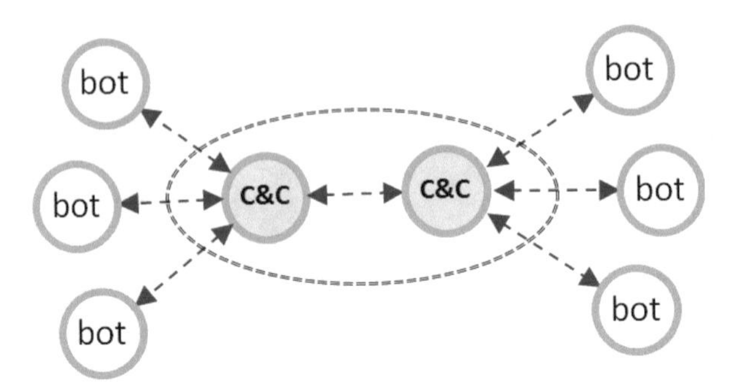

Fig. 5 Central structure (multi C&C)

2.4.2 Decentralized Topology

This type of topology is precisely the opposite of the central type. In P2P, there is no botmaster and commands transfer from one bot to another. Each bot has an address list of adjacent bots, with which sends and receives commands; in this way commands spread all over the network. Despite the difficulty in creating a P2P topology, they are far harder to detect than central types. In fact, in P2P type we could say C&C server is hidden [28].

Decentralized topologies are of two models:

- *Distributed Model*: In this type, many servers control a group of bots each. These C&C servers are directly connected to each other and are located in various geographical spots. Advantages of this model are flexibility, traffic distribution, and accessibility; so if a server is down, its bots will be distributed among other servers (Fig. 6).
- *Random Model*: This model was proposed by E. Cooke [34]. In this model, bots are not permanently connected with botmaster and/or other bots; rather they wait for communication signal from botmaster. In this way, botmaster regularly scans the network to discover existing bots and sends commands to the discovered ones (Fig. 7).

2.5 Hybrid Topology (Unstructured Topology)

This type is a mixture of above models and covers each one's shortcomings (Fig. 8).

In Table 2, advantages and shortcomings of above models have been analyzed.

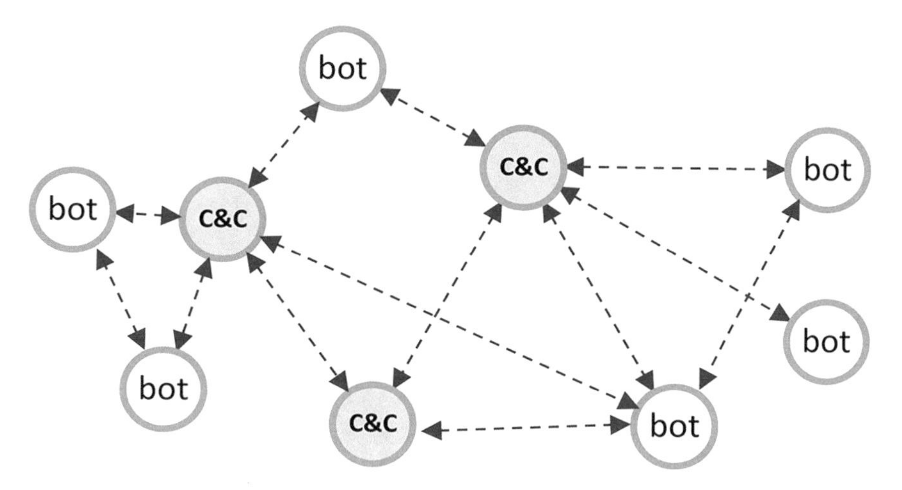

Fig. 6 Distributed model

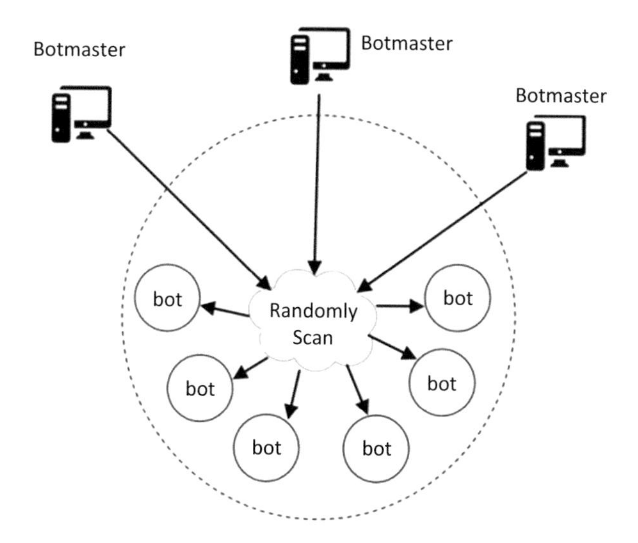

Fig. 7 Random topology

2.6 Botnets Communicative Protocols

A botmaster sends commands to bots through one of IRC, HTTP, DNS, Bluetooth and other protocols [28].

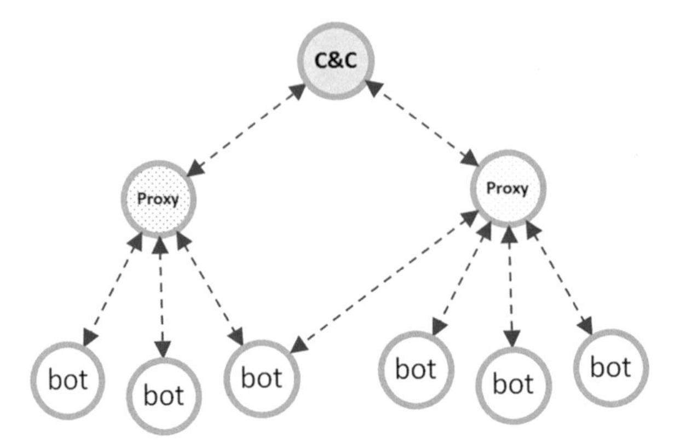

Fig. 8 Hybrid topology

Table 2 C&C structures and basic properties [18]

Topology	Complexity	Detectability	Message latency	Survivability
Centralized	Low	Medium	Low	Low
Decentralized (P2P)	Medium	Low	Medium	Medium
Untrusted	Low	High	High	High

2.6.1 IRC Protocol

This protocol is used to send text messages over the internet. The first version of IRC was designed in 1993 by the name RFC 1459 (Illustrated in Fig. 9); this version works on a client-server platform with four popular messages mentioned further [35]:

- *Join message*: It is used to start listening to a specific channel.
- *Ping message*: It is used to test live machines in a network.
- *Pong message*: It is a reply to a ping message.
- *Error message*: It is a report of a serious or fatal error to operators.

2.6.2 HTTP Protocol

This protocol is designed for websites, used to hide malicious activities by the botmaster. Since malicious commands disguise in ordinary HTTP messages, they are difficult to detect [23].

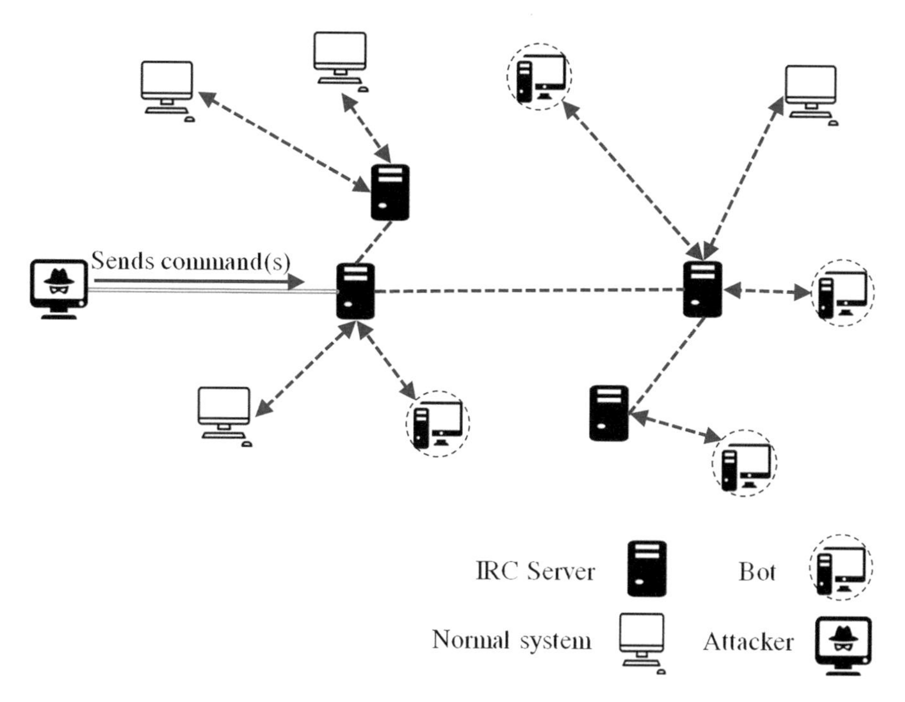

Fig. 9 IRC structure

2.6.3 POP3 Protocol

This protocol is designed to support offline mail processing and works best for those who use single computers all the time [19].

2.6.4 Other Protocols

There are different protocols such as IMAP that can be used by botnets, but botnet designers do not use them, due to their complexity in comparison with previously mentioned protocols.

2.7 The Application of Botnets

So far we got acquainted with structure of botnets in detail. In this section, we will mention some uses of botnets [36]:

- *Click fraud*: Nowadays, ads play an important role in business. Website owners are paid based on the number of clicks on ads. The attacker can increase the number of clicks on ads using infected systems.
- *DDoS attacks*: In this type of attack, a large amount of traffic and supposedly regular requests are sent to the target. Source of this traffic are clients that have been infected by the attacker. In these attacks, the target will not be able to process this massive amount of traffic and will be out of service.
- *Information theft*: After infecting systems, the attacker can access the user's sensitive data such as bank account passwords, emails, etc.
- *Sending spam emails*: The attacker could send spam emails via infected systems.
- *Bitcoin mining*: The more systems get infected by the attacker, the more bitcoins are mined.
- *Phishing attacks*: For this attack to done, some servers are needed to invariably fake host websites. Given that, some bots are connected to the internet, and the attacker could use them as servers to host these websites.

2.8 The Life Cycle of Botnets

To beat botnets, it is necessary to assess their life cycle. It includes four stages that each will be discussed further and shown in Fig. 10 [17]:

- Forming step
- Command and control stage
- Attack stage
- Post-attack stage

2.8.1 Forming Step

In this step, the attacker breaks into the victim system using its weak points or social engineering and installs the malware. Next, the malware runs and converts the system into a bot.

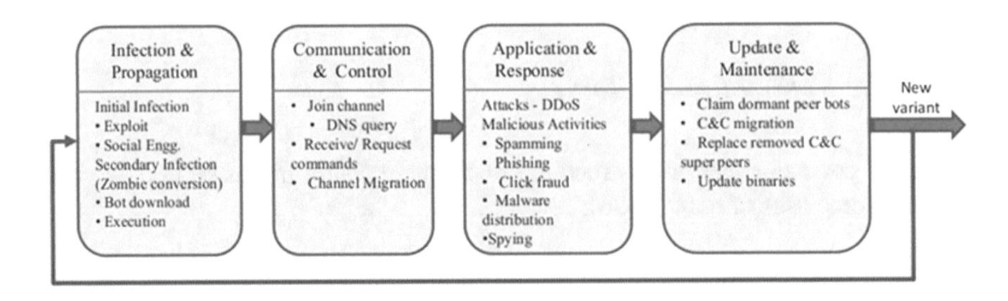

Fig. 10 The life cycle of botnet life cycle [17]

2.8.2 Command and Control Step (C&C)

After the system becomes a bot, it tries connecting to the server via C&C. In fact, in this step, the botmaster can remotely send commands to the bots.

2.8.3 Attack Step

In this phase, the bots run the malicious commands.

2.8.4 Post-Attack Step

In this stage, essential commands are sent to the bots to update their binary codes to stay hidden. Updating binary codes has another use which is preparation for future attacks. Bots need to gain new abilities to prepare for new attacks; in some cases, the most essential reason for updating is to find a new C&C server which is very vital for a botnet to survive.

3 Botnet Detection Methods

To detect a botnet, it is necessary to diagnose one of its composing elements such as the botmaster, C&C channels, and bots. The tree diagram below (Fig. 11) categorizes botnet detection methods.

Botnet detection is the most critical research field in network researching. Numerous papers were written in this field. In [17, 24, 37] different categories have been proposed for botnet detection. We will review some of these methods.

Botnet detection approaches split into two major categories: Honeynet and IDS.

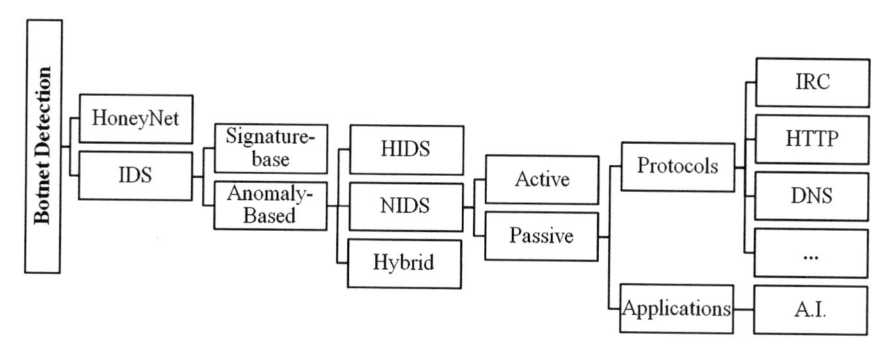

Fig. 11 Taxonomy of botnet detection techniques

3.1 Honeynet

It is a network of honeypots which is used as a trap in the network. If honeypots get infected by malware, it can store and report that bot's interactive and behavioral features [38]. In fact, honeypot is a technology that is valuable due to its illegal and prohibited use. Many studies such as [39] have been done on honeypot.

3.2 IDS

This method has two parts: signature-based and anomaly-based classes.

3.2.1 Signature-Based

In this method through analysis of input/output packets, particular patterns are discovered and compared with store patterns in the database to detect bot existence. This method is not applicable in Zero-day botnets that have no signatures available. For instance, Snort software uses this method [40, 41].

3.2.2 Anomaly-Based Detection

This method works along with anomaly in the network traffic e.g. high network latency, traffic on unusual ports and high volume of traffic. This approach is classified into three main groups: host-based, network-based, and hybrid [42].

- *Host-based method (HIDS)*: In this method, the detection of illegal activities is done via analysis of a computer's internal data like log-files [43, 44].
- *Network-based method (NIDS)*: It is vice-versa of host-based method, in this way that traffic of different hosts is analyzed. This method is categorized into two groups: Active and Passive monitoring [45].
  - *Active Monitoring*: In active monitoring, the extra packets are injected into the network to analyze the reaction of the network and discover malicious activities.
  - *Passive Monitoring*: In passive monitoring, traffic is monitored and analyzed. This approach has two subcategories: Applications and Protocols. In the *applications approach*, artificial intelligence (A.I.) and machine learning techniques are used. In *protocols approach*, IRC, DNS and other especial protocols that are used by botnets are analyzed.
- *Hybrid method*: In a hybrid method, a combination of HIDS and NIDS methods is applied to enjoy the advantages of both ways [46].

In the rest of this section, we will review some of the researches about botnet detection:

In [47], comprehensive research of capabilities and development cycle of botnets is proposed. In this research, it is tried to analyze various methods of botnet detection and related challenges which should be taken into consideration in detection algorithm selection. One of the challenges mentioned in the article is the hidden connection with C&C server that might be undetectable for the algorithm. Another serious challenge analyzed in the article is forming of the botnets, an important stage that must be considered when selecting an algorithm due to the normal traffic flow botnets use in forming stage.

In [28], Stevanovic and Pedersen proposed a flow-based detection method for p2p botnets. In their proposing system, pre-processing and classifier components are used. In pre-processing component, the network traffic flow is analyzed and its features are extracted. Then in classifier component, the traffic flow is classified into malicious and benevolent groups using machine learning algorithms.

In [48] Shanthi et al. proposed a detection method based on machine learning, monitoring and analyzing traffic and its features. Process detection is written in four stages. First, traffic flow is monitored via open source tools. Second, normal packets such as DHCP packet are excluded to decrease the traffic flow. Third, features are extracted then j48 Decision tree and Naive Bayes use network training in final stage. In the proposed system, Decision tree had better efficient than Naïve Bayes.

In [49], Kheir et al. proposed a system by the name of Mentor that permits removal of domain names from C&C servers blacklist. In fact, Mentor distinguishes legal domains from illegal ones through statistical features, website content crawling, and domain names. The authors carried out the procedure based on supervised machine learning that led to 99.02% experimental results for legal domain names.

In [50], Abraham et al. proposed an intrusion detection system on the anomaly platform to enhance network security and decrease human error. The most important differentiating factor in this method is traffic tracing. Given that content analysis of the traffic in trace mode is more comfortable than packet mode, the authors extracted temporal, and size features from traces to detect botnet traffic, and then applied them in various learning algorithms such as logistic regression, Naive Bayes, SVM, random jungle, and neural networks. Ninety nine percent was the best rate among algorithms, which produced by random jungle.

In [51], Yin et al. proposed a deep learning algorithm with a generative adversarial network (GAN). Generative Adversarial Networks (GANs) are a framework in which two neural networks compete with each other. Finally, the competition between two networks benefits both in terms of training and performance. Given that most of detection methods like machine learning consider limited features of network flow and face problems detecting new botnets. Thus, Chan et al. tried to cover all network flow features to improve network detection.

In [52], Hoang and Nguyen applied machine learning techniques and DNS query data to detect botnets. The proposing algorithm includes five stages. (1) DNS data collection, (2) Extraction of domain names, (3) Extraction of features by pre-processing, (4) Training, (5) Algorithm running. The authors used n-gram algorithm and features like vowel distribution characteristics of domain names in training. After running tests and applying different algorithms, Random forest algorithm provided the best results in the proposing method.

In [53], Costa et al. proposed an anomaly and host-based method to detect mobile botnets. The authors used a machine learning algorithm to detect anomalies in statistical features which were extracted from the system calls of applications. Based on assessments, Random forest algorithm has good results.

Kirubavathi and Anitha [54] proposed a five-phase method of structural analysis to detect android botnets. First, there is a collection of android software including application programs and botnets to detect C&C structure and malicious activities. Second, in order to extract features of each program and their permissions, manifest file in each APK files is extracted. Third and fourth stages include analysis of requested permissions features to produce unique patterns. Finally, machine learning algorithm is run. After the tests, it was indicated that support vector machine produced the best results with 98.97% accuracy among Naïve Bayesian and REPTree algorithms.

The authors of [55] applied the ensemble learning point of view to improve accuracy of botnet detection. In fact, they designed a detection system for p2p traffic flow including three machine learning algorithms (Naive Bayes, Bayesian Network, and C4.5). PCA and CSE algorithms were used to reduce features, and lastly two Stacking and Voting techniques were applied. In the proposing algorithm Stacking had better performance than Voting.

Haddadi and Zinshir in [56] analyzed five different botnet detection methods. The methods were divided into signature-based and machine learning methods. BotHunter and Snort in signature-based methods, Tranalyzer-2, Packet Payload and flow aggregation/function in machine learning methods were analyzed. In machine learning methods network flow-based and packet-based features were analyzed. In the article, efficiency of algorithms was assessed using binary and multi classifiers on 25 datasets. After the tests, C4.5 algorithm indicated the best classification accuracy.

In [57], an ensemble leaning intrusion detection has been proposed. In above-mentioned paper, authors developed an ensemble learning method, AdaBoost, based on established statistical flow features to mitigate malicious events. AdaBoost used three machine learning algorithms named Decision Tree, Naive Bayes, and Artificial Neural Network. Their strategy consists of three stages (1-extracting features 2-selecting features 3-run AdaBoost), in which to extract features and evaluate performance of AdaBoost, authors used two UNSW-NB15 and NIMS datasets. Finally, the experiments demonstrated high detection rate and low false-positive rate among other recent methods.

4 Challenges of Botnet Detection

Botmasters apply different methods to keep their bots alive, and these methods are code obfuscation, P2P structure, traffic encryption, flash crowd mimic. These methods are challenges we need to pay attention to while introducing detection algorithms.

4.1 Encrypted Traffic

Early versions of botnet were based on IRC protocol. In these botnets, the content traffic is plain text which is easily distinguished from normal traffic content. Nowadays, new botnets use cryptographic algorithms to challenge botnet detection [20, 58].

> *Signature-based* detection techniques face difficulties with encrypted traffic; hence, behavior-based techniques are used instead [17].

4.2 Code Obfuscation

In early botnet versions, in which C&C IP addresses were hard-coded, pentesters detect a bot, and through reserve, engineering discover C&C addresses. Then they block the addresses and defend the attacks. But in new generation botnets, hackers use various techniques to obfuscate the bot's binary codes, so as to challenge reverse engineering [12].

> The most important advantage of hard-coded IP address is high speed and ordinary skill needed in programming.

4.3 Dynamic Domain Name System (Dynamic DNS)

We know DNS task is to translate domain names to IP addresses which is a normal procedure. DNS-based botnets work in opposite fashion from hard-coded IP addresses in a way that they put agent addresses in DNS servers instead of binary codes. Thus, each agent connecting to another (generally C&C), must send a request to DNS server to receive the address. This process is depicted in Fig. 12. The attackers hide the identity of agents using this technique. In [59, 60] proposing methods were introduced to detect above-mentioned botnets. Lately, attackers apply dynamic DNS technique to complicate detection process. In this technique, dynamic addresses are used instead of static IPs for C&C. Actually, in this method, IP addresses related to C&C domain names change in periods.

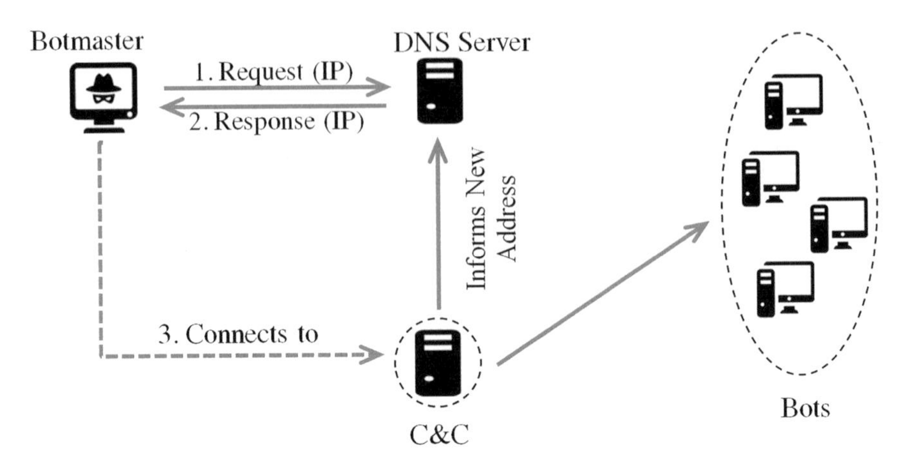

Fig. 12 The structure of DDNS

4.4 Domain Generation Algorithm (DGA)

This algorithm is used to generate domain names. It is possible to alter C&C addresses periodically and keep them away from block list. In this algorithm, which Fig. 13 illustrates, pseudo-code is an example of these addresses and time is normally used as seed so as to generate dynamic addresses [27].

It is worthy to mention that each DGA-based botnet uses a customized algorithm. For instance, Tropping botnet from current date and trending words on Twitter as seed are used for domain generation [61].

Main weakness of this method is that DGA algorithm codes are stored in binary bots, so by detecting a bot and reverse engineering, DGA algorithm is discovered then generated names are put blacklist. A better approach is reserving domain names that will be generated in a particular time after DGA discovery. In this way, it is possible to take over the control of the botnet [62]. Different methods like [63, 64] were proposed to detect this type of botnets. An alternative approach for DGA is Fast-Flux technique which will be explained below.

4.5 Fast-Flux

The possibility of fast geographical spot alteration of an internet-based service, e.g. email service from one or more online computers to another group of computers to reduce delay or detection interference is called Fast-flux (Fig. 14). Attackers use Fast-Flux feature to interfere detection of C&C spot and increase robust botnet infrastructure. By traffic monitoring and assessing domains, it is possible

```
suffix =  ["anj", "ebf", "arm", "pra", "aym", "unj",
    "ulj", "uag", "esp", "kot", "onv", "edc"]

def generate_daily_domain():
    t = GetLocalTime()
    p = 8
    return generate_domain(t, p)

def scramble_date(t, p):
    return (((t.month ^ t.day) + t.day) * p) +
        t.day + t.year

def generate_domain(t, p):
    if t.year < 2007:
        t.year = 2007
    s = scramble_date(t, p)
    c1 = (((t.year >> 2) & 0x3fc0) + s) % 25 + 'a'
    c2 = (t.month + s) % 10 + 'a'
    c3 = ((t.year & 0xff) + s) % 25 + 'a'
    if t.day * 2 < '0' || t.day * 2 > '9':
        c4 = (t.day * 2) % 25 + 'a'
    else:
        c4 = t.day % 10 + '1'
    return c1 + 'h' + c2 + c3 + 'x' + c4 +
        suffix[t.month - 1]
```

Fig. 13 Torpig daily domain generation algorithm

to encounter this approach and detect Fast-Flux. Different methods like [59, 65] are proposed for this technique.

4.6 *Stepping-Stone*

The attacker can use a number of proxy systems between himself and C&C server to outrun the detection (Fig. 3). In this technique, cyber-security experts detect a bot and follow its connections to reach a stepping stone layer; if proxy systems be located in different. Countries or continues, detection of the attacker would be very difficult [23].

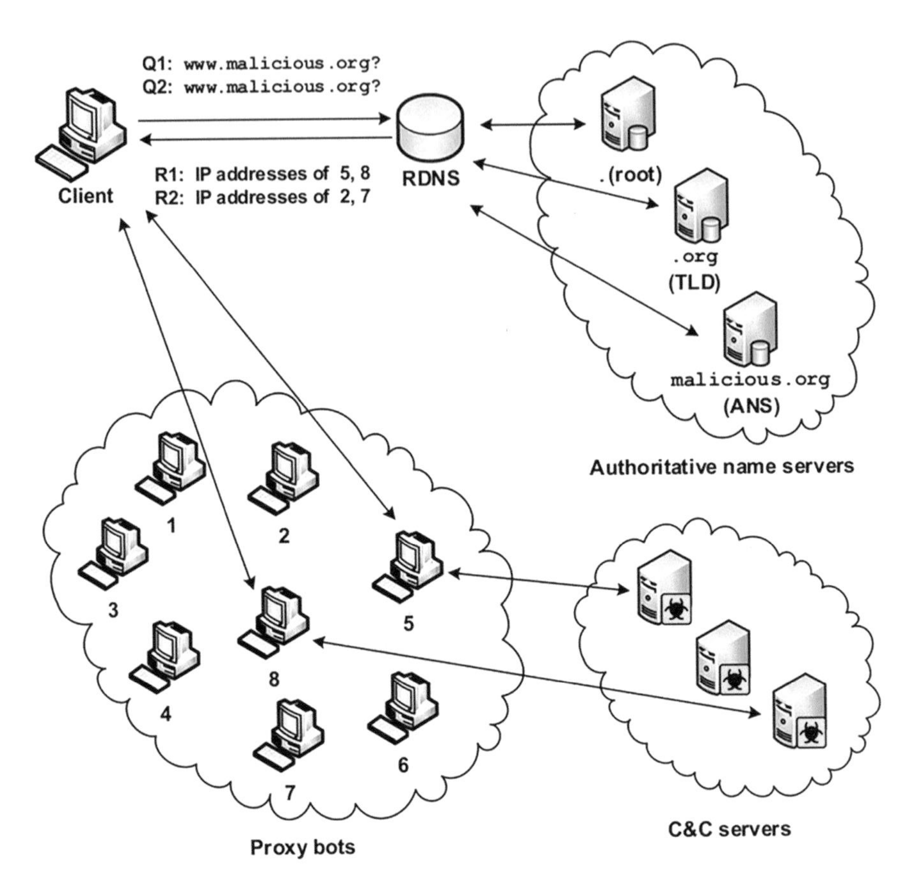

Fig. 14 Structure of fast-flux [59]

4.7 Flash-Crowd Mimic

When many legal users send requests to the server simultaneously, it can cause packet loss and congestion. This problem is normal and, e.g. can happen in any sudden surges of users to read a particular piece of news. The same scenario occurs in DDoS attacks with this difference that requests are sent crowd pattern to hide the attackers [66].

5 Summary

We categorized this chapter into three sections, analyzing one of the most important and newest malware named botnet.

In the first section, we analyzed botnets in terms of history, structure, communication protocols, and life cycle, which help pentesters to detect botnets in the networks.

In the second section, we examined botnet detection methods, lead us to a general classification including IDS and Honeynet methods. We covered these methods in detail, and we reviewed several related works in the botnet detection field, including a variety of methods.

Considering botnet detection methods faces several problems, in the third part we thoroughly analyzed essential challenges, such as Code Obfuscation, Dynamic DNS, Domain Generation Algorithm (DGA), Fast-Flux, Stepping-Stone, and Flash-crowd mimic, which should be taken into consideration by researchers to propose an efficient detection method.

References

1. H. Karimipour, A. Dehghantanha, R.M. Parizi, C. Kim-Kwang Raymond, H. Leung, A deep and scalable unsupervised machine learning system for cyber-attack detection in large-scale smart grids. IEEE Access **7**, 80778–80788 (2019)
2. H. HaddadPajouh, A. Dehghantanha, R.M. Parizi, M. Aledhari, H. Karimipour, A survey on internet of things security: requirements, challenges, and solutions. Internet Things (2019). https://doi.org/10.1016/j.iot.2019.100129
3. Kaspersky Lab, https://www.usa.kaspersky.com. Accessed May 2019
4. AV-TEST Institute, https://www.Av-test.org. Accessed Oct 2019
5. J. Demarest, Taking down botnets: public and private efforts to disrupt and dismantle cybercriminal networks, in U.S. senate, Statement before the Subcommittee on Crime and Terrorism, Washington, 2014
6. I. Ghafir, S. Jakub, V. Prenosil, A survey on botnet command and control traffic detection. Int. J. Adv. Comput. Netw. Secur **5**(2), 75–80 (2015)
7. M. Begli, F. Derakhshan, H. Karimipour, A layered intrusion detection system for critical infrastructure using machine learning, in *2019 IEEE 7th International Conference on Smart Energy Grid Engineering (SEGE)* (IEEE, 2019), pp. 120–124
8. S. Mohammadi, H. Mirvaziri, M. Ghazizadeh-Ahsaee, H. Karimipour, Cyber intrusion detection by combined feature selection algorithm. J. Inf. Secur. Appl. **44**, 80–88 (2018)
9. F. Amiri, M. Rezaei Yousefi, C. Lucas, A. Shakery, N. Yazdani, Multivariate mutual information feature selection for intrusion detection, in *IEEE Canada Electrical Power and Energy Conf. (EPEC), Toronto, Canada* (IEEE, 2018)
10. S. Chakkaravarthy, D. Sangeetha, V. Vaidehi, A survey on malware analysis and mitigation techniques. Comput. Sci. Rev. **32**, 1–23 (2019)
11. A. NamavarJahromi, S. Hashemi, A. Dehghantanha, K.-K.R. Choo, An improved two-hidden-layer extreme learning machine for malware hunting. Comput. Secur. **89**, 101655 (2019)
12. M.F.A. Razak, N.B. Anuar, R. Salleh, A. Fir, The rise of "malware": bibliometric analysis of malware study. J. Netw. Comput. Appl. **75**, 58–76 (2016)
13. V. Subrahmanian, M. Ovelgönne, T. Dumitras, B.A. Prakash, *The Global Cyber-Vulnerability Report* (Springer, Cham, 2015)
14. R. HosseiniNejad, H. HaddadPajouh, A. Dehghan Tanha, A cyber kill chain based analysis of remote access trojans, in *Handbook of Big Data and IoT Security* (Springer, Cham, 2019)
15. S.S. Silva, R.M. Silva, R.C. Pinto, R.M. Salles, Botnets: a survey. Comput. Netw. **57**(2), 372–403 (2013)

16. E. Bertino, N. Islam, Botnets and internet of things security. Computer **50**(2), 76–79 (2017)
17. R.S. Rawat, E.S. Pilli, R.C. Joshi, Survey of peer-to-peer botnets and detection frameworks. Int. J. Netw. Secur. **20**(3), 547–557 (2018)
18. A. Kumar Tyagi, G. Aghila, A wide scale survey on botnet. Int. J. Comput. Appl. **34**(9), 10–23 (2011)
19. W.Z. Khan, M.K. Khan, F.T.B. Muhaya, M.Y. Aalsalem, H.C. Chao, A comprehensive study of email spam botnet detection. IEEE Commun. Surv. Tutor. **17**(4), 2271–2295 (2015)
20. N. Kaur, M. Singh, Botnet and botnet detection techniques in cyber realm, in *2016 International Conference on Inventive Computation Technologies (ICICT)*, vol. 3 (IEEE, 2016), pp. 1–7
21. T.S. Hyslip, J.M. Pittman, A survey of botnet detection techniques by command and control infrastructure. J. Digit. Forensic Secur. Law **10**(1), 7–26 (2015)
22. R. Hadianto, T.W. Purboyo, A survey paper on botnet attacks and defenses in software defined networking. Int. J. Appl. Eng. Res. **1**(13), 483–489 (2018)
23. G. Vormayr, T. Zseby, J. Fabini, Botnet communication patterns. IEEE Commun. Surv. Tutor. **19**(4), 2768–2796 (2017)
24. A. KARIM, R.B. SALLEH, M. SHIRAZ, S.A.A. SHAH, Botnet detection techniques: review, future trends, and issues. J. Zhejiang. Univ. Sci. C **15**(11), 943–983 (2014)
25. A. Habibi Lashkari, S.G. Ghalebandi, M.R. Moradhaseli, A wide survey on botnet, in *Proceedings of the International Conference on Digital Information and Communication Technology and Applications* (Springer, Cham, 2011), pp. 445–454
26. A.K. Sood, S. Zeadally, R.J. Enbody, An empirical study of HTTP-based financial botnets. IEEE Trans. Dependable Secur. Comput. **13**(2), 236–251 (2016)
27. B. Stone-Gross, M. Cova, L. Cavallaro, B. Gilbert, M. Szydlowski, R. Kemmerer, C. Kruegel, G. Vigna, Your botnet is my botnet: analysis of a botnet takeover, in *CCS'09, November 9–13, 2009, Chicago, Illinois, USA* (ACM, 2009)
28. M. Stevanovic, J.M. Pedersen, An efficient flow-based botnet detection using supervised machine learning, in *International Conference on Computing, Networking and Communications (ICNC), Honolulu, HI, USA* (IEEE, 2014)
29. SymantecSecurity, http://www.symantec.com. Accessed May 2019
30. F. Haddadi, A.N. Zincir-Heywood, Benchmarking the effect of flow exporters and protocol filters on botnet traffic classification. IEEE Syst. J. **10**(4), 1–12 (2014)
31. Microsoft, https://microsoft.com. Accessed May 2019
32. S. Ryu, B. Yang, A comparative study of machine learning algorithms and their ensembles for botnet detection. J. Comput. Commun. **6**(5), 119–129 (2018)
33. S. Khattak, N. Rasheed Ramay, K. Riaz Khan, A.A. Syed, S.A. Khayam, A taxonomy of botnet behavior, detection, and defense. Commun. Surv. Tutor. **16**(2), 898–924 (2013)
34. E. Cooke, F. Jahanian, D. McPherson, The zombie roundup: understanding detecting and disrupting botnets, in *Proc. Steps to Reducing Unwanted Traffic on the Internet Workshop (SRUTI'05), Cambridge, MA* (ACM, 2005)
35. https://tools.ietf.org/html/rfc1459. Accessed May 2019
36. S. Amina, R. Vera, T. Dargahi, A. Dehghantanha, A bibliometric analysis of botnet detection techniques, in *Handbook of Big Data and IoT Security* (Springer, Cham, 2019), pp. 345–365
37. H. Ostap, R. Antkiewicz, A concept of clustering-based method for botnet detection, in *International Conference on Mathematical Methods, Models, and Architectures for Computer Network Security* (Springer, Cham, 2017), pp. 223–234
38. K.-C. Lu, I.-H. Liu, M.-W. Sun, J.-S. Li, A survey on SCADA security and honeypot in industrial control system, in *International Conference of Reliable Information and Communication Technology* (Springer, Cham, 2018), pp. 598–604
39. M. Zuzcak, T. Sochor, *Behavioral Analysis of Bot Activity in Infected Systems Using Honeypots* (Springer, Cham, 2017)
40. E. Modiri Dovom, A. Azmoodeh, A. Dehghantanha, D. Ellis Newton, R.M. Parizi, H. Karimipour, Fuzzy pattern tree for edge attack detection and categorization in IoT. J. Syst. Archit. **9**, 1–7 (2018)

41. A. Namavarjahromi, J. Sakhnini, H. Karimipour, A. Dehghantanha, An unsupervised feature selection approach for effective cyber-physical attack detection and identification, in *29th Annual International Conf. on Computer Science and Software Engineering, Toronto, Canada* (ACM, 2019)
42. H. Karimipour, H. Leung, Relaxation-based anomaly detection in cyber-physical systems using ensemble kalman filter. IET Cyber-phys. Syst. Theor. Appl. **3**, 29–38 (2019)
43. C.V. Martinez, B. Vogel-Heuser, A host intrusion detection system architecture for embedded industrial devices. J. Frankl. Inst. (2019). https://doi.org/10.1016/j.jfranklin.2019.03.037
44. S.K. Gautam, H. Om, Computational neural network regression model for host based intrusion detection system. Perspect. Sci. **8**, 93–95 (2016)
45. K. Alieyan, A. ALmomani, A. Manasrah, M.M. Kadhum, A survey of botnet detection based on DNS. Neural Comput. Appl. **28**, 1541–1558 (2017)
46. A. Patelab, M. Taghavi, K. Bakhtiyari, J.C. Júniorc, An intrusion detection and prevention system in cloud computing: a systematic review. J. Netw. Comput. Appl. **36**, 25–41 (2013)
47. M. Bailey, E. Cooke, F. Jahanian, Y. Xu, M. Karir, A survey of botnet technology and defenses, in *2009 Cybersecurity Applications & Technology Conference for Homeland Security* (IEEE, 2009), pp. 299–304
48. K. Shanthi, D. Seenivasan, Detection of botnet by analyzing network traffic flow characteristics using open source tools, in *2015 IEEE 9th International Conference on Intelligent Systems and Control (ISCO)* (IEEE, 2015), pp. 1–5
49. N. Kheir, F. Tran, P. Caron, N. Deschamps, Mentor: positive DNS reputation to skim-off benign domains in botnet C&C blacklists, in *ICT Systems Security and Privacy Protection* (Springer, Berlin/Heidelberg, 2014)
50. A. Brendan, A. Mandya, R. Bapat, F. Alali, D.E. Brown, M. Veeraraghavan, A comparison of machine learning approaches to detect botnet traffic, in *2018 International Joint Conference on Neural Networks (IJCNN)* (IEEE, 2018), pp. 1–8
51. C. Yin, Towards accurate node-based detection of P2P botnets. Sci. World J. **2014**(425491), 1–10 (2014)
52. X.D. Hoang, Q.C. Nguyen, Botnet detection based on machine learning techniques using DNS query data. Future Internet **10**(5), 43 (2018)
53. V.G.T.d. Costa, S. Barbon, R.S. Miani, J.J.P.C. Rodrigues, B.B. Zarpelão, Detecting mobile botnets through machine learning and system calls analysis, in *IEEE International Conference on Communications (ICC)* (IEEE, 2017)
54. G. Kirubavathi, R. Anitha, Structural analysis and detection of android botnets using machine learning. Int. J. Inf. Secur. **17**(2), 153–167 (2018)
55. J.M. Reddy, C. Hota, P2p traffic classification using ensemble learning, in *Proceedings of the 5th IBM Collaborative Academia Research Exchange Workshop* (ACM, 2013)
56. F. Haddadi, A.N. Zincir-Heywood, Botnet behaviour analysis: how would a data analytics-based system with minimum a priori information perform? Int. J. Netw. Manag. **27**(4), e1977 (2017)
57. N. Moustafa, B. Turnbull, C. Kim-Kwang Raymond, An ensemble intrusion detection technique based on proposed statistical flow features for protecting network traffic of internet of things. IEEE Internet Things J. **6**(3), 4815–4830 (2018)
58. S.-C. Su, Y.-R. Chen, S.-C. Tsai, Y.-B. Lin, Detecting P2P botnet in software defined networks. Secur. Commun. Netw. **2018**, 1–13 (2018)
59. M. Stevanovic, J.M. Pedersen, A. D'Alconzo, S. Ruehrup, A method for identifying compromised clients based on DNS traffic analysis. Int. J. Inf. Secur. **16**(2), 115–132 (2017)
60. K. Alieyan, A. Almomani, R. Abdullah, M. Anbar, A rule-based approach to detect botnets based on DNS, in *2018 8th IEEE International Conference on Control System, Computing and Engineering (ICCSCE)* (IEEE, 2018), pp. 115–120
61. C.-D. Chang, H.-T. Lin, On similarities of string and query sequence for DGA botnet detection, in *2018 International Conference on Information Networking (ICOIN)* (IEEE, 2018), pp. 104–109

62. S.T. Ali, P. McCorry, P.H.-J. Lee, F. Hao, ZombieCoin 2.0: managing next-generation botnets using Bitcoin. Int. J. Inf. Secur. **17**(4), 411–422 (2018)
63. T.-S. Wang, H.-T. Lin, W.-T. Cheng, C.-Y. Chen, DBod: clustering and detecting DGA-based botnets using DNS traffic analysis. Comput. Secur. **64**, 1–15 (2017)
64. A. Satoh, Y. Nakamura, D. Nobayashi, T. Ikenaga, Estimating the randomness of domain names. IEEE Commun. Lett. **22**(7), 1378–1381 (2018)
65. A. Almomani, Fast-flux hunter: a system for filtering online fast-flux botnet. Neural Comput. Appl. **29**(7), 483–493 (2018)
66. S. Yu, W. Zhou, W. Jia, S. Guo, Y. Xiang, F. Tang, Discriminating DDoS attacks from flash crowds using flow correlation coefficient. IEEE Trans. Parallel Distrib. Syst. **23**(6), 1073–1080 (2012)

Learning Based Anomaly Detection in Critical Cyber-Physical Systems

Farnaz Seyyed Mozaffari, Hadis Karimipour, and Reza M. Parizi

1 Introduction

Cyber-physical system (CPS) is the result of the efficient combination of cyber systems and the physical world as an integrated structure for vital tasks which are originated from advancements in digital electronics [1]. In such systems, physical procedures and computational resources are integrated through communication links for remote monitoring and control [2, 3].

Even though CPSs develop power network interaction with the consumer and other parties, many challenges have emerged, including security, reliability, stability, maintainability, safety, predictability, etc. [4, 5]. Security is one of the most important challenges in the cyber-physical systems due to the integration of many components, which have made them susceptible to be attacked on both the physical and cyber sides [6]. The system operation can be interrupted by malicious attack and may cause data leakage in vital components [5, 7, 8]. Cyber-physical systems are facing the tsunami of generated data on different components that are too large and complex for real-time processing. Cloud computing techniques, along with analytics methods such as machine learning, which the latter refers to learning and making predictions from available data by a system, can help the generated information to be secure whilst being processed, analyzed and stored [9–11].

Confidentiality, Integrity, and Availability (CIA) triad is one of the important concepts of information security [5]. These three concepts are the core objective in

F. S. Mozaffari · H. Karimipour (✉)
School of Engineering, University of Guelph, Guelph, ON, Canada
e-mail: fseyyedm@uoguelph.ca; hkarimi@uoguelph.ca

R. M. Parizi
College of Computing and Software Engineering, Kennesaw State University, Kennesaw, GA, USA
e-mail: rparizi1@kennesaw.edu

H. Karimipour et al. (eds.), *Security of Cyber-Physical Systems*,
https://doi.org/10.1007/978-3-030-45541-5_6

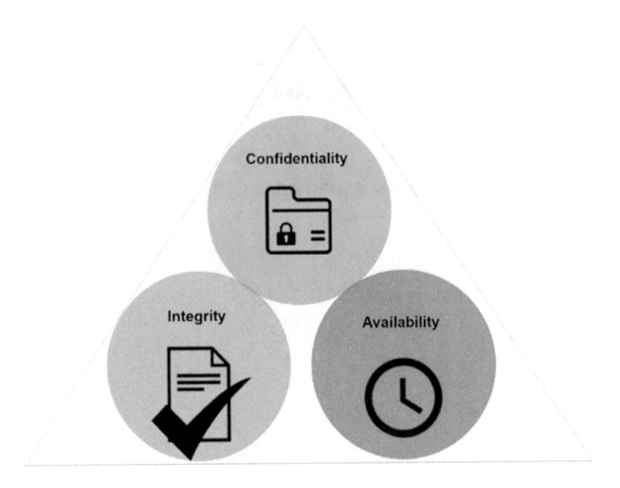

Fig. 1 CIA triad

cyber-physical systems' security, and in order to have a secure CPS, this essential goal should be achieved. The CIA triad model is shown in Fig. 1. Confidentiality is obtained by preventing disclosure of data to unauthorized systems or individuals and can be achieved by encrypting information. For example, a patient's health records in a healthcare cyber-physical system should be encrypted during any communication among health record devices in order to prevent unauthorized party obtains personal information. Confidentiality in CPS is crucial since users' privacy should be maintained, and it can be achieved by preventing attackers from eavesdropping the communication channel between the sensors and controller, as well as between the controller and actuator [12]. Different type of attacks can target confidentiality of CPS, such as the man-in-the-middle and eavesdropping. Integrity refers to no modification of resources without authorization, which is achieved when the receiver receives the true data sent by the sender. A breach of integrity occurs when important data are modified or deleted by an unauthorized individual for a nefarious purpose. Furthermore, receiving false data while believing that it is true, indicates a violation of integrity. Integrity can be achieved in CPS by preventing modification of information received by sensors, actuator or controllers [13]. Spoofing and data injection can target the integrity of CPS. Availability is obtained when both physical controls and communication channels are available at the requested time. Denial of Service (DOS) and buffer overflow are examples of attacks that can affect availability principle and prevents the user from having access to desired resources in a demanded time.

This paper provides a comprehensive survey on machine-learning methods for anomaly detection in cyber-physical systems. In addition, we present the results of a case study on the comparison of various machine learning algorithms in this area. Anomaly detection is referred to as detecting patterns from which is not fit into predictable behavior [14–16]. Web of Science is used as the search engine

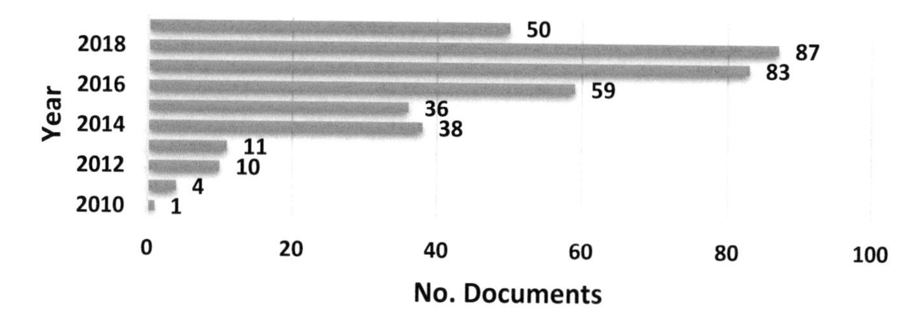

Fig. 2 The number of publications in the last 10 years

in order to categorize the publications about this topic in the last 10 years. First, the related keywords are inputted for extracting publications. Then, non-relevant and non-English publications were removed. As a result, in primitive search, 379 publications were founded related to this subject.

Figure 2 shows 87 documents are published in 2018 while there was only one publication in 2010. Considering the fact that the study was conducted in August 2019, it is predictable to see the number of publications to go even higher for 2019 compared to 2018.

The bibliometric analysis was also performed in order to have a better understanding of which type of intrusion detection had the greatest number of publications. As can be seen in Fig. 3, anomaly detection was the most common topic, which was discussed by the researcher in the area of security of cyber-physical systems.

Figure 4 demonstrates that among different types of machine-learning algorithms, neural networks, accounting 52%, is studied the most in the area of anomaly detection in cyber-physical systems. Furthermore, support vector machine and Bayesian network accounting for 20% and 12% respectively were the second largest number of publications after the neural network.

2 Background and Literature Review

National Science Foundation (NSF) introduced the term of cyber-physical system (CPS) in 2006 for the first time [17], and it has been defined from a different perspective in various articles. For instance, Lou et al. [18] describe it as "embedded systems together with their physical environment", while Edward [19] described it from a design viewpoint: "integration of computation with physical processes". On the other hand, Rajkumar [20] provided a general perspective for CPS: "an engineered system whose operations are monitored and controlled by a computing core". In simple words, cyber refers to a virtual environment, and physical refers to physical systems, and the term of a cyber-physical system is the connection of

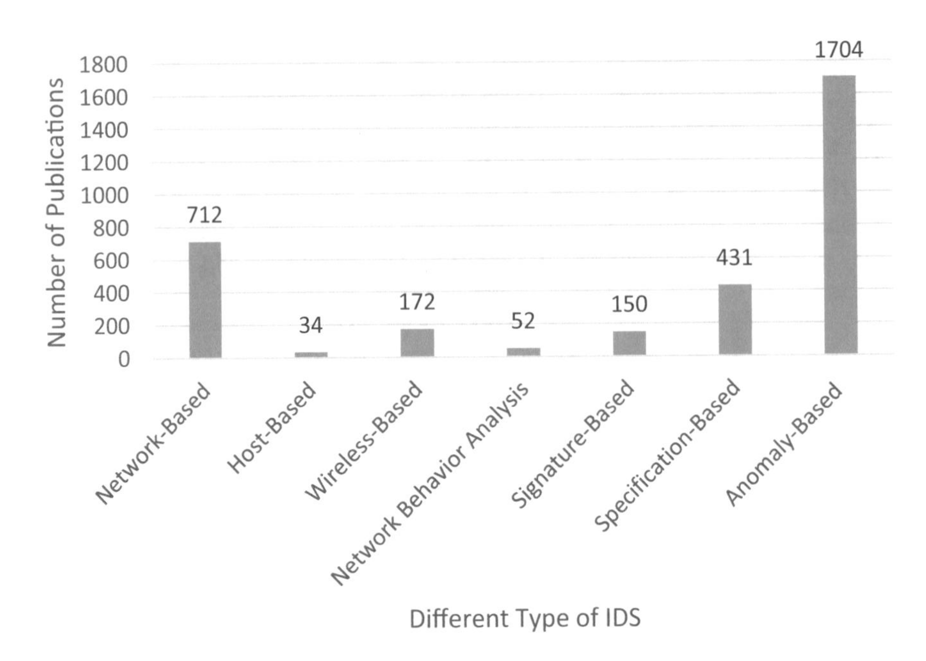

Fig. 3 The number of publications in different Type of IDS

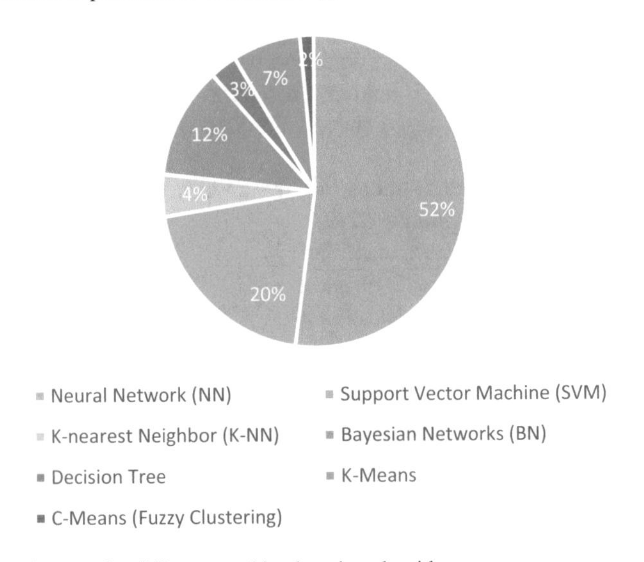

Fig. 4 Publication rate for different machine learning algorithms

these two, with feedback loops where physical components can affect the cyber part and vice versa [21]. As an intellectual viewpoint, CPS should be considered as an integration of both components and understanding each part separately is not sufficient; instead, their interaction should be taken into account. Typically, cyber-physical systems should consist of three parts including, embedded systems, sensors, and communication systems. The real-time data is collected from the physical environment through sensors and stored for processing. Then, the real-time signal commands will be given to the actuator section in order to achieve the desired task.

Some characteristics and features of CPSs are explained as follows: [22–24].

- They should be integrated closely.
- The system resources such as network bandwidth are limited.
- CPSs are strictly constrained by real-time, while each component may have different timing scale. This has led the CPSs complex at multiple temporal and spatial scales.
- Since these systems have a complex structure, they should have the adaptive capability.
- The network section has an extreme scale; for example, it should include WLAN, Bluetooth, GSM and etc.
- The advanced feedback control technologies are required for CPSs, which has led them to have high degrees of control loops.
- Operations in CPSs must be dependable, and since they are used in large-scale systems, their security and reliability play a vital role.

The current cyber-physical systems with their complex architecture use a widely inconsiderable number of large-scale and critical industries. However, they are facing various challenges, which should be addressed [25]. For instance, since these systems are used in vital applications, their safety and reliability play an important role. They should be designed in such a way that no threads can affect their efficiency. Moreover, due to unpredictability situations of the physical world, it is difficult to design the components of the CPS. In other words, these systems should be designed in such a way that they are robust to unexpected conditions from the physical world. Besides, some CPSs are extremely small, and as a result, they are more vulnerable to various existing threats. Furthermore, CPSs should operate at high speed, and the lack of timing in computing has resulted in facing some challenges in designing the operating systems and networking. In addition, it is nearly impossible to predict the execution time of the codes in software, and this has led to some architectural problems. In order to address this issue and make CPSs more predictable, designers should choose alternative processor architecture [26]. Much research has conducted on security challenges in CPS. For example, Sanislav et al. [27] discuss system-level requirements and challenges for designing CPS and Shi [28] give an outline of CPS challenges and features.

Baheti et al. describe system-level aspects of CPS from scientific and social impact view and suggests a research direction for designing CPS [29]. Lee [26] shows that more programming language models are required for CPS design and [30] introduced two new approaches, cyberizing the physical and physicalizing the cyber, for CPS systems.

Moreover, new security concerns have been introduced due to tight coupling between communication technology and physical systems, and this has made novel challenges that require new approaches to the field of cyber security [31]. For instance, prior security researches related to smart grid infrastructure are either inapplicable or insufficiently scalable, and new approaches are required to achieve smart grids secure [32]. Furthermore, there are a considerable number of critical infrastructures that use CPS, and they need appropriate methods for detecting and preventing attacks. The attack on the Ukraine power plant in 2016 [33] and the attack on nuclear power plants [34] areas examples that show the existence of vital threats to the CPS and the huge need to improve its security. In order to have secure IT systems, only well encryption of data is enough, while this is not enough in CPS. The behavior of CPS should be observed continuously due to the fact that CPS has more complex architecture design and they face more critical threats and vulnerabilities that can have a fundamental impact on the behavior of the system [35, 36].

2.1 CPS Attack Classification

There is an attack classification method that was proposed by Yampolskiy et al. [37], and it categorizes any infected items in CPS into two different elements: Influenced element and victim element. In this classification method, the influenced element de-scribes the component that has been manipulated by the attack, and it can be either a component in an integral part of CPS or a physical or cyber environment. The victim element should not necessarily be the same component, and it can be at different layers or levels of abstraction. The main difference between the influenced element and victim element is as follows: The first one is directly manipulated by a malicious attack, while the latter is manipulated by existing interaction in CPS. These two elements are independent of each other, and they can belong either to the cyber or physical domain. The attacks can be categorized into four classes based on the domain of these components: Cyber-to-Cyber (C2C), Physical-to-Physical (P2P), Cyber-to-Physical (C2P) and Physical-to-Cyber (P2C). Figure 5 shows this classification.

2.2 Intrusion Detection System

One vital component of the security of cyber-physical systems is the Intrusion detection system, which gathers data and detects an attack by doing analyzation on collected data [38, 39]. In order to have secured CPS, most IDS technologies provide four common capabilities: gathering, logging, detection, and prevention. Gathering refers to collecting data from the observed operation, and logging refers to validating alerts in case of detected attacks [40]. Detection and prevention methodologies are the most important stages in CPS threat mitigation.

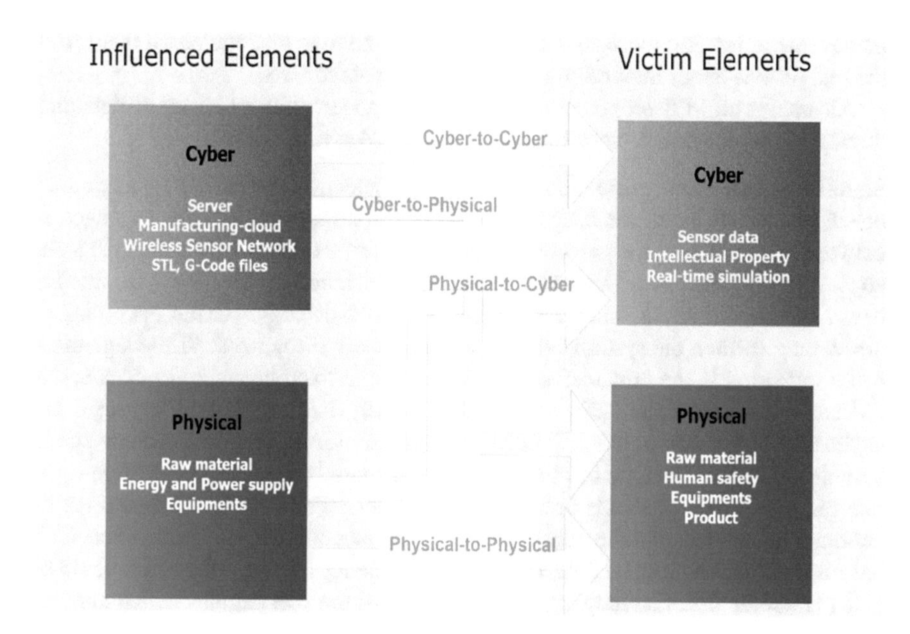

Fig. 5 Attack classification based on infected elements in CPS

Historically, IDS were categorized as either being passive or active. A passive IDS will give an alert to the system if it finds any entity that has been compromised, while active IDS could also respond to attack and take actions like blocking or shutting down the infected resources. National Institute of Standards and Technology divides IDS into these four categories [41]:

1. Network-Based IDS, which detects and corresponds to attacks related to network protocols and traffic
2. Host-based, which monitors the behavior of certain hosts like traffic, configuration behavior, and application behavior. One advantage of host-based IDS is the ability to detect malicious traffic or network packet which were originated from the host itself.
3. Wireless-based, which is similar to network-based IDS, but it monitors wireless network traffic
4. Network Behavior Analysis, which captures network traffic flow in order to identify suspicious behavior or attack

One drawback of IDS is the limitation in supplying absolute accurate detection [40]. The degree of accuracy of IDS can be assessed by having two indicators of False Positive (FP) and False Negative (FN). False-positive occurs when IDS incorrectly detect and attack, which has not happened, and false negative refers to failing in identifying malicious activity. Due to some security consideration, security administrator prefers to find an approach to increase FPs than decrease FNs. This means that indicating more suspicious activities and then distinguishing them from the real ones is better than failing in detecting threats to CPS.

Alcaraz et al. [42] proposed three types for modern IDS in terms of detecting threats: Signature-based, Specification-based and Anomaly-based.

Signature-Based Detection A signature-based detection system is a type of attack detection, which looks for a specific type of patterns, such as byte sequence in network traffic. It uses the gathered signatures in order to be able to detect the same attack. The efficiency of this system can be determined by the speed of creating new signatures. One advantage of signature-based detection is the easy task of developing it when the system behavior is completely recognized. While signature-based detection is the best method for detecting the fixed behavioral attack types, its inability to identify and preventing the previously unknown attacks is considered as the main drawback of this approach. It is difficult for a signature-based detection system to identify the type of attacks that are using advanced exploit technologies like encrypted data channels and payload encoders. Thus, these systems must be continually updated, and the new signatures must be collected and maintained. This has resulted in reducing the performance and efficiency of the system. In order to be able to increase the efficiency, systems should use more IDS engines with numerous processors [43].

Specification-Based Detection Specification-based detection relies on manually developing the specification of captured behaviors of the system. The main advantage of specification-based is the decreased rate of false-positive alarms while their downside is the time consuming of this process [44].

Anomaly-Based Detection Anomaly detection is the process of finding any unusual pattern of data that can change the behavior of the system. The anomalies might or might not have harmful impacts, but detecting anomalies play a vital role in critical infrastructures such as power systems [45]. Anomaly-based detection is discussed more in section four [46].

3 Different Type of Attacks to CPS

In order to be able to have secured CPS, we should be aware of different threats, which CPS might face. The detection and prevention of the attacks depend on having strong knowledge on possible threats to CPS. There are different types of attacks which might occur to CPS, but the major ones are compromised-key attacks, a man-in-the-middle attack (MITM) and denial of service (DoS).

3.1 False Data Injection (FDI) Attack

False data injection attacks are one of the common types of malicious cyber-attacks that can produce serious threats to the system by manipulating the measurements of the system without posing any anomalies. This type of attack might happen when the attacker has the knowledge of power network information and have access to even a small part of smart resources [47].

3.2 Compromised-Key Attack

Although it is assumed that data in every secured system is encrypted and decrypted via keys, compromised key attack is the stealing and using of the encrypted key in order to be able to have access to secured data. In this type of attack, the malicious attacker can easily decrypt and modify data and have access to secured communication while the receiver and the sender are not usually aware of the attack. Furthermore, by using the compromised key, the attacker is able to compute additional keys in the system and gain access to the whole system. In CPS, a compromised-key attack might happen by capturing the sensors and doing reverse engineering jobs or by pretending to be a valid sensor node in order to find out the keys.

3.3 Man-in-the-Middle Attack (MITM)

MITM is the type of attack in which the malicious actor has access to the communication channel in which two parties are transmitting data in it and can possibly alter the information. In MITM attack to CPS, the malicious attacker tries to send either false positive or false negative to the operator and causing the system to take an undesirable action, or prevent it to take any action while it is required [12]. There are several types of man-in-the-middle attacks, including IP Spoofing, DNS Spoofing, SSL hijacking, Email hijacking and etc.

3.4 Denial-of Service

Denial of Service attack is another kind of attack method in which the attacker does not steal or alter the data; instead, it makes the system's resources unavailable. DoS attack consists of two general techniques: Flooding and Crashing. Flood attack happens when too much traffic is sent to the server, which results in slowing down and eventually stopping the whole system and Crash attack occurs when the perpetrator exploits vulnerabilities to the system and result in making the system crash. Another type of DoS attack is Distributed Denial of Service attack (DDoS). In DDoS attack, the target system is attacked from different locations and makes the detection and prevention of the attack difficult tasks, since a considerable number of machines from different locations are constantly targeting the system.

3.5 *Eavesdropping*

Eavesdropping is considered a passive attack, in which that intruder does not interfere with the operation of the system and only observes the operation. For instance, in CPS, the attacker can monitor the data, which is transferring through communication channels. Eavesdropping violates confidentiality in information security.

3.6 *Code Injection*

Code injection is another type of attack in which the intruder tries to insert malicious software codes to the system in order to interrupt the systems' operation. SQL injection, remote file injection and cross-site scripting are different types of code injection. SQL injection involves inserting SQL statements in the queries and causing in failing to receive input data, while cross-site scripting occurs when malicious code is added into a web application. When malicious code is downloaded from the internet, and the code is extending by itself, remote file injection happened.

4 Anomaly Detection

Anomaly detection is considered as a promising technique that can protect cyber-physical systems from any kind of vulnerabilities and attack. Therefore, it has been used in a wide range of applications such as smartphone malware detection [48, 49], medical anomaly detection [15] and fraud detection.

Anomaly detection consists of two different phases. In the first part, the correct pre-defined behavior of the system is modeled by using machine learning and data analytics algorithms, and the performance of the system is calculated. In the second phase, anomalies can be detected by comparing the current behavior with the normal model. In this part, the system is monitored at runtime, and any deviation from the normal expected behavior can be discovered. The main feature of anomaly-based detection is the ability to detect any unknown vulnerability, which there is no prior knowledge about them and use it as a second line of defense in the system.

One problem with this technique is the high rate of false-positive detection in situations that normal behavior and attack data are similar to each other or when the normal behavior (training dataset) is not stable and should change over time [50]. Another drawback of anomaly detection is the difficulty in analyzing and collecting the normal behavior of all the components and defining all possible data pattern which might occur in the system. Once the correct behavior is defined, secured system memory is required for maintaining the gathered data in order to detect correctly.

Security administrators face so many challenges in anomaly detection. For instance, the sensors in CPS might receive some noises when they are collecting data, and distinguishing between actual anomalies and the normal patterns of the system, in this case, is a difficult task. Moreover, some anomalies appear normal, and detecting them is a challenging job. Based on the type of processing, anomaly detection can be classified into three categories: Statistical-based, Knowledge-based and machine learning-based. The fundamentals and features of each category are described below [51].

4.1 Statistical-Based

In the statistical-based method, the behavior of each component is represented by different viewpoints, such as login session variables, timers, and resources. The network traffic activity is captured and the stochastic behavior of the system that includes traffic rate, the number of packets and the number of different IP addresses is formed. The system can identify the malicious activity by using statistical properties like mean and variance of current behavior from normal behavior. Statistical-based uses two different datasets during the anomaly detection: One of them corresponds to the currently observed profile, and the other corresponds to a previously trained profile. The earliest statistical approaches by using Gaussian random variables Denning and Neumann was modeled in 1985 [52]. Later, Ye et al. [53] proposed a multivariate model that considers the correlation between more than one metric in the system.

The main characteristic of this type of anomaly detection is decreased false rate among the other methods [51]. In other words, they can provide more accurate detection of malicious behavior over a long period of time. Another advantage of this type of detection is that general knowledge about the normal activity of the system is not required; instead, they have the ability to learn from observation.

However, statistical-based has some drawbacks which should be pointed out. First, the intruder still can attack the system by generating network traffic in such a way that it looks normal to the system. Second, all the models in the system cannot be modeled in stochastic methods. Moreover, statistical-based relies on the assumption of a quasi-stationary process, which is not always realistic [54].

4.2 Knowledge-Based

This type of detection gathers knowledge about the different types of attacks and uses them for detecting malicious behavior. If the CPS does not have any prior knowledge about the particular attack, then it is not able to detect it. In other words, the system should have a significant amount of knowledge (system data such as protocols, specifications, network traffic and etc.) about several attack types in order to behave correctly. The main advantage of knowledge-based detection is the less false-positive rates, of hence more accurate results than the other types of anomaly detection. Whilst the drawback is the time-consuming operation of gathering knowledge/data.

The knowledge-based detection technique can be classified into three categories of Finite State Machines, Expert systems, and signature analysis. The finite state machine was first proposed by Porras and Kemmerer [55] and is a sequence of states and transitions, in which the states graphically show different malicious activities that might occur to the system. Expert systems are a set of rules and methods that, in case of attack, the general approach can be generated and concluded from this set. Signature analysis is the same as the expert system, but the method in acquiring is different.

4.3 Machine Learning-Based

Machine learning is a collection of algorithms and statistical models, which are used by systems to find a pattern and make predictions and decisions based on the previous data in order to perform a task [56]. There are several types of machine learning algorithms that have different purposes, and the type of machine-learning algorithm can be chosen based on the type of input and output data. There are three main categories of machine learning algorithms: Supervised Learning, unsupervised learning, semi-supervised learning. When data is labeled, and the purpose of using machine learning is to find a model that explains the dataset, the problem is called supervised learning, while when the main objective of the program is to find a pattern for unlabeled data, the problem is considered unsupervised. In semi-supervised learning, only a portion of the data is labeled. Machine learning can be used in so many applications such as CPSs when the prediction is one of the vital operations. By using machine learning algorithms, supervised or unsupervised learning anomaly detectors can be built, which can lead CPSs to be secured.

The machine learning approaches usually separates data into different categories: Training and testing. Training data, which commonly has a bigger portion, is used for learning and providing a model for the system. However, testing data, which is completely independent of the training dataset, is used to assess the performance of the algorithm. In anomaly-based detection, the normal behavioral pattern is described and modeled by using the training set. Then, the model is applied to the testing dataset in order to classify it as either normal or anomalous. In addition, some machine learning methods separate datasets into three categories instead of two, including the validation dataset. The validation dataset is used in the machine-learning algorithms that their training set provides various models. For instance, the number of layers and nodes in artificial neural networks can be varied and deciding the best parameters that have less estimation of error and more efficient to be built depends on the performance on the validation dataset.

One important part of any anomaly detection is evaluating the performance of machine-learning algorithms. Classification accuracy is the most intuitive method in this evaluation, which measures the performance of the model by computing the ratio number of correct predictions to a total number of observations. The main drawback of this metric is that it works properly only when the dataset has equal values for false positive and false negative. F1 score is another metric in measuring the accuracy in uneven class distribution, which computes the balance between Precision and Recall. Precision is the ratio of correctly predicted positive observation to total positive observation (Both true positives and false positives), while Recall is the ratio of correct positive prediction to a total number of predictions in the same class (True positives and false negatives of the same class). As a result, F1-score can compute the performance by taking both false positive and false negatives into account. In multi-label machine-learning algorithms, F1-score is usually used to evaluate the classification performance. Therefore, by maximizing the F1-score in multi-label classification, the performance of the algorithm can be considerably improved.

Different authors used machine learning methods to study anomaly detection [57]. For instance [15, 58] used machine learning algorithms for anomaly detection. Junejo and Jonathan [43] proposed an approach by modeling the physical system of CPS and detecting anomaly attacks when any changes in the behavior of the system have occurred. Moreover, Patric [59] successfully applied a machine-learning algorithm to detect anomaly attacks in the industrial gas system and it showed that in case of complex attacks, the performance of the detector decreases to near 25 percent. Furthermore, neural networks are used in attack detection in [60]. Moreover, Beaver et al. [61] applied different machine learning algorithms in the context of SCADA communication.

5 Machine-Learning Techniques

The three types of supervised learning algorithms, which are used the most in anomaly-detection in cyber-physical systems, are the neural network, support vector machine, and naïve Bayes. The following subsections will explain each of these algorithms with their mathematical approaches for classifying the dataset.

5.1 Artificial Neural Network (ANN)

The neural network is a supervised learning algorithm that can learn the classification function by training the dataset. Given a set of features $(x_1, \ldots, x_n)$ and an output y, the ANN architecture is made of input layer consists of a set of nodes $\{x_i \mid x_1, \ldots, x_n\}$ representing the input feature and hidden layers that process information based on specific weights $(w_1, \ldots, w_n)$ by the non-linear function of $w_1x_1 + w_2x_2 + \ldots + w_nx_n$. Therefore, tuning ANNs requires a number of parameters for the number of hidden nodes, layers, and iterations.

5.2 Support Vector Machine (SVM)

Support Vector Machine is a classifier algorithm with the goal of prediction, that classifies data by mapping them into a set of hyper-planes in high dimensions. New examples in SVM are predicted to be in a class based on their location in that space relative to the hyperplanes [62]. SVM can be applied to three kinds of outcomes: Bernoulli (binary), multinomial and continuous [63]. In order to simplify the computation, different kernels can be used. For instance, the Gaussian kernel, which is the most popular choice of nonlinear kernel, classifies data based on statistical variances and its mathematical equation is defined as follows [63]:

$$K(x_i, x_{i'}) = \exp\left\{-\gamma \sum_{j=1}^{p} \left(x_{ij} - x_{i'j}\right)^2\right\}$$

where γ is an additional tuning parameter called kernel coefficient, which regulates how close observations should be in order to contribute to the classification decision. The exponent becomes extremely negative and $K(x_i, x_{i'})$ becomes close to zero, when the training and testing observations are far from each other.

5.3 Naïve Bayes (NB)

Naïve Bayes was first adopted into the field of machine-learning in 1992 [64] and is a probabilistic supervised classifier, which is based on the Bayes theorem with the naïve assumption of conditional independence between every pair of features. The following relationship states Bayes theorem:

$$P(y|x_1, \ldots, x_n) = \frac{P(x_1, \ldots, x_n|y)\, P(y)}{P(x_1, \ldots, x_n)}$$

where y is the given class variable and $x_1, \ldots, x_n$ are dependent feature vectors.

Since it is assumed that, the classifier is using the naïve conditional independence:

$$P(x_i|y, x_1, \ldots x_{i-1}, x_{i+1}, \ldots, x_n) = P(x_i|y)$$

Therefore, from the above equations, the Naïve Bayes relationship is simplified to:

$$P(y|x_1, \ldots, x_n) = \frac{\left[\prod_{i=1}^{n} P(x_i|y)\right] P(y)}{P(x_1, \ldots, x_n)}$$

Each distribution in the Naïve Bayes classifier can be independently estimated as a one-dimensional distribution. As a result, compared to other sophisticated machine-learning classifiers, Naïve Bayes learners perform at extremely higher speed.

6 Case Study

In this study, three different types of supervised machine-learning algorithms are used in order to classify the behavior of the smart grid as either normal or malicious. Artificial neural networks (ANN), support vector machine (SVM), and Naïve Bayesian (NB) classifiers are implemented and evaluated in terms of classification prediction accuracy and F1-score.

6.1 Dataset

The data used in this study was drawn by fifteen datasets provided by the Mississippi State University power system attack dataset [62]. There were near 5200 data in each set and near 78,000 data in total, including samples of measurements in power systems for three different types of outputs: attack situations, natural operations, and no events. Attack situations are simulated with the assumption that the system is manipulated by the attacker, who gained access to the system for performing malicious activities. Different attack scenarios are modeled as follows [62]:

- Short-circuit-fault: When the attacker makes a short in power line
- False Data Injection: When the attacker manipulates the system parameters and causes a blackout
- Remote tripping command: When the attacker sends commands to a rely and causes a breaker to open

- Relay setting change: When the attacker modifies the configurations of a relay and causes the relay to be disabled and prevented from tripping

6.2 System Model

The Cyber-physical framework used in this study is a combination of control systems interacting with different smart electronic components and network, which monitors modules such as SNORT and Syslog systems. The model evaluated for this purpose is depicted in Fig. 6 [62].

As can be seen in the figure above, this model consists of two power generators (G), four LEDs that can control the four breakers (BR), and turn them on or off. These LEDs use a distance protection scheme, and in case of anomaly detection, they switch the breakers on. Moreover, there are two transmission lines from bus B1 to bus B2 and bus B2 to bus B3. The attack scenarios are generated with the assumption that the attacker has access to the substation networks and has the ability to pose threats by performing commands from the substation switch. In this analysis,

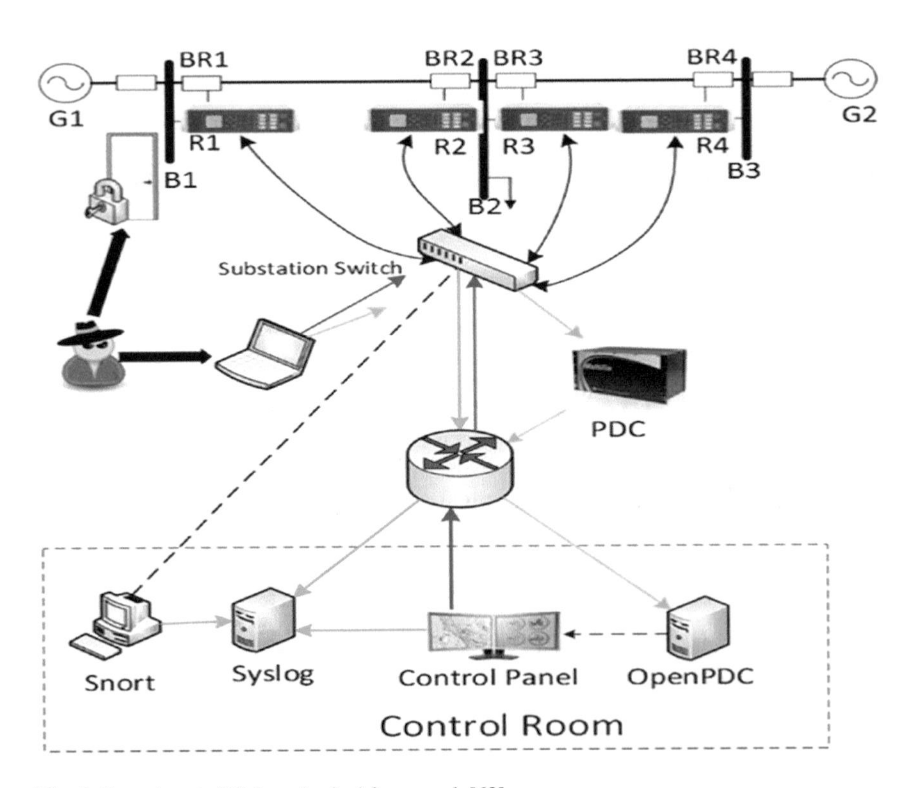

Fig. 6 Experimented Cyber-physical framework [62]

multiple attack scenarios are included in order to gain confidence that this system can perform properly and can detect anomaly detections in various situations [62].

6.3 Analytic Approach

The implementation of this experiment required four main steps:

1. Preprocessing of data: In order to have the highest efficiency, the data was analyzed, and among the 128 features, the useless ones were removed from the dataset. Besides, since the purpose of these implementations is to detect attack events, the outputs must be converted to binary format. Therefore, the three different statues of "Attack", "Natural" and "No event" were labeled to 1,0 and 0, respectively.
2. Splitting the data into training and testing set: The data for this study were split into two categories: 80% was used for the training phase, and the remaining 20% was used for the testing phase.
3. Training the model: Each of machine-learning algorithms can be implemented with different models and parameters. K-fold cross-validation is the method that finds the best parameters and the best model in each classifier. Although larger K might have more accurate results, in this process, three-fold cross-validation ensured the most efficient computational time.
4. Evaluating the performance: The classification algorithms in this study were evaluated based on the accuracy and f1-sore. Accuracy is the percentage of correct classification relative to the total number pf classification on testing data. The accuracy of each algorithm was computed and compared based on the results of the below equation: [14]

$$\text{Accuracy} = \frac{\text{Number of Correct Predictions}}{\text{Number of Testing Data}}$$

Besides, the performance of each algorithm is analyzed based on the True Positives (TP), the True Negatives (TN), the False Positives (FP), and the False Negatives (FN). TP refers to an outcome, where the model correctly predicts the positive classes and TN refers to the prediction of the negative outcomes correctly. While, FP is an outcome, where the model classifies the positive class incorrectly and FN is the outcome, where the model incorrectly predicts the negative classes. From these metrics, precision, recall, and F-1 score can be defined:

$$\text{Precision} = \frac{\text{TP}}{\text{TP} + \text{FP}}$$

$$\text{Recall} = \frac{\text{TP}}{\text{TP} + \text{FN}}$$

$$F1 - \text{Score} = \frac{2 \times \text{precison} \times \text{Recall}}{\text{Precision} + \text{Recall}}$$

6.3.1 Classification Algorithms

ANN The ANN algorithm was implemented with a varying learning rate of α (from 10^{-10} to 10^{10}), hidden layers (from one layer to six layers), and nodes (from 8 to 2048 nodes).

SVM This algorithm was tuned by Gaussian kernel and was cross-validated with varying penalty parameters, C and Kernel coefficient, λ in range of 0.001 to 1000.

Naïve Bayes This probabilistic classifier was implemented by two different kernels (Gaussian and Bernoulli Kernel) and was tuned with varying α in range of 0 to 4.

6.4 *Results and Discussion*

Parameter optimization for each of these three algorithms was performed through cross-validation with various parameters in order to achieve the highest accuracy. Figure 7 illustrates the accuracy of SVM classification with different parameters on the Mississippi State University power system attack dataset. Since C indicates the trade-off between training error and the complexity of the model, a large value of C might cause losing the generalization of properties and overfitting as well as increasing the training computational time. In contrast, a small value of C might cause the optimizer to misclassify soma data and result in less accuracy. As it can be seen in Fig. 7, the best possible C leading to the highest accuracy in the given dataset was 1000.

NB classifier with Gaussian Kernel did not perform successfully, and the highest accuracy achieved was 31%, while this classifier with Bernoulli kernel performed better, and the accuracy reached 60%. The results for these two kernels in Naïve Bayes classifier can be seen in Fig. 8. Since Naive Bayes classifier has conditional independence assumption, it did not perform properly and was a bad estimator for this study. This classifier assumes that the features are independent of each other, while multiple features in this study depend on one other classes.

Optimal parameters for these classifiers were selected based on the maximum accuracy rate achieved for a given dataset. These parameters and the corresponding accuracies are stated in Table 1.

The training and validation accuracy of deep learning neural network with the proposed architecture was provided by TensorBoard and can be visualized in Fig. 9.

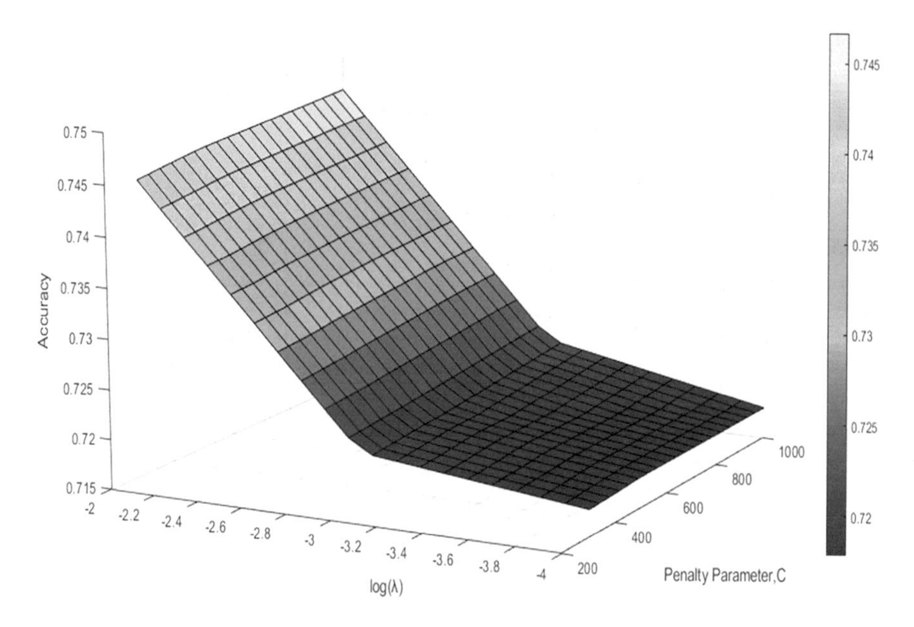

Fig. 7 Accuracy of SVM with varying parameters of C and λ

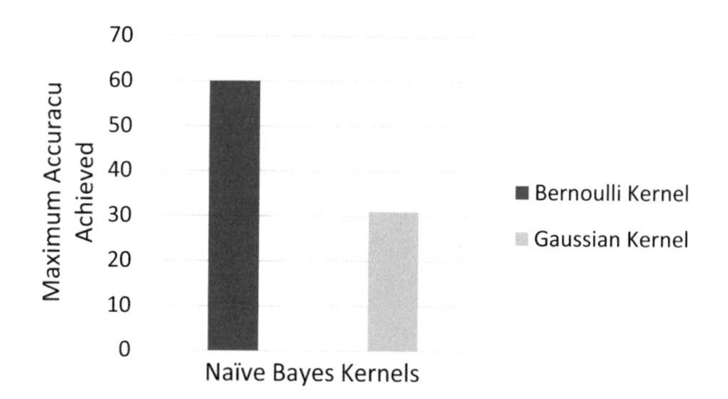

Fig. 8 Accuracy of NB with different kernels

The model was trained for 350 epochs and the accuracy on both datasets was increased until it was not raising anymore. Since the gap between the accuracy of these two datasets is not huge, it can be concluded that the parameters were chosen wisely and the model did not face over fitting nor underfitting.

The results show that ANN with the proposed architecture was successful at anomaly detection, and it achieved the highest accuracy among these three classifiers. On the other hand, although the NB classifier had less computational time and it performed faster than other algorithms, it had low performance.

Table 1 Optimal parameters for each algorithm with the corresponding accuracy

Algorithms	Optimal parameters	Precision (%)	Recall (%)	F1-score (%)	Accuracy	Computational time (s)
ANN	Optimizer: Adam layer size: (256,128,128)	96	96	96	94	131.9
SVM	C = 1000, $\lambda = 0.01$	77	92	84	74	1.49
Naïve Bayes	Kernel: Bernoulli $\alpha = 0.5$	74	66	70	60	0.0099

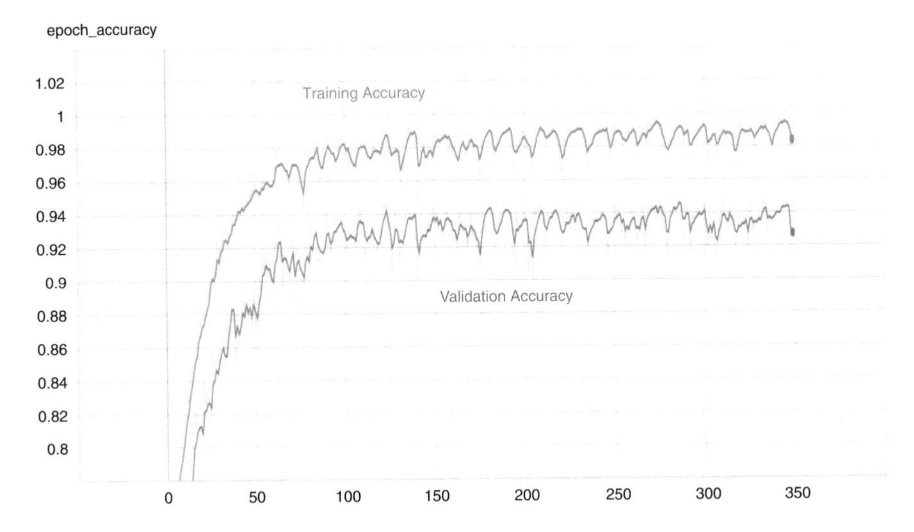

Fig. 9 Accuracy of proposed ANN architecture

7 Conclusion

This paper provides a timely review on the security of cyber-physical systems since there are a considerable number of threats and vulnerabilities to such systems. Special emphasis is placed on providing a general framework for anomaly detection after discussing major challenges and issues of cyber-physical systems. Machine-learning is explained as a promising approach in anomaly-based detection, and different research direction on anomaly detection within the scope of machine-learning is discussed.

Additionally, the effectiveness results of three different supervised machine-learning classifiers, including neural network, support vector machine, and naïve Bayes in classifying power systems behaviors and detecting the future attacks, are presented with respect to the accuracy and f1-score. Based on the results of applying machine-learning methods for anomaly detection, it can be concluded that machine-learning and deep learning are viable approaches in detecting attacks in power system operators and among different algorithms that were used in this study, Neural Network provided the highest performance and outperform the other algorithms.

References

1. V. Gunes, S. Peter, T. Givargis, F. Vahid, A survey on concepts, applications, and challenges in cyber-physical systems. KSII Trans. Internet Inform. Syst. **8**(12), 4242–4268 (2014)
2. J. Goh, S. Adepu, M. Tan, Z.S. Lee, Anomaly detection in cyber physical systems using recurrent neural networks, in *Proceedings of IEEE International Symposium on High Assurance Systems Engineering* (2017), pp. 140–145
3. A. Jones, Z. Kong, C. Belta, Anomaly detection in cyber-physical systems: A formal methods approach, in *Proceedings of the IEEE Conference on Decision and Control* (2014), pp. 848–853
4. D. Serpanos, The cyber-physical systems revolution. Computer (Long. Beach. Calif). **51**(3), 70–73 (2018)
5. C.K. Keerthi, M.A. Jabbar, B. Seetharamulu, Cyber physical systems (CPS): Security issues, challenges and solutions, in *IEEE International Conference on Computational Intelligence and Computing Research*, ICCIC 2017 (2017), pp. 1–4
6. H. Karimipour, H. Leung, Relaxation-based anomaly detection in cyber-physical systems using ensemble kalman filter. IET Cyber-Phys. Syst. Theor. Appl. **5**(1), 49–58 (2019)
7. E.M. Dovom, A. Azmoodeh, A. Dehghantanha, D.E. Newton, R.M. Parizi, H. Karimipour, Fuzzy pattern tree for edge malware detection and categorization in IoT. J. Syst. Archit. **97**, 1–7 (2019)
8. H. Karimipour, A. Dehghantanha, R.M. Parizi, K.K.R. Choo, H. Leung, A deep and scalable unsupervised machine learning system for cyber-attack detection in large-scale smart grids. IEEE Access **7**, 80778–80788 (2019)
9. C.W. Tsai, C.F. Lai, M.C. Chiang, L.T. Yang, Data mining for internet of things: A survey. IEEE Commun. Surv. Tutorial. **16**(1), 77–97 (2014)
10. R. Altawy, A.M. Youssef, Security tradeoffs in cyber physical systems: A case study survey on implantable medical devices. IEEE Access **4**, 959–979 (2016)
11. H. Karimipour, V. Dinavahi, Robust massively parallel dynamic state estimation of power systems against cyber-attack. IEEE Access **6**, 2984–2995 (2017)
12. E.K. Wang, Y. Ye, X. Xu, S. M. Yiu, L.C.K. Hui, K.P. Chow, Security issues and challenges for cyber physical system, in *Proceedings—2010 IEEE/ACM International Conference on Green Computing and Communications, GreenCom 2010, 2010 IEEE/ACM International Conference on Cyber, Physical and Social Computing, CPSCom 2010* (2010), pp. 733–738
13. J. Madden, B. McMillin, A. Sinha, Environmental obfuscation of a cyber physical system—vehicle example, in *Proceedings—International Computer Software and Applications Conference* (2010), pp. 176–181
14. S. Mohammadi, H. Mirvaziri, M. Ghazizadeh-Ahsaee, H. Karimipour, Cyber intrusion detection by combined feature selection algorithm. J. Inform. Secur. Appl. **44**, 80–88 (2019)
15. V. Chandola, A. Banerjee, V. Kumar, Anomaly detection: A survey. *ACM Comput. Surv.***41**(3), 1–58 (2009)
16. H. Karimipour, S. Geris, A. Dehghantanha, H. Leung, Intelligent anomaly detection for large-scale smart grids, in *2019 IEEE Canadian Conference of Electrical and Computer Engineering, CCECE 2019* (2019)
17. E.A. Lee, S.A. Seshia, *Introduction to Embedded Systems: A Cyber-Physical Approach* (MIT Press, Cambridge, 2017)

18. Y. Luo, K. Chakrabarty, T. Y. Ho, A cyberphysical synthesis approach for error recovery in digital microfluidic biochips, in *Proceedingsof Design, Automation and Test in Europe* (2012), pp. 1239–1244
19. E.A. Lee, Cyber-physical systems—Are computing foundations adequate? *NSF Work. Cyber-Phys. Syst.***1**, 1–9 (2006)
20. R. Rajkumar, I. Lee, L. Sha, J. Stankovic, *Cyber-Physical Systems: The Next Computing Revolution Conference*, Anaheim, California, USA (2010)
21. S. Geris, H. Karimipour, Joint state estimation and cyber-attack detection based on feature grouping, in *Proceedings of 2019 the 7th International Conference on Smart Energy Grid Engineering, SEGE 2019* (2019), pp. 26–30
22. J. Li, H. Gao, and B. Yu, Concepts, features, challenges, and research progresses of CPSs. *Development Report of China Computer Science* (2009), pp. 1–17
23. B.H. Krogh, Cyber-physical systems: The need for new models and design paradigms. Comput. Eng. **1**, 1–1
24. B.X. Huang, Cyber physical systems: A survey. *Present Report* (2008)
25. H. HaddadPajouh, A. Dehghantanha, R.M. Parizi, M. Aledhari, H. Karimipour, A survey on internet of things security: Requirements, challenges, and solutions. *J. Internet Things***1**, 100129 (2019)
26. E.A. Lee, Cyber physical systems: Design challenges, in *Proceedings—11th IEEE Symposium on Object/Component/Service-Oriented Real-Time Distributed Computing, ISORC 2008* (2008), pp. 363–369
27. T. Sanislav, L. Miclea, Cyber-physical systems—Concept, challenges and research areas. *J. Contr. Eng. Appl. Informt.* (2012). Accessed 28 Oct 2019. http://www.ceai.srait.ro/index.php?journal=ceai&page=article&op=view&path%5B%5D=1292&path%5B%5D=968
28. J. Shi, J. Wan, H. Yan, H. Suo, A survey of cyber-physical systems, in *2011 International Conference on Wireless Communications and Signal Processing, WCSP 2011* (2011)
29. R. Baheti, H. Gill, Cyber-physical systems (2011). Accessed 28 Oct 2019. https://scholar.google.ca/scholar?q=Baheti,+Radhakisan,+and+Helen+Gill.+%22Cyber-physical+systems.%22&hl=en&as_sdt=0&as_vis=1&oi=scholart
30. E.A. Lee, CPS foundations, in *Proceedings of Design Automation Conference* (2010), pp. 737–742
31. Y. Mo et al., Cyber-physical security of a smart grid infrastructure. Proc. IEEE **100**(1), 195–209 (2012)
32. J. Sakhnini, H. Karimipour, A. Dehghantanha, R.M. Parizi, G. Srivastava, Security aspects of Internet of Things aided smart grids: A bibliometric survey. J. Internet Things **1**, 100111 (2019)
33. R.M. Lee, M.J. Assante, T. Conway, *Analysis of the Cyber Attack on the Ukrainian Power Grid Defense Use Case* (E-ISAC, Washington, DC, 2016)
34. N. Falliere, L.O. Murchu, and E. Chien Security response contents (2011)
35. A.A. Cárdenas, S. Amin, S. Sastry, Secure control: Towards survivable cyber-physical systems, in *Proceedings—International Conference on Distributed Computing Systems* (2008), pp. 495–500
36. A.A. Cárdenas, S. Amin, S. Sastry, Research challenges for the security of control systems
37. M. Yampolskiy, P. Horvath, X.D. Koutsoukos, Y. Xue, J. Sztipanovits, Taxonomy for description of cross-domain attacks on CPS, in *HiCoNS 2013—Proceedings of the 2nd ACM International Conference on High Confidence Networked Systems, Part of CPSWeek 2013*, (2013), pp. 135–142
38. M. Begli, F. Derakhshan, H. Karimipour, A layered intrusion detection system for critical infrastructure using machine learning, in *Proceedings of 2019 the 7th International Conference on Smart Energy Grid Engineering, SEGE 2019*, (2019), pp. 120–124
39. S. Mohammadi, V. Desai, H. Karimipour, Multivariate Mutual Information-based Feature Selection for Cyber Intrusion Detection, in *2018 IEEE Electrical Power and Energy Conference, EPEC 2018* (2018)

40. H.J. Liao, C.H. Richard Lin, Y.C. Lin, K.Y. Tung, Intrusion detection system: A comprehensive review. J. Netw. Comput. Appl. **36**(1), 16–24 (2013)
41. K. Scarfone, P. Mell, Guide to intrusion detection and prevention systems (IDPS). Natl. Inst. Stand. Technol. **800**(94), 127 (2007)
42. C. Alcaraz, L. Cazorla, G. Fernandez, Context-awareness using anomaly-based detectors for smart grid domains, in *Lecture Notes in Computer Science (including subseries Lecture Notes in Artificial Intelligence and Lecture Notes in Bioinformatics)*, vol. 8924, (2015), pp. 17–34
43. K.N. Junejo, J. Goh, Behaviour-based attack detection and classification in cyber physical systems using machine learning, in *CPSS 2016—Proceedings of the 2nd ACM International Workshop on Cyber-Physical System Security, Co-located with Asia CCS 2016*, (2016), pp. 34–43
44. R. Sekar et al., Specification-based anomaly detection: A new approach for detecting network intrusions, in *Proceedings of the ACM Conference on Computer and Communications Security*, (2002), pp. 265–274
45. J. Sakhnini, H. Karimipour, A. Dehghantanha, Smart grid cyber attacks detection using supervised learning and heuristic feature selection, in *Proceeding of IEEE SEGE 2019* (2019), pp. 108–112
46. S. Agrawal, A. Jitendra, Survey on anomaly detection using data mining techniques. Proc. Comput. Sci. **60**, 708–713 (2015)
47. H. Zhong, D. Du, C. Li, X. Li, A novel sparse false data injection attack method in smart grids with incomplete power network information. Complexity **2018**, 1–16 (2018)
48. G. Dini, F. Martinelli, A. Saracino, D. Sgandurra, MADAM: A multi-level anomaly detector for android malware, in *Lecture Notes in Computer Science (including subseries Lecture Notes in Artificial Intelligence and Lecture Notes in Bioinformatics)*, vol. 7531 (LNCS, 2012), pp. 240–253
49. I. Burguera, U. Zurutuza, S. Nadjm-Tehrani, Crowdroid: Behavior-based malware detection system for android, in *Proceedings of the ACM Conference on Computer and Communications Security* (2011), pp. 15–25
50. I. Žliobaitė, Learning under concept drift: an overview (2010)
51. S. Jose, D. Malathi, B. Reddy, D. Jayaseeli, A survey on anomaly based host intrusion detection system. J. Phys. Conf. Ser. **1000**(1), 12049 (2018)
52. D.E. Denning, P.G. Neumann, Requirements and model for IDES-a real-time intrusion detection expert system. SRI Int. **333** (1985)
53. N. Ye, S.M. Emran, Q. Chen, S. Vilbert, Multivariate statistical analysis of audit trails for host-based intrusion detection. IEEE Trans. Comput. **51**(7), 810–820 (2002)
54. P. García-Teodoro, J. Díaz-Verdejo, G. Maciá-Fernández, E. Vázquez, Anomaly-based network intrusion detection: Techniques, systems and challenges. Comput. Secur. **28**(1–2), 18–28 (2009)
55. P. A. Porras, A. Valdes Live traffic analysis of TCP/IP gateways
56. A.N. Jahromi, J. Sakhnini, H. Karimpour, A. Dehghantanha, A deep unsupervised representation learning approach for effective cyber-physical attack detection and identification on highly imbalanced data, in *Proceedings of the 29th Annual International Conference on Computer Science and Software Engineering*. IBM Corp., (2019), pp. 14–23
57. A. Namavar Jahromi et al., An improved two-hidden-layer extreme learning machine for malware hunting. Comput. Secur. **89**, 101655 (2020)
58. C.F. Tsai, Y.F. Hsu, C.Y. Lin, W.Y. Lin, Intrusion detection by machine learning: A review. Expert Syst. Appl. **36**(10), 11994–12000 (2009)
59. P. Nader, P. Honeine, P. Beauseroy, Lp-norms in one-class classification for intrusion detection in SCADA systems. IEEE Trans. Ind. Inf **10**(4), 2308–2317 (2014)
60. W. Gao, T. Morris, B. Reaves, D. Richey, On SCADA control system command and response injection and intrusion detection, in *General Members Meeting and eCrime Researchers Summit, eCrime 2010* (2010)

61. A. Almalawi, A. Fahad, Z. Tari, A. Alamri, R. Alghamdi, A.Y. Zomaya, An efficient data-driven clustering technique to detect attacks in SCADA systems. IEEE Trans. Inf. Forensics Secur. **11**(5), 893–906 (May 2016)
62. R.C. Borges Hink, J.M. Beaver, M.A. Buckner, T. Morris, U. Adhikari, S. Pan, Machine learning for power system disturbance and cyber-attack discrimination, in *7th International Symposium on Resilient Control Systems, ISRCS 2014* (2014)
63. N.G. Guenther, M. Schonlau, Support vector machines. State J. **16**(4), 917–937 (2016)
64. P. Langley, W. Iba, and K. Thompson, An analysis of Bayesian classifers. Proc. Tenth Natl. Conf. Artif. Intell.**415**, 223–228 1992

Data-Driven Anomaly Detection in Modern Power Systems

Mucun Sun and Jie Zhang

1 Introduction

Date-driven methods have been witnessed to have significant impacts on power system operations. Large amounts of data are being collected by smart grid devices, such as advanced metering infrastructures and phasor measurement units, which facilitates deep applications of big data and data-driven techniques to power systems.

The power grid is a complex cyber-physical system incorporating a vast volume of distributed devices, which by nature results in a large attack surface. Malicious attackers can compromise power devices, communication and control facilities, or market interfaces, leading to local outages, equipment damages, grid instabilities, or individual financial gains.

Cyber attacks in the recent literature can be generally classified into three groups: (1) false data injection attacks, that compromise data integrity of the measurements; (2) physical response attacks, where the attackers leverage the vulnerabilities and easy access of grid edge devices to create certain power usage patterns, with the objective to destabilize the system operation or marketing; (3) market interface attacks, where attackers disturb the bid/offer process of market participants, with the objective to gain financial profits or generate social chaos with sharp price changes and even widespread outages.

One of the promising cyber defense approaches, i.e., non-intrusive to the operational system and adds additional protection, is the physical response-based anomaly detection. For cyber-physical systems, evaluating physical performance from sensor data is a common practice, but using these data to detect cyber attacks

M. Sun · J. Zhang (✉)
The University of Texas at Dallas, Richardson, TX, USA
e-mail: Mucun.Sun@utdallas.edu; jiezhang@utdallas.edu

H. Karimipour et al. (eds.), *Security of Cyber-Physical Systems*,
https://doi.org/10.1007/978-3-030-45541-5_7

is under-developed. A few anomaly detection technologies have been presented in the literature with power systems applications. For example, Wang et al. [17] presented a power consumption anomaly detection method based on long short-term memory (LSTM) point forecasts. In [9], Krishna et al. adopted the Principal Component Analysis (PCA) and density-based spatial clustering on noise pattern to detect the anomalies which are deviations from the normal electricity consumption behavior. Kim et al. [11] presented a framework which utilized spatial and temporal correlation between multiple solar farms to defend against data integrity attacks and learns the inter-farm/intra-farm correlation between measurements to perform anomaly detection.

To better understand data-driven anomaly detection applications in power systems, this chapter first reviews state-of-the-art anomaly detection methods in electricity markets, especially on locational marginal price (LMP) anomaly detection. Then, two data-driven LMP anomaly detection methods are developed, including a deterministic anomaly detection method and a probabilistic anomaly detection method.

2 Anomaly Detection Methods Review

Anomaly detection methods can be categorized based on different criteria, such as detection task (e.g., electricity markets, automated surveillance systems, and transportation systems) and method type. Figure 1 shows a general classification of anomaly detection methods, which are divided into clustering-based methods and regression-based methods. Clustering-based methods classify data into different clusters based on its intrinsic pattern and detect abnormal behavior, while regression-based methods correlate the anomaly behavior with predictors.

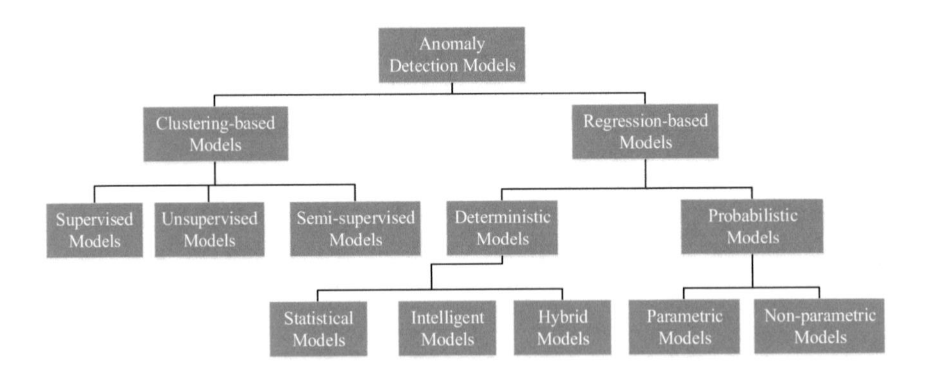

Fig. 1 Anomaly detection method categorization

2.1 Clustering-Based Anomaly Detection Methods

Generally, clustering-based anomaly detection methods can be further divided into three categories: supervised models, unsupervised models, and semi-supervised models. Supervised clustering-based models refer to methods utilizing labelled data to train detection models. Unsupervised clustering-based models are label-free, which requires no training data and detects the anomaly deviation in measurements. Semi-supervised anomaly detection techniques construct a model representing normal behavior from a given normal training data set, and then test the likelihood of a test instance to be generated by the learnt model. Conventional supervised clustering-based models include: support vector machine (SVM), neural network, linear and logistics regression, random forest, and classification trees. Conventional unsupervised clustering-based models include: k-means, k-nearest neighbors (KNN), information theory, and hierarchical clustering. Buzau et al. [2] used an extreme gradient boosted trees to detect electricity losses, which used all the information from the smart meters (i.e., energy consumption, alarms, and electrical magnitudes) to train a supervised model. An unsupervised anomaly detection algorithm was proposed by Mohammadpourfard et al. [12] to identify cyber attacks in power systems that are caused by sustainable energy sources or system reconfigurations. Zhou et al. [22] proposed a semi-supervised anomaly detection method through a density-based spatial clustering of applications with noise (DBSCAN) algorithm, where limited amount of labelled data is utilized to guide the anomaly detection process.

2.2 Regression-Based Anomaly Detection Methods

Generally, regression-based anomaly detection methods can be further divided into two categories: deterministic forecasting-based models and probabilistic forecasting-based models. Deterministic forecasting-based models refer to the utilization of deterministic forecasting models to identify anomaly behavior. Probabilistic forecasting-based models refer to models using prediction intervals or scenarios to quantitatively describe anomaly events. Conventional deterministic forecasting-based models include: (1) statistical models, such as the autoregressive model and the autoregressive integrated moving average (ARIMA) model; (2) machine learning models, such as artificial neural networks (ANNs), support vector regression (SVR), random forest (RF), and gradient boosting machine (GBM); and (3) hybrid models, such as ANN-ARIMA and SVR-ARIMA. For example, Li et al. [10] proposed an ARIMA-based anomaly detection model in real-time water consumption to detect meter stilting. Fontugne et al. [7] adopted a Fourier transform-based empirical mode decomposition model to detect the abnormal electricity consumption behavior. Esmalifalak et al. [5] proposed a support vector machine-based techniques for stealthy attack detection in smart grid. Wang et al.

[18] adopted a deep autoencoder for phasor measurement unit data manipulation anomaly detection. For hybrid models, more deterministic anomaly detection models can be found in a review paper [8].

The anomaly detection methods discussed above are mainly deterministic models, and thus insufficient to characterize uncertainties in cyber attacks. Probabilistic approaches that provide quantitative uncertainty information associated with cyber attacks are therefore expected to better assist power system operations. Probabilistic anomaly detection methods usually take the form of prediction intervals, quantiles, or predictive distributions to quantify the abnormal behavior in power systems. Generally, probabilistic anomaly detection methods can be classified into parametric and nonparametric approaches [14]. Parametric approaches generally require low computational cost since a prior assumption of the predictive distribution shape is made before parameter estimation. Nonparametric approaches are distribution free, and their predictive distributions are inferred through observations or scenarios. Only a few recent studies have been performed in the literature on probabilistic anomaly detection. For example, Zhang et al. [21] proposed a probabilistic anomaly detection method for a wind farm based on bootstrap and ensemble neural networks.

3 Deterministic and Probabilistic LMP Anomaly Detection

3.1 Deterministic Anomaly Detection

A deterministic forecasting-based anomaly detection methodology is developed for electricity price, named the machine learning based multi-model error distribution analysis (M3-EA), as shown in Fig. 2. The M3-EA deterministic model has two steps. In the first step, a two-layer forecasting method is adopted to generate deterministic forecasts. The first layer consists of several machine learning models, including ANN, SVM, GBM, and RF, which are built based on historical data. These models forecast locational marginal price as the output. A blending model is developed in the second layer to combine the forecasts produced by different algorithms from the first layer, and to generate deterministic forecasts. This blending model is expected to integrate the advantages of different algorithms by canceling or smoothing the local forecasting errors. The mathematical description is shown as:

$$y_i = f_i(x_1, x_2, \ldots, x_p) \tag{1}$$

$$\hat{y} = \Phi(y_1, y_2, \ldots, y_m) \tag{2}$$

where $f_i(\cdot)$ is the ith algorithm, y_i is the LMP forecasted by $f_i(\cdot)$, and $\hat{y}$ is the second-layer blending algorithm.

Once LMP deterministic forecasts are generated by the M3 deterministic forecasting model, a Gaussian mixture model [15] is adopted in the second anomaly

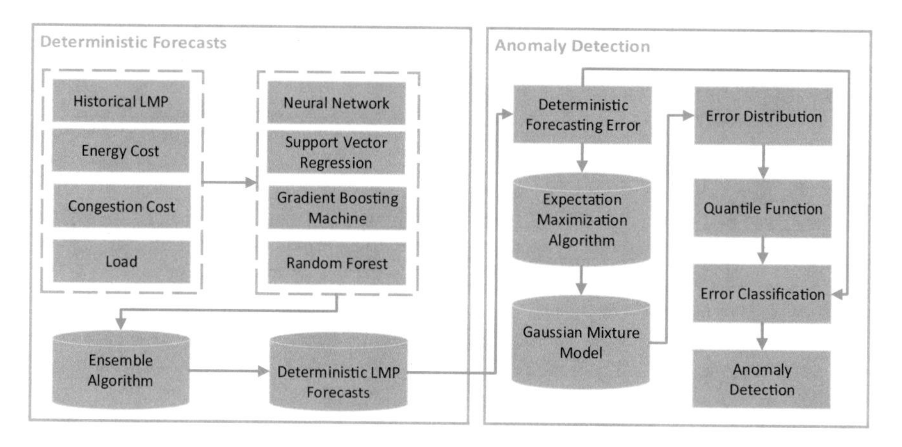

Fig. 2 Framework of the M3-EA deterministic anomaly detection model

detection step to model the distribution shape of the LMP forecasting error from the first step, and to generate empirical percentiles. The probabilistic density function (PDF) of GMM is formulated as follows:

$$f_G(x|N_G, \omega_i, \mu_i, \sigma_i) = \sum_{i=1}^{N_G} \omega_i g_i(x|\mu_i, \sigma_i) \tag{3}$$

where N_G is the number of mixture components, $U(\mu_i \in U)$ is the expected value vector, $\Sigma(\sigma_i \in \Sigma)$ is the standard deviation vector, and the $\Omega(\omega_i \in \Omega)$ is the weight vector. Each component $g(x; \mu_i, \sigma_i)$ follows a normal distribution, which is expressed as:

$$g(x|\mu, \sigma) = \frac{1}{\sqrt{2\pi\sigma^2}} e^{-\frac{(x-\mu)^2}{2\sigma^2}} \tag{4}$$

The GMM distribution has two constraints: (1) the integral of Eq. (5) equals unity, and (2) the summation of weight parameters equals unity as well, which are expressed as follows:

$$\int_{-\infty}^{+\infty} f_G(x|N_G, \omega_i, \mu_i, \sigma_i)\, dx = \int_{-\infty}^{+\infty} \sum_{i=1}^{N_G} \omega_i g_i(x|\mu_i, \sigma_i)\, dx = 1 \tag{5}$$

$$\sum_{i=1}^{N_G} \omega_i = 1 \tag{6}$$

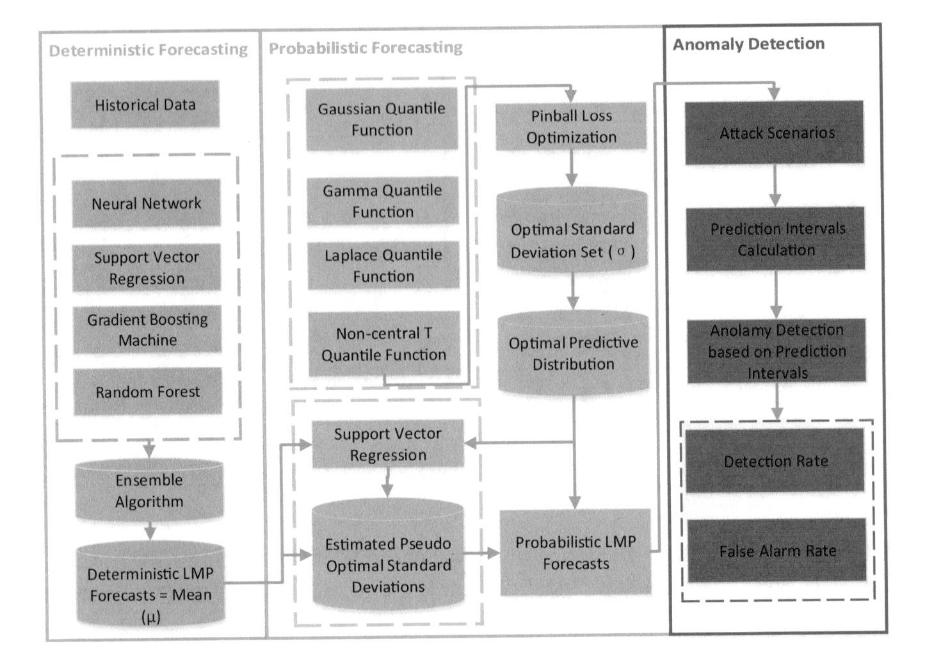

Fig. 3 Framework of the M3 probabilistic anomaly detection model

The parameters of GMM are estimated through the expectation maximization (EM) algorithm. The goal of EM is to maximize the likelihood function with respect to the parameters. More details about EM can be found in [1].

Once the PDF of error distribution is estimated, the percentiles can be calculated. The anomaly behaviors can be detected when the forecasting error falls out of a certain percentile range.

3.2 *Probabilistic Anomaly Detection*

In addition to the deterministic forecasting-based anomaly detection model, an M3 probabilistic anomaly detection framework is also developed in this chapter. The overall framework of the developed probabilistic anomaly detection methodology is illustrated in Fig. 3, which consists of three major steps: (1) deterministic LMP forecasting, (2) probabilistic LMP forecasting, and (3) anomaly detection. In the first step, the M3 model is adopted to generate deterministic forecasts.

After obtaining deterministic forecasts, a multi-distribution database is formulated to model possible shapes of the predictive distribution. Four widely used predictive distribution types are considered: Gaussian, Gamma, Laplace, and non-central t distributions. PDFs of the four distribution types are summarized in [16].

PDFs of the four distributions can be represented by the mean μ and standard deviation σ as $f(x|\mu, \sigma)$, and the corresponding cumulative distribution functions (CDFs) are deduced and denoted as $F(x|\mu, \sigma)$. A set of unknown parameters in the predictive distribution are determined by minimizing the pinball loss [6], using the genetic algorithm (GA) [3]. Note that the optimal distribution parameter is adaptive and dynamically updated based on the deterministic forecast value at each time stamp. The optimal adaptive predictive distribution parameters are first determined offline with the historical training data. Then a SVR surrogate model is developed to represent the optimized distribution parameter as a function of the deterministic forecast. The mathematical description of the SVR surrogate model is shown as:

$$\hat{\sigma} = f(\hat{x}_i) \tag{7}$$

where $\hat{x}_i$ is a point forecast, and $f(\cdot)$ is the SVR surrogate model of the optimal standard deviation of the predictive distribution. At the online forecasting stage, the surrogate model is used together with deterministic forecasts to adaptively predict the unknown distribution parameters and thereby generate probabilistic forecasts. Details about the M3 probabilistic forecasting method can be found in [14].

Once probabilistic forecasts are generated, prediction intervals (PIs) can be calculated. Anomaly behaviors can be detected when observations fall out of a certain PI range.

4 Results and Discussion

The performance of the deterministic anomaly detection and probabilistic anomaly detection models are evaluated in this section. Three evaluation metrics are used to assess the deterministic forecasting accuracy, which are the normalized root mean squared error (nRMSE), normalized mean absolute error (nMAE), and mean absolute percentage error (MAPE). The mathematical expressions of the three metrics are expressed as:

$$nRMSE = \frac{1}{x_{max}} \sqrt{\frac{\sum_{i=1}^{n} (\hat{x}_i - x_i)^2}{n}} \tag{8}$$

$$nMAE = \frac{1}{n} \sum_{i=1}^{n} \left| \frac{\hat{x}_i - x_i}{x_{max}} \right| \tag{9}$$

$$MAPE = \frac{100\%}{n} \sum_{i=1}^{n} \left| \frac{x_i - \hat{x}_i}{x_i} \right| \tag{10}$$

where $\hat{x}_i$ is the forecasted LMP, x_i is the LMP observation, x_{max} is the maximum LMP observation, and n is the sample size. For these metrics, a smaller value indicates better forecasting performance.

In order to carry out the performance of detection, we calculate the detection rate (DR) and false alarm rate (FAR) of the anomaly detection. The mathematical expressions of the two metrics are expressed as:

$$DR = \frac{TP}{TP + FP} \qquad FAR = \frac{FP}{FP + TN} \tag{11}$$

where TP, FP, and TN denote true positive, false positive, and true negative, respectively. DR measures the positive predictive value, and FAR measures the probability of falsely rejecting the null hypothesis for a particular test. For the DR metric, a value approaching 1 indicates strong detection performance; while for the FAR metric, a value closer to 0 indicates better performance.

4.1 Case Study I: Deterministic LMP Anomaly Detection Based on M3-EA

The M3-EA anomaly detection method is applied to the real-time LMP data from PJM with a 5-min resolution, which includes 5 zones (as shown in Table 1) with energy cost, congestion cost, and load forecasts information. The ratio between the number of training to testing samples is 3:1.

A false data injection attack (FDIA) is implemented in the case study. FDIA is one of the most common cyber attacks in power systems. Motivated by financial benefits, attackers can create false load estimation or transmission congestion limits to mislead real-time electricity pricing algorithms in producing biased LMPs. Attackers then take advantage of the biased LMP to gain monetary profits using bids and offers via a market interface. In this study, to generate FDIA, we shift load from morning (i.e., 8:00 a.m.–10:00 a.m.) to arbitrary a same time length in the afternoon or evening as a fake load profile. In this way, abnormal LMPs are generated.

To show the effectiveness of the M3-EA anomaly detection model, a persistence method (PS)-based error distribution analysis model is adopted as a baseline. In this study, the nRMSE, nMAE, and MAPE of the deterministic LMP forecasting through

Table 1 Data summary at selected zones

Case No.	Zone name	Duration
C1	AEP	07-01-2019 to 10-01-2019
C2	MIDATL	07-01-2019 to 10-01-2019
C3	COMED	07-01-2019 to 10-01-2019
C4	DEOK	07-01-2019 to 10-01-2019
C5	DOM	07-01-2019 to 10-01-2019

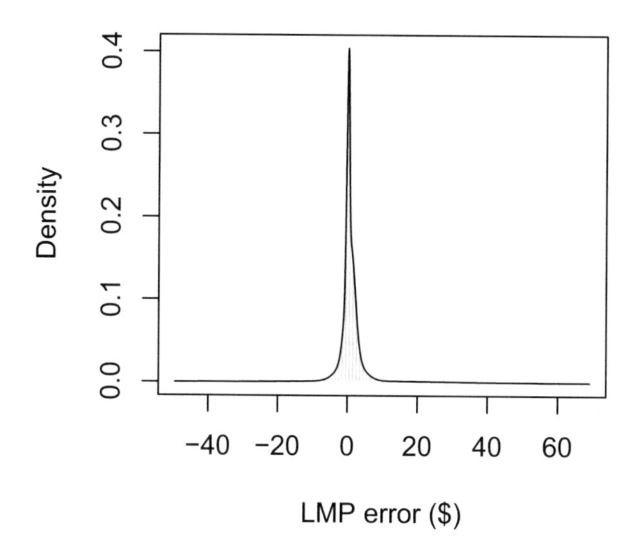

Fig. 4 PDF of LMP forecasting error

Table 2 Detection rate and false alarm rate of anomaly detection

Metric	M3-EA	PS
DR	0.72	0.54
FAR	0.29	0.30

M3 are 3.68%, 1.43%, and 5.77%, respectively, which are better than those of the baseline PS method (i.e., 9.60%, 3.62%, and 9.33%).

Figure 4 shows the PDF of the LMP forecasting error. Based on the error distribution, percentiles of the errors can be calculated, where the anomaly behavior is detected based on the 95% threshold. The detection evaluation metrics are summarized in Table 2. Overall, the proposed anomaly detection method has a higher DR (over 72%) and lower FAR than PS.

Figure 5 summarizes the temporal statistics of the number of false alarms. It is seen that the proposed M3-EA method has relatively more false alarms during peak hours (i.e., 11 a.m.–7 p.m.) than off-peak hours, which indicates that anomaly detection becomes more difficult during peak hours. To reduce false alarms during peak hours, the deterministic anomaly detection threshold should be differentiated between peak hours and off-peak hours. For example, the current anomaly detection threshold (i.e., 95% percentile) of error distribution should be relaxed to a lower percentile for peak hours.

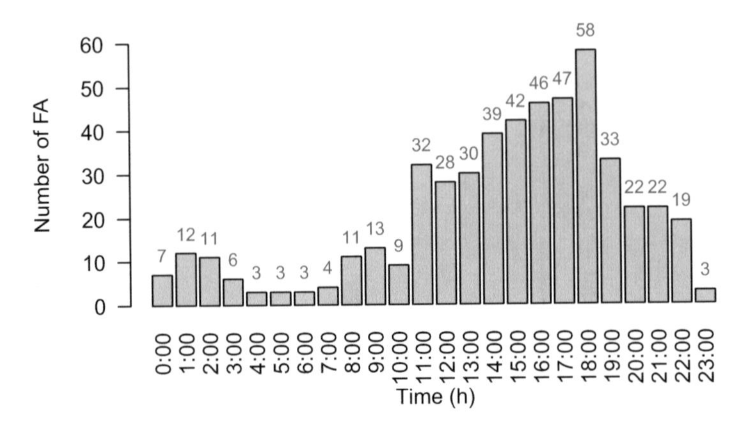

Fig. 5 Temporal statistics of FAs (Time-of-Day)

4.2 *Case Study II: Probabilistic LMP Anomaly Detection*

For data preparation, an electricity market simulator is first built based on MATPOWER [23] using a combined model of day-ahead economic dispatch (ED) and real-time incremental economic dispatch (IED). We then run the simulator on the IEEE 14-bus system with 11 loads selected from PJM load profiles. Day-ahead hourly load forecasts and 5-min. real-time load forecasts from Sept. 19th to Oct. 17th are used for ED and IED, respectively. The simulated 5 min. real-time LMP, energy cost, congestion cost, and load forecasts are used for training and testing. The first 3/4 of the data is used as training data, in which the first 11/12 is used to train deterministic forecasting models and the remaining 1/12 of the training data is used to build the surrogate models of the optimal standard deviations and weight parameters. The effectiveness of the forecasts is validated by the remaining 1/4 of the data.

Two kinds of attacks are implemented in the case study: Load Redistribution Attack (LRA) and Price Responsive Attack (PRA). LRA was first introduced by Yuan et al. [20], where only the measurements related to certain load buses power injection are attacked. LRA redistributes the load by increasing/decreasing certain loads at certain buses while keeping the total load unchanged [4].

PRA is inspired by the time-responsive load [19] and manipulation of demand attack (MAD) [13], which changes the load behavior to damage the power grid. The motivation of our PRA is that the quick growth of smart grid foresees the wide usage of smart electronics, which can change the load behavior based on the current LMP. For example, the controller of smart appliances can switch to a full-power mode when the price is low, and keep in an energy saving mode when the price is high. Unlike the power grid infrastructures that are well-protected, smart electronics are located at the user end, which are much less secure to defend cyber attacks. The PRA is designed by injecting false price signals to load controllers so as to inverse the controller logic, and use more power when LMP is high. The increasing load could cause a high opportunity of line congestions in the power grid. By increasing the load demand at such a critical time period, it may even change the LMP due to increased congestions.

The evaluation metrics are summarized in Table 3. Overall, the proposed anomaly detection method has a high DR (over 80%) and a lower FAR (below 26%), which shows the effectiveness of the probabilistic anomaly detection algorithm. In addition, it is shown that the DR of LRA is higher than that of PRA. This is mainly due to the larger LMP magnitude change under LRA. To show the effectiveness of the M3 probabilistic anomaly detection model, a quantile regression (QR) method is adopted as a baseline. It is seen from Table 3 that overall the M3 probabilistic anomaly detection method performs better than the QR method.

To better visualize probabilistic anomaly detection results, PIs of a selected time period under LRA and PRA are illustrated in Figs. 6 and 7, respectively. It is observed that at most part of the no attack periods, the LMP reasonably lies within the PIs. When the observation in the attack period falls out of the 70% PI, it is defined as a truth positive detection. It is seen that the PRA frequency in Fig. 7 is higher than that of LRA in Fig. 6, and the magnitude change of LMP under LRA is higher than that under PRA. The width of the PI varies with the LMP variability. When the LMP fluctuates more frequently, the PI tends to be wider, and thereby the uncertainty under PRA is relatively higher.

Table 3 Detection rate and false alarm rate of anomaly detection

Model	Metric	LRA	PRA
M3	DR	0.83	0.80
	FAR	0.22	0.26
QR	DR	0.68	0.53
	FAR	0.31	0.30

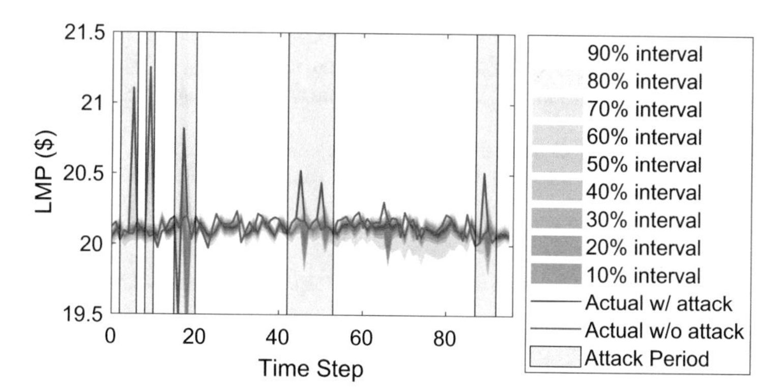

Fig. 6 PIs of LMP under the LRA attack

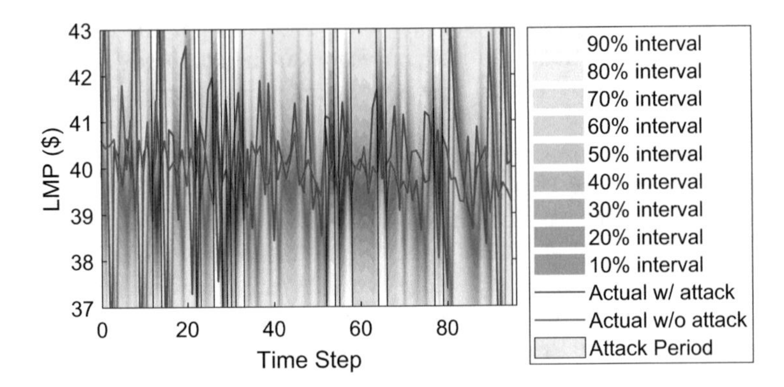

Fig. 7 PIs of LMP under the PRA attack

5 Conclusion

In this chapter, two data-driven anomaly detection methods were introduced, including a deterministic anomaly detection method and a probabilistic anomaly detection method. The developed methods were applied to detect attacks in LMP. Specifically, a machine learning-based multi-model method was developed to provide very-short-term deterministic and probabilistic LMP forecasts, which are used for anomaly detection. Results based on publicly available and simulated datasets showed that the developed data-driven anomaly detection methods outperformed corresponding benchmarks.

References

1. C. Biernacki, G. Celeux, G. Govaert, Choosing starting values for the EM algorithm for getting the highest likelihood in multivariate Gaussian mixture models. Comput. Stat. Data Anal. **41**(3–4), 561–575 (2003)
2. M.M. Buzau, J. Tejedor-Aguilera, P. Cruz-Romero, A. Gómez-Expósito, Detection of non-technical losses using smart meter data and supervised learning. IEEE Trans. Smart Grid **10**(3), 2661–2670 (2018)
3. L. Davis, *Handbook of Genetic Algorithms* (Elsevier, Amsterdam, 1991)
4. R. Deng, G. Xiao, R. Lu, H. Liang, A.V. Vasilakos, False data injection on state estimation in power systems—attacks, impacts, and defense: a survey. IEEE Trans. Ind. Inf. **13**(2), 411–423 (2016)
5. M. Esmalifalak, L. Liu, N. Nguyen, R. Zheng, Z. Han, Detecting stealthy false data injection using machine learning in smart grid. IEEE Syst. J. **11**(3), 1644–1652 (2014)
6. C. Feng, M. Sun, J. Zhang, Reinforced deterministic and probabilistic load forecasting via Q-learning dynamic model selection. IEEE Trans. Smart Grid **11**, 1377–1386 (2019)
7. R. Fontugne, N. Tremblay, P. Borgnat, P. Flandrin, H. Esaki, Mining anomalous electricity consumption using ensemble empirical mode decomposition, in *2013 IEEE International Conference on Acoustics, Speech and Signal Processing* (IEEE, Piscataway, 2013), pp. 5238–5242

8. M. Gupta, J. Gao, C. C. Aggarwal, J. Han, Outlier detection for temporal data: a survey. IEEE Trans. Knowl. Data Eng. **26**(9), 2250–2267 (2013)
9. V.B. Krishna, G.A. Weaver, W.H. Sanders, PCA-based method for detecting integrity attacks on advanced metering infrastructure, in *International Conference on Quantitative Evaluation of Systems* (Springer, Cham, 2015), pp. 70–85
10. H. Li, D. Fang, S. Mahatma, A. Hampapur, Usage analysis for smart meter management, in *2011 8th International Conference and Expo on Emerging Technologies for a Smarter World* (IEEE, Piscataway, 2011), pp. 1–6
11. K.G. Lore, D.M. Shila, L. Ren, Detecting data integrity attacks on correlated solar farms using multi-layer data driven algorithm, in *2018 IEEE Conference on Communications and Network Security (CNS)* (IEEE, Piscataway, 2018), pp. 1–9
12. M. Mohammadpourfard, A. Sami, Y. Weng, Identification of false data injection attacks with considering the impact of wind generation and topology reconfigurations. IEEE Trans. Sustain. Energy **9**(3), 1349–1364 (2017)
13. S. Soltan, P. Mittal, V. Poor, Protecting the grid against MAD attacks. IEEE Trans. Netw. Sci. Eng. (2019). DOI: 10.1109/TNSE.2019.2922131 (Early Access)
14. M. Sun, C. Feng, E.K. Chartan, B.M. Hodge, J. Zhang, A two-step short-term probabilistic wind forecasting methodology based on predictive distribution optimization. Appl. Energy, **238**, 1497–1505 (2019)
15. M. Sun, C. Feng, J. Zhang, Conditional aggregated probabilistic wind power forecasting based on spatio-temporal correlation. Appl. Energy, **256**, 113842 (2019)
16. M. Sun, C. Feng, J. Zhang, Multi-distribution ensemble probabilistic wind power forecasting. Renew. Energy **148**, 135–149 (2020)
17. X. Wang, T. Zhao, H. Liu, R. He, Power consumption predicting and anomaly detection based on long short-term memory neural network, in *2019 IEEE 4th International Conference on Cloud Computing and Big Data Analysis (ICCCBDA)* (IEEE, Piscataway, 2019), pp. 487–491
18. J. Wang, D. Shi, Y. Li, J. Chen, H. Ding, X. Duan, Distributed framework for detecting PMU data manipulation attacks with deep autoencoders. IEEE Trans. Smart Grid **10**, 4401–4410 (2019)
19. L. Xie, Y. Mo, B. Sinopoli, Integrity data attacks in power market operations. IEEE Trans. Smart Grid **2**(4), 659–666 (2011)
20. Y. Yuan, Z. Li, K. Ren, Modeling load redistribution attacks in power systems. IEEE Trans. Smart Grid **2**(2), 382–390 (2011)
21. Y. Zhang, M. Li, Z.Y. Dong, K. Meng, A probabilistic anomaly detection approach for data-driven wind turbine condition monitoring. CSEE J. Power Energy Syst. **5**, 149–158 (2019)
22. Y. Zhou, W. Hu, Y. Min, L. Zheng, B. Liu, R. Yu, Y. Dong, A semi-supervised anomaly detection method for wind farm power data preprocessing, in *2017 IEEE Power & Energy Society General Meeting* (IEEE, Piscataway, 2017), pp. 1–5
23. R.D. Zimmerman, C.E. Murillo-Sánchez, R.J. Thomas, MATPOWER: steady-state operations, planning, and analysis tools for power systems research and education. IEEE Trans. Power Syst. **26**(1), 12–19 (2010)

AI-Enabled Security Monitoring in Smart Cyber Physical Grids

Hossein Mohammadi Rouzbahani, Zahra Faraji, Mohammad Amiri-Zarandi, and Hadis Karimipour

1 Introduction

Power systems are growing sharply due to the increasing demand for electrical energy that is expected to grow by 30% by 2035 [1]. Smart grids as the new generation of integrated power systems have been utilized in a wide area [2] These systems use smart tools for making two-way communication across the network to integrate renewable energy sources and improve reliability, stability, and security [3]. Figure 1 shows a schematic of the smart grids. There are two major challenges that the new generation of the power system are facing; resource limitation and cyber threat.

Due to the expansion of the power systems and given that the cost of traditional fuel, pollution, etc. wind energy as a renewable resource has been taken into consideration [4]. It is worth noting that wind power generation is increasing sharply to fulfill the demand of electricity all around the world [5]. Figure 2 shows the capacity of wind energy that have been installed in different countries. One of the major challenges in dealing with wind energy is related to the prediction wind specifications like speed, turbulence, temperature,atmospheric pressure, etc. which are very important for forecasting the output energy [6]. Since the output of wind turbines is highly regarded to weather conditions, reliable and persist short-term forecasting plays a vital role in using wind energy in regard to scheduling, dynamic control of turbines, demand-side management, etc.

H. M. Rouzbahani · Z. Faraji · H. Karimipour (✉)
School of Engineering, University of Guelph, Guelph, ON, Canada
e-mail: hmoham15@uoguelph.ca; mamiriza@uoguelph.ca; hkarimi@uoguelph.ca

M. Amiri-Zarandi
School of Computer Science, University of Guelph, Canada
e-mail: zfaraji@basu.ac.ir

H. Karimipour et al. (eds.), *Security of Cyber-Physical Systems*,
https://doi.org/10.1007/978-3-030-45541-5_8

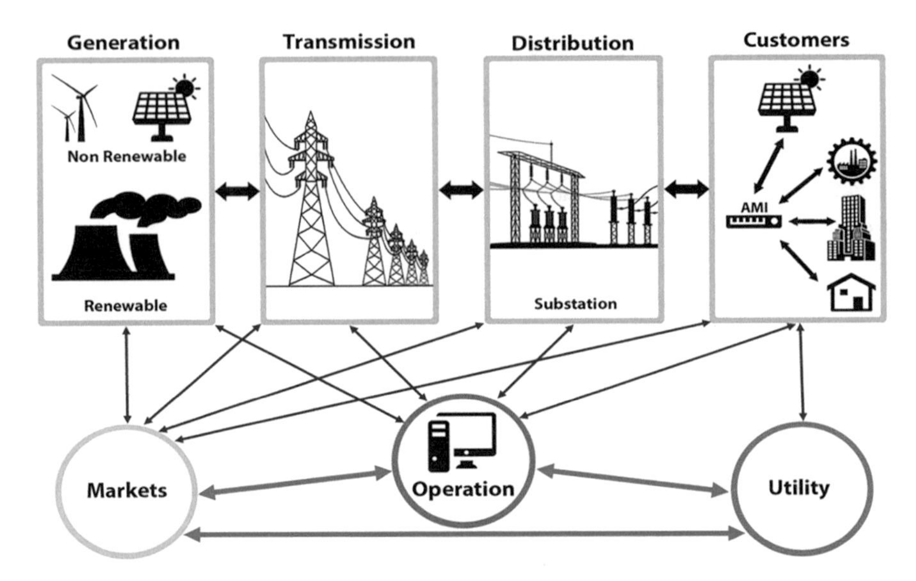

Fig. 1 The structure of smart grids

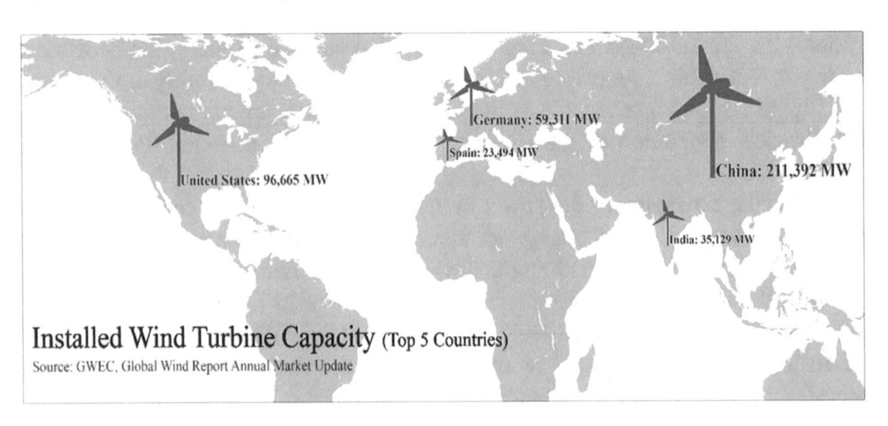

Fig. 2 Installed capacity by five top countries

Alongside the benefits of smart grids, two-way communication between utilities, control centers, and customers can make these networks vulnerable to cyber-attacks. The most frequent power grids incidents in 2013 and 2014 were related to cybersecurity [3] which shows this is a critical issue since attacker access to the network can lead to pernicious consequences.

Figure 3 shows there are three major layers in smart grids that could be targeted for attackers [7]. An attack against the power and energy system layer can target the control station and some of the equipment. Communication protocol and network attacks are two main types of attacks against the communication layer while the Computer/IT layer might be attacked by malware or software.

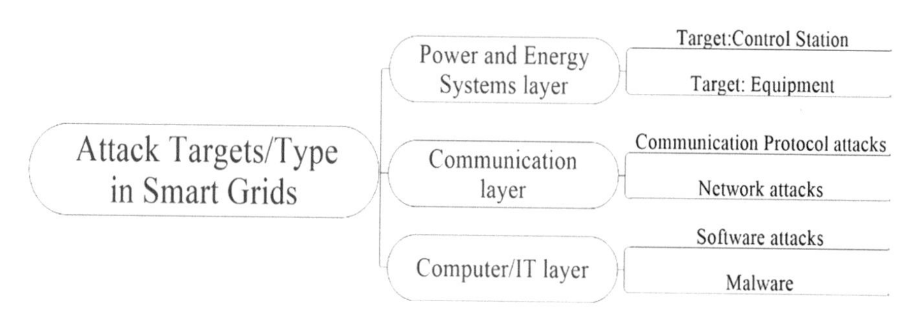

Fig. 3 Cyber-attacks in smart grids

Machine Learning can be used for spatial, temporal prediction as a purely data-driven model that can give good results compared to traditional methods [8]. There are three major challenges; dealing with big data which is required high-level processors [9], reliability of predictions that is very important to tackle market competition and finally the third challenge is the security that originated from two-way communication systems in smart grids which are needed to up-date data continuously [10].

There are different methods of attack detection in the smart grids that have been proposed in the literature whereas some of the new attacks cannot be detected by the traditional methods therefore, it is essential to use intelligent methods like using machine learning. Many papers have compared the results of different methods of learning and suggested one of those methods. Mohammad et al. [8] show that a semi-supervised anomaly detection algorithm has better performance in comparison to a Support Vector Machine (SVM) on the IEEE 118-bus case study. Ozay et al. [11] have concluded that the performance of a single-layer perceptron, KNN and SVM is acceptable for attack detection. They used IEEE 9-bus, 30-bus, 57-bus and 118-bus scenarios in their study. A deep learning method has been studied on the IEEE 9-bus, 14-bus, 30-bus, and 118-bus scenarios in [12]. In [13] the authors used reinforcement learning to attack detection on the NSL-KDD dataset. Finally, Jahromi et al. [14] suggested a deep unsupervised representation learning approach for cyber-physical attack detection on the power system dataset from ORNL Laboratories.

To find out the importance of the application of machine learning in a wind energy system a bibliometric analysis seems necessary. This analysys allowed scholars to recognize the various valuable criteria of research activities while contains information on various applications of science, history, and sociology to research evaluation. Until the emergence of the Scopus and Google Scholar in 2004, the only famous tool for bibliometric analysis was Web of Science (WoS) which provides many comprehensive features for doing this analysis [15].

After selecting the search engine (WoS) some related keywords are selected to start the extraction of results. The query to collect the data for bibliometric analysis was as follows: TS = ((wind power OR wind energy OR wind turbine OR wind

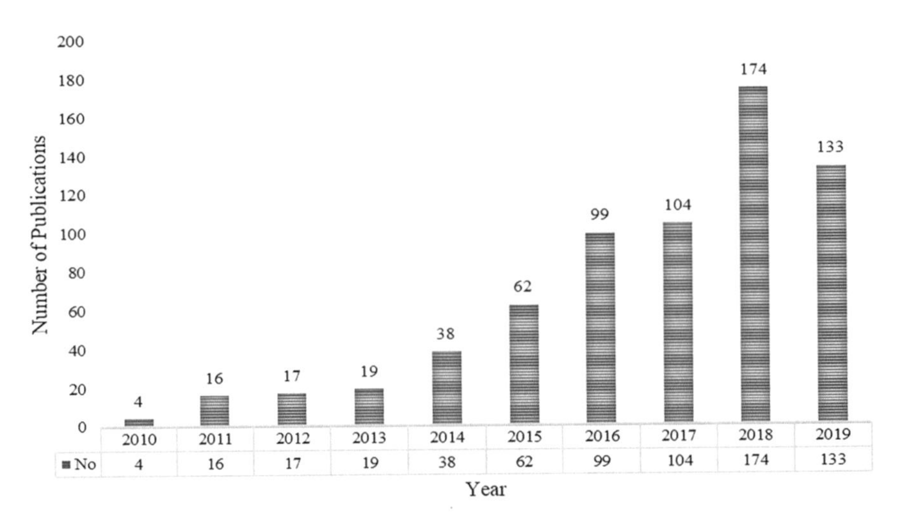

Fig. 4 Number of Publications in different years

farms OR Offshore wind power OR onshore wind power OR Offshore wind energy OR onshore wind energy) AND (forecasting OR prediction OR estimation OR predicting OR estimating) AND (machine learning OR ML OR supervised learning OR unsupervised learning OR semi-supervised learning OR reinforcement learning OR deep learning OR data mining AND data analytic AND big data)).

As a result, 677 records including journals, books, conferences, and patterns were detected over a period of 10 years. Figure 4 shows the number of publications per year between 2010 and 2019 which demonstrates a significant increasing trend. However, in 2019, as of 19th September 2019, the number of publications is 133 it is more highly likely that it would be more than 2018 in the remaining of 2019.

Another factor that can prove the quality of the topic is the analysis which is a manner of presenting the evidence of content and concept in the articles to demonstrate the increasing quantity of research activities. Figure 5 shows the citations received by the publications over the last 10 years. As the number of citations increases, there is an increase in the number of publications. The average annually number of citations is 197 between 2010 and 2019.

Eventually, these two trends indicates the importance of the subject as a top tire subject.

This chapter aim to study the application of ML in both major challenges as follows:

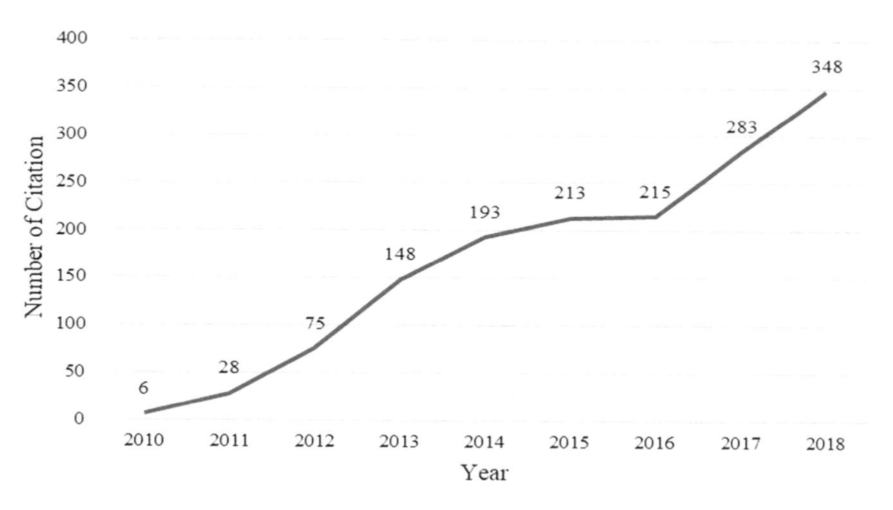

Fig. 5 Number of citation in different Years

1.1 Application of ML in Wind Energy System

Two different methods of learning have been applied in this section; LSTM as an artificial recurrent neural network (RNN) [16] has been tested regarding wind nature as time series forecasting problem that follows a pattern for a certain period like a day, month or year. Also, a neural network has been examined as a broadly distributed processor with a parallel structure that can handle a large amount of input.

The dataset that has been used for this study contains historical data of wind from the National Renewable Energy Laboratory (NREL) throughout a period of 6 years which contains like time, date, air temperature, wind direction, the power generated by the system, pressure, and wind speed.

An algorithm should be trained by different learning models then their performances would be compared to choose the most accurate one to do two different tasks. It should be noted that the first mission is the prediction of the weather condition in order to estimate the generated energy as a second task.

There are two useful pieces of software that have been used to train algorithms named [17] and Python [18] to apply NN and LSTM respectively, then the performances would be compared based on MAE.

1.2 Application of ML for Attack Detection

In this part, we are going to predict attack in a basic model of smart grids similar to what was performed in [14] and our goal is to improve the performance of the results from some aspects. The evaluated smart grid is consisted of generators, transmission lines, breakers and an Intelligent Electronic Device (IED) which has been installed on each breaker that collects data, receives the command and finally controls the breaker.

To pick the best method to predict an attack regarding the dataset, five different algorithms are selected and then trained. The reason for using these algorithms which are supervised learning comes from the nature of our dataset which contains different labels. Different techniques like subsampling and oversampling are used to deal with imbalanced data and turn it into a balanced dataset.

The outline of this paper is as follows. We present an introduction to the wind energy system in Sect. 2. Thereafter in Sect. 3, Machine Learning application for attack detection and load prediction is introduced briefly. Two different case studies and results are presented in Sect. 4. Finally, Sect. 5 is the conclusion of the study.

2 Wind Energy System as a CPS

There is various kind of renewable energy like Wind, Solar, Biomass, Geothermal and hydropower [19]. Although the biggest share of renewable energy belongs to hydropower by more than 80 percent but also regarding climate change and the importance of water resources, wind, geothermal and solar has been taken into consideration more than ever [20].

Recently wind energy portion in the power system is increasing while most of the wind turbines have been designed to operate by constant-speed. Due to the stochastic nature of the wind including speed fluctuation and undesirable variations, most of the turbines can not work on this situation but the sharp development of wind turbine designing methods and strategies causes a massive increase in complexity and scale of wind energy systems [21].

Also, considering the wing energy system as a CPS inable these systems to apply innovative approaches like using machine learning methods and applying two way communication to have better performances [22]. Computing machines and communication through the internet depend on embedded systems and global networks respectively which are two majors factors of development in information and Communication Technologies (ICTs) [23]. A wind energy system can be considered as a CPS whereas cyber and physical components have been integrated to do forecastic and predicting in real time.

Tower, blades, rotor, generator, gearbox, wiring system, nacelle, and foundations are the physical parts of the wind energy system while the cyber layer contains collaboration between various hardware and software. This collaboration may cause tree different challenges for wind energy systems as a CPS including security, safety, and sustainability [24, 25].

In this chapter, security has been taken into consideration whereas real-time monitoring and control of the system is the key to the prediction of output energy but presently the main issue is the prediction of output energy. Kinetic energy (KE) of air is wind energy and total wind energy (E) can be calculated as follows:

$$E = \frac{1}{2}mv^2 = \frac{1}{2}(Avt\rho)\,v = \frac{1}{2}At\rho v^3 \quad (1)$$

Where ρ is the density of air, v is wind velocity, A is effective area and t represent time. Eventually, Eq. (2) shows the output power which is a fraction of E and time.

$$P = \frac{E}{t} \quad (2)$$

Finally, at the end of 2018, the worldwide total installed wind power generating electricity amounted to 591,549 MW was increased up to 9.7% compared to the prior year.

3 Machine Learning

Machine Learning is a combination of various areas of science including artificial intelligence, statistics, mathematics, etc. [26] which focused on the development of high-performance algorithms to make prediction on data. The main goal of ML is to create an algorithm that can work and update automatically without human interference or guidance.

Machine learning examines the huge amount of data and frequently produces outputs to find a beneficial opportunity or a hazardous threat. There are three major approaches for ML; supervised, unsupervised and reinforcement learning [27]. Supervised learning needs training with labeled data to train an algorithm while in unsupervised learning method, unlabelled datasets are used to predict well. Reinforcement learning train an algorithm based on the feedback received through the interactions. Figure 6 shows a taxonomy of machine learning [28].

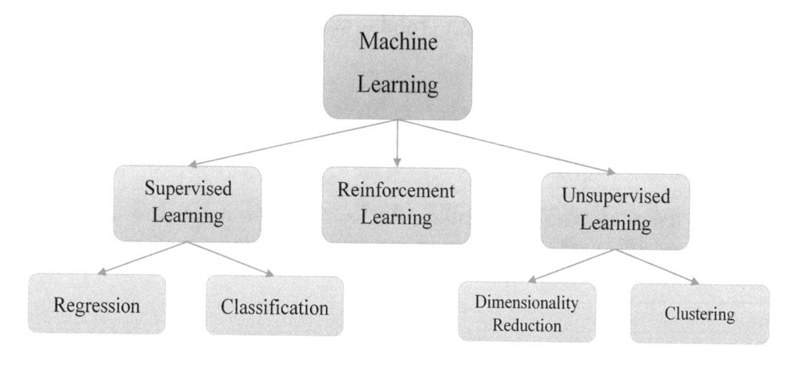

Fig. 6 A taxonomy for machine learning

Characteristics of machine learning can be defined as follows:

1. Visualization relation between data automatically
2. Improve the knowledge of the complex model through train and test
3. Using the least assumption considering the efficiency
4. Accurate data analysis

Many factores must be considered to choose a proper learning method to build a high performance algorithm. First of all, the type of dataset can specify the method regarding the labels. Afterward, the number of features and the environment of the dataset should be considered. There are many diverse machine learning approaches that can be used based on different criteria's which are essential in a study. In this chapter two groups of techniques have been applied as follows:

3.1 Machine Learning to Attack Detection

In this section several sorts of supervising learning have been applied to find out the best algorithm for attack detection in smart grids. All F-score, Accuracy, and runtime should be considered simultaneously to evaluate methods. Figure 7 presents an algorithm that shows this process.

3.1.1 Decision Tree

The decision tree is one of the supervised learning approaches that predict the label of a target example based on a tree data structure [29].

3.1.2 Random Forest

Random forest is a tree-based classifier with high accuracy that comes from the ensemble learning method. The most popular class is selected based on the vote of the trees [30].

3.1.3 K-Nearest Neighbor

As a distance metric learning algorithm, the k-nearest neighbor is one of the oldest methods of classification that work based on Euclidean or Manhattan distance between feature values [31].

Algorithm 1: Attack Detection Using Machine Learning

Input: Dataset containing Attack, Normal and Natural examples
Training Phase:
Split data for Train and Test
Pre-processing
Do the Feature Selection;
Do the Over/Sub Sampling;
Model Training
Train SVC on Imbalanced data (I-SVC)
Train SVC on Oversampled data (O-SVC)
Train Linear-SVC on Imbalanced data (I-LSVC)
Train Linear-SVC on Oversampled data (O-LSVC)
Train Decision Tree on Imbalanced data (I-DT)
Train Random Forest-15 on Imbalanced data (I-RF15)
Train Random Forest-50 on Imbalanced data (I-RF50)
Train Bootstrapping on Imbalanced data (I-BS)
Train Gradient Boosting on Imbalanced data (I-GB)
Train Gradient Boosting on Oversampled data (O-GB)
Train XG-Boosting on Imbalanced data (I-XGB)
Train XG-Boosting on Oversampled data (O-XGB)
Train Adaboost on Imbalanced data (I-Ada)
Train Adaboost on Oversampled data (O-Ada)
Train KNN on Imbalanced data (I-KNN)
Train KNN on Oversampled data (O-KNN)
Train KNN on Subsampled data (S-KNN)
Train KNN-Manhattan on Oversampled data (O-KNNM)
Testing Phase:
Extract the selected features
Pass the test sample trough all Training methods;

Output: Attack/Normal/Natural

Fig. 7 Algorithm of Attack Detection

3.1.4 Support Vector Classifier (SVC)

This algorithm tries to separate the training patterns by a hyperplane while getting the maximal margin [32]. Typically, when we are facing batch input data this algorithm can be useful [33].

3.1.5 Gradient Boosting

The main idea of gradient boosting is a strong learner can be made from an ensemble of weak learners and it should be noted that the performance of a weak learner is better than a random chance. The most popular boosting methods are gradient boost [34], Adaptive Boost [35], and XG-boost [36].

3.2 *Machine Learning to Prediction*

Regarding our dataset that is a time series, two different methods have been tested to predict output energy; LSTM and ANN.

3.2.1 Long Short-Term Memory (LSTM)

As a type of artificial recurrent neural network, Long Short-Term memory is the best time-series pattern methods which were proposed in 1997 [37] LSTM not only processes a single data such as image, but it can also process a sequence of data like a video or speech. Consequently this method can arguably the most commercial Artificial intelligence (AI) achievement [38].

LSTM network is most appropriate for problems like processing, classifying, and decision-making or making the prediction of any time-series data. This methos is composed of a few cells, such as input, output, and a forget gate. The memory cell is the basic unit of LSTM which is a fix-weight unit that can be known as a self connection part [39]. There is an input gate that protects the error flow within the memory cell an also. Also, an output gate protects other units from perturbation by currently inappropriate memory contents. Finally, access to the error flow will be controlled by the gates. it should be noted that the gates have been learned before.

LSTM network is most appropriate for problems like processing, classifying, and decision-making or making the prediction of any time series data because there may be any data missing or lag in the dataset of any event of time series pattern [40].

There are many types of LSTM architecture [41]. The most common one is a composition of a cell and three regulators, usually known as gates, where the flow of the information takes place. The input gate controls the flow of new values, whereas the output gate controls the value present in the cell, which is used for the computational purpose and the forget gate control values that remain in the cell. The activation role of LSTM is the logistic function of the sigmoid [42].

Figure 5 shows the structure of the LSTM. The cell of LSTM is composed of three gates- input, forget, and output. This structure is efficient in solving any type of problem like gradient explosion, missing data, etc. The dependent connection between the data in the input sequence is caught by the cell and the forget gate controls the extent of the values which remain in the cell. The input gate controls the extent of the values, which flow into the cell while the values of the cell are calculated by means of the output gate [43].

Different elements are shown in Fig. 8. A multiplication sign means it is multiplying the elements, and con presents vector merge. Also, f_t, i_t, g_t, o_t are represent the input, output, and forgotten gate respectively. Equations (3)–(8)

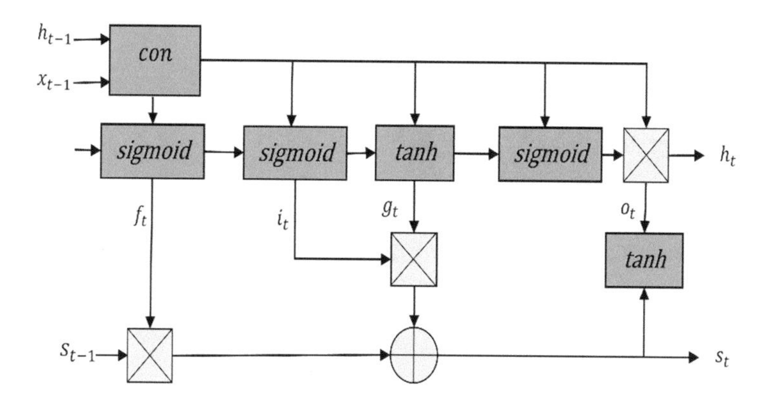

Fig. 8 The structure of LSTM

demonstrate the formulas which are related to the structure of LSTM. Where w_f, w_i, w_g, w_o are the equivalent weight matrix linking the input signal $[h_{t-1}, x_t]$ and σ represent the sigmoid activation remembers preceding values over arbitrary time intervals.

$$f_t = \sigma\left(w_f\,[h_{t-1}, x_t] + b_f\right) \tag{3}$$

$$i_t = \sigma\,(w_i\,[h_{t-1}, x_t] + b_i) \tag{4}$$

$$g_t = \tanh\left(w_g\,[h_{t-1}, x_t] + b_g\right) \tag{5}$$

$$o_t = \sigma\,(w_o\,[h_{t-1}, x_t] + b_o) \tag{6}$$

$$s_t = (s_{t-t} * f_t) + g_{t*i_t}\Big) \tag{7}$$

$$h_t = \tanh\,(s_t) * v_t \tag{8}$$

3.2.2 Neural Network (NN)

Neural Networks are broadly known as an alternative method to deal with complex issues. A wide range of problems in different areas like engineering analysis, forecasting, robotics, pattern recognition, predicting, optimisation, etc. have been

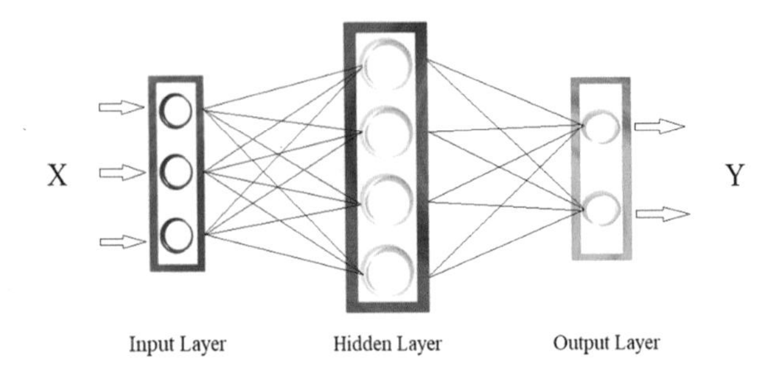

Fig. 9 Multi Layer NN

answered by utilizing NN [27]. NNs imitate the learning process of the human brain which is modeled like a black box without needing additional information on the system.

Figure 9 shows a diagram of a typical multilayer feed forward neural network which displays neural network typically includes three different layers including an input layer, some hidden layers and an output layer [44]. The simple form of network is the connection of every single neuron to other neurons of a previous layer by adjustable synaptic weights. Information is normally kept as a set of related weights [45]. In the process of training by using an appropriate learning method the connection weights are changing in some orderly manner. In the learning mode, the weights are adjusted in order to produce the desired output by an input that is presented to the network [46]. Before training weights are random and have no meaning whereas after training, they contain significant information [41].

Equation (9) describe the relationship between the inputs X_i (the ith input to the neuron) and the output Y_i as follows [39]:

$$Y_i = f_i\left(\sum_{j=1}^{n} w_{ij} X_j + b_i\right) \tag{9}$$

Where, w_{ij} is the connection weight between the input and output node, b_i is the bias of the node, and f_i is known as the activation function.The mean squared error (MSE) [43] of the network is displayed by Eq. (10)

$$MSE = \frac{1}{N}\sum_{i=1}^{N}(T_i - Y_i)^2 \tag{10}$$

Where Y_i is the prediction of the network output, N is the number of the training set, and T_i is the target.

3.3 *Forecasting Based on Machine Learning*

One of the computer science domains is Machine Learning that tries to enable learning ability to computers or other smart devices without being expressly worked. The main purpose of machine learning is developing methods and algorithms to describe the manner of the dataset and with respect to historical data and features, learn a structure to predict the output. Machine learning algorithms are successfully employed to explain the behavior of the dataset, model input features regarding the attended output, and forecast output features regarding its historical records [28].

Forecasting is an on growing technique that can be used everywhere such as weather patterns, economic events, etc. Accurate weather forecasting is hard with traditional methods, although machine learning forecasting might be inaccurate due to various nature of weather, different capability such as detecting environmental changes and adapt to the new environment can make it easy to the learning algorithms [44].

There are many precious and advanced machine learning techniques including NNs, artificial intelligence, deep learning, etc. that can be used for forecasting. Forecasting can be considered as a beneficial application of machine learning [45]. Regarding power demand is changing every minute for customers, prediction can be a critical issue which can be used by managers of energy companies to organize power-reduction policy. A developed company with forecast help can create tactics of how to optimize the exacting operation and power storage system [46].

4 Case Study and Results

4.1 *Case 1: Output Power Prediction in the Wind Energy System*

Accurate and reliable wind speed forecasts are a significant challenge due to the stochastic nature of wind. As mentioned before, the quantity and quality of generated power by a wind turbine are highly dependent on wind speed, pressure, temperature and direction which are highly non-linear. This makes it difficult for any machine-learning model to figure out a pattern and give an accurate prediction. We tried to interpret this problem as time series forecasting problem as wind follows a pattern for a certain period like a day, month or year. Fortunately, wind speed follows a certain pattern over a while.

Historical wind energy data over a period of six years (2007–2012) have been taken from NREL to do this study. After preprocessing there are different feautures like timestamp, air temperature (C), pressure (atm), wind direction (deg), wind speed (m/s) and output power (kW). Table 1 shows a sample of the dataset.

Table 1 Sample of the dataset

Date	Time	Temperature	Pressure	Speed	Wind direction	Output power
2007-01-01	00:00:00	10.92	0.979103	9.014	229	33688.1
2007-01-01	01:00:00	9.91	0.979566	9.428	232	37261.9
2007-01-01	02:00:00	8.57	0.979937	8.745	236	30502.9
2007-01-01	03:00:00	7.87	0.980053	8.383	247	28419.2
2007-01-01	04:00:00	7.25	0.978544	8.254	256	27370.3
2007-01-01	05:00:00	6.57	0.985487	7.587	265	25805.9
2007-01-01	06:00:00	5.89	0.985789	7.421	271	11546.8

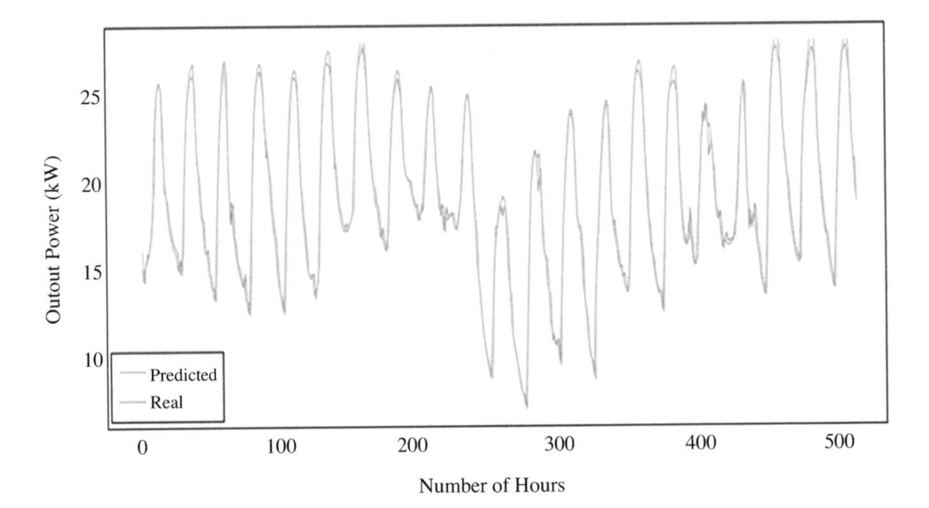

Fig. 10 The comparison of real and the predicted output power (LSTM)

Regarding the nature of the dataset which has sampled hourly, the LSTM can be an appropriate method for making a prediction. After that, the NN will be applied and finally, we make a comparison between the results of these two methods.

Estimation models are useful if we get the weather information about the present day or the future publicly using machine learning with certain accuracy. Then this model can be used to be the perfect estimation of power generated by the system. Multivariate time series forecasting with LSTM with Keras library has been applied. Creating the baseline model and initial experimentations specified that eight look backs are a great number, which provides significant results in the prediction.

The dataset has been divided into train and test parts with a share of 70% and 30% respectively. The impressive result is observed where the mean absolute error is 2.54 kW and the accuracy is 96.9%. Unfortunately, the runtime is too high which took around 2 min. In the following, we will see that the NN is pretty faster than LSTM. Figure 10 shows the comparison of true and the predicted value of output power. Also Fig. 11 demonstrates the error plot of the LSTM model.

A neural network is composed of various neurons that are inspired by the biological nervous system. Many hidden neurons may worsen the performance of

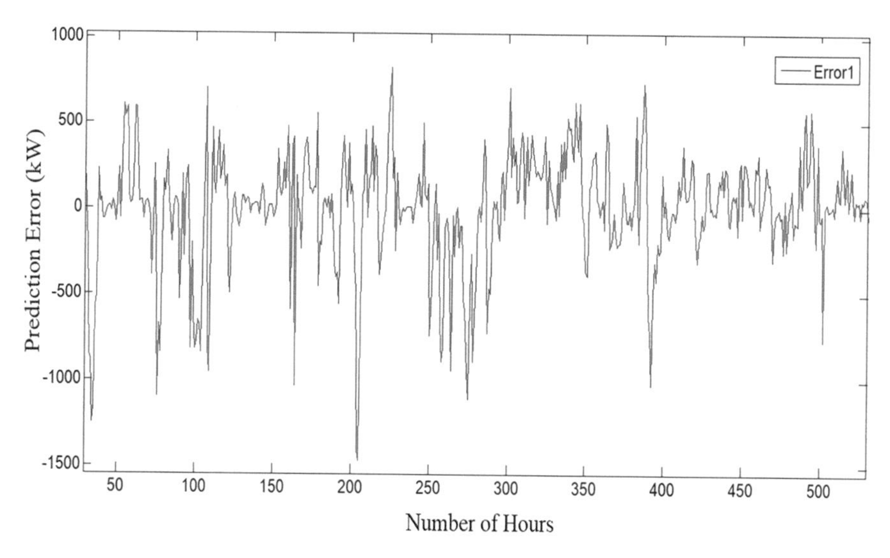

Fig. 11 Error plot of LSTM model

the network as it requires huge storage for the variables. However, if fewer numbers of neurons are used then the network will not be able to adjust the weights and prejudices correctly during training, resulting in overfitting. As we know, the neural network approach requires large input and output samples and a proper number of hidden layers. A three layer neural network as an ML approach, have been trained, validated and tested by using Matlab.

After applying NN we are able to make a comparison between two considered models. Data has been divided into two parts. The first part contains statistics like velocity, the direction of the wind and the other part holds has information on turbine power generation. By using a separate NN model for each turbine ensures better output, fast and accurate training and it will reduce the size and complexity of the model. The dataset has been devided into 3 different parts. The share of trainin and validation are 70% and 15% respectively while 15% is belonged to test.

The wind features and weather conditions have been imported as input data while the target value is the power generated by a wind turbine that must be predicted. When the input and output values are given to the network, it tries to find out the relationship among them. Afterward it corrects the error rate by using the backpropagation algorithm which caused to lower error rate. Then the network is ready to find the Power output in kW when any input data is given with a minimalized error.

Figure 12 shows the result of the neural network which demonstrates the model fits very well. A comparison between real and predicted value for output power is shown in Fig. 13. The Mean absolute error and accuracy are 4.7 kW and 92.90% respectively and the runtime is more than 50% better than LSTM. Also, Fig. 14 shows the error rate of NN in this case.

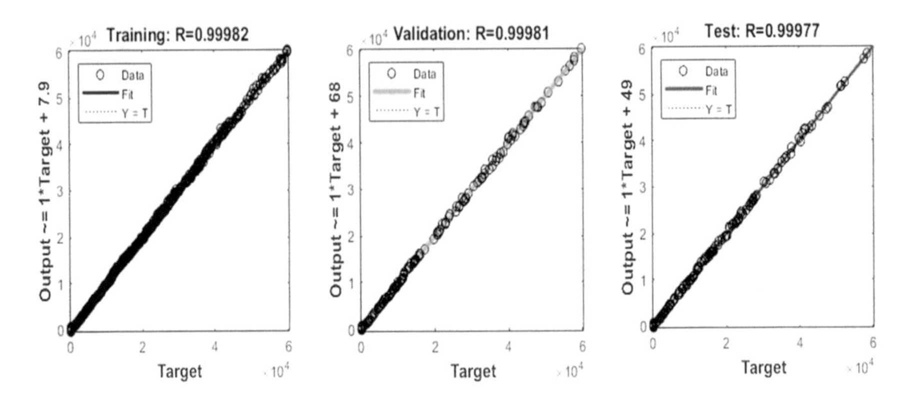

Fig. 12 Result of the neural network

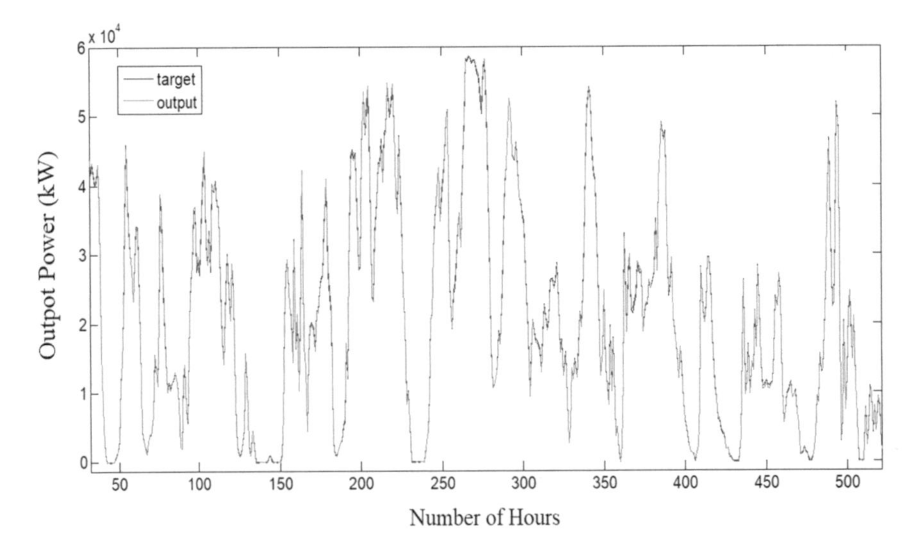

Fig. 13 The comparison of real and the predicted output power (NN)

Finally, a comparison between results for both methods is presented in Table 2. As we can see, LSTM is more accurate but the complexity is high which caused high runtime. The NN is more simple and fast and it would be beneficial to give up some accuracy to make prediction faster.

4.2 Case 2: Attack Prediction in Smart Grids

The dataset that is used in this section is generated in Oak Ridge National Laboratories (ORNL). The data set includes 78,377 individual examples that sampled

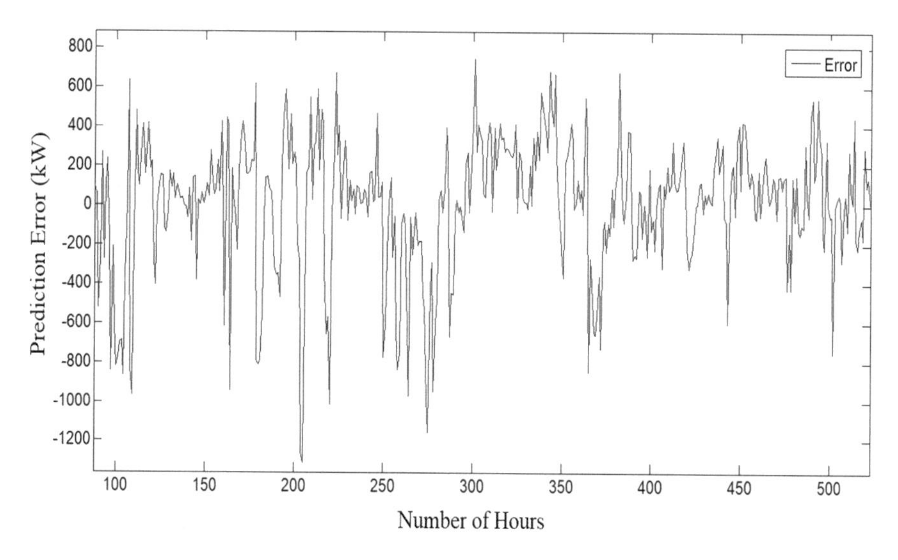

Fig. 14 Error plot of NN

Table 2 Relust comparison

Method	Accuracy (%)	MAE (kW)	Minimum run-time (s)	Complexity
LSTM	96.6	2.54	120	High
NN	92.9	4.7	60	Low

throughout 128 features. Three different statuses can be considered for the power systems as labels including attack, normal and natural event.

In this project, different classification algorithms have been used on the dataset. For all of them, the results are obtained for both imbalanced and balanced data and some techniques like featured selection have been driven (except for tree-based algorithms). In addition, to have more stable measuring, cross-validation is utilized in the deployment of the machine learning techniques. Finally, the results are evaluated based on evaluation factors while we get a glimpse of training time (Mean Fit time) and Test time (Mean Score time) as follows:

SVC Both Linear and non-linear SVC have been tested on the dataset and it should be noted that RBF kernel is used in non-linear ones. As you can see in Table 4, the SVC is very slow and it cannot deal with imbalanced data. Although if we balance data by oversampling, the evaluation factors would be better, the Training time is still long for non-linear SVC.

Decision Tree The techniques like scaling, feature selection, and balancing are not useful and necessary when tree-based algorithms (Decision tree, Random forest and boosting) are used.

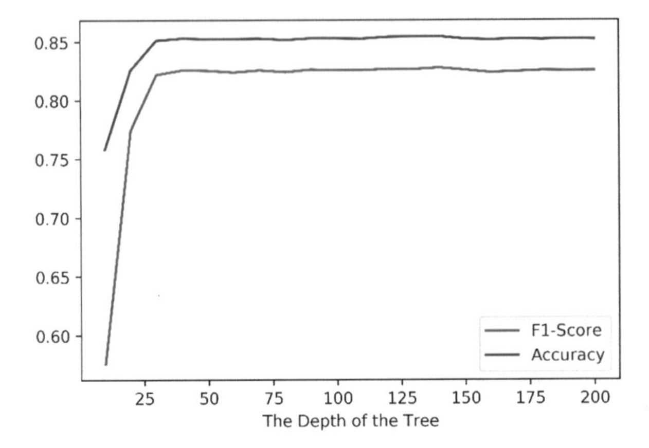

Fig. 15 F-score and accuracy for decision tree with different depths

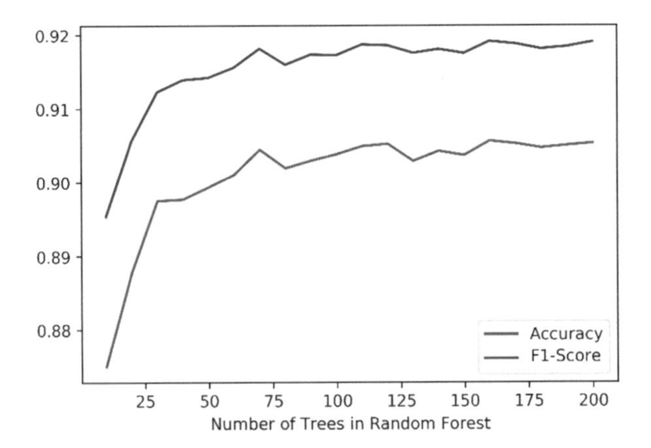

Fig. 16 F-Score and accuracy for random forest method with the different numbers of trees

As shown in Table 4, the decision tree is so faster than SVC and F-Score shows this algorithm is more accurate as well. For the next step, the performance of the decision trees with different depths is evaluated.

As is shown in Fig. 15, the trees with a depth of more than 40 do not have a significantly better F-score. The reason is that the number of features that can effectively enhance the prediction performance in decision trees is limited and, in this case, it is approximately 40.

Random Forest First we used 15 trees to achieve equal training time with the decision tree where the result shows F-Score for the random forest is higher than the decision tree. After that, we used different numbers of trees and the performance of the method is demonstrated in Fig. 16.

Table 3 Evaluation factors for KNN

Method	K	Precision	Recall	F-score	Accuracy
KNN	1	0.971	0.971	0.971	0.901
KNN	11	0.885	0.883	0.881	0.821

As is shown is Fig. 16 increasing the number of trees enhances the F-score, especially in smaller numbers. But, using more than 50 trees does not change the performance considerably, while it leads to longer test time. Therefore, it seems in this case 50 is a reasonable value for the number of trees in the random forest technique.

Bootstrapping Decision tree is the base classifier that is used in bootstrapping in our study. We used 10 bootstraps in this technique and the result shows that the F-score is not good enough for this method.

Boosting Gradient boosting, Adaboost, and XG-boost are the methods that are utilized in this section. In all these methods, a decision stump is selected as the weak learner. The results demonstrate that none of these algorithms can reach a good performance in comparison to the decision tree and random forest.

KNN In this part, L1-Norm is used for feature selection alongside scaling. Table 3 presents a comparison among different evaluation factors for KNN and also it should be noted that the best performance is obtained when $K = 1$.This is reasonable because in the dataset the classes are to close and the margin is small. The result shows that the best training time belongs to KNN while the test time is acceptable as well.

To deal with imbalanced data both subsampling and oversampling have been used. The training time has strongly improved after subsampling while oversampling caused the best performance among all of the algorithms that we tested in this project especially when the Manhattan distance is applied.

As it is demonstrated in Table 4, the KNN method provides better predictions in comparison to the other evaluated methods. But, the main issue in this method is the long test time which is required for this learning technique. As the decisions in the smart grids should be taken in the shortest possible time, this method is not well-matched with the requirement of the considered systems. All in all, we believe that the random forest is the most useful learning method among considered techniques to detect attacks in the smart grids.

5 Conclusion

As mentioned before, the demand for electrical energy is increasing continuously which forced the utility companies to expand the power stations. It is very expensive and it might cause environment and air polushion. Also, the expansion of traditional

Table 4 Relust comparison

Method	Balancing	NOF	NOE	Fit time (s)	Score time (s)	F-score
SVC	Imbalance	91	78,377	521.13	76.78	0.342
SVC	Oversampled	117	166,989	2436.44	510.42	0.685
Linear SVC	Imbalance	93	78,377	38.61	0.190	0.333
Linear SVC	Oversampled	113	166,989	147.57	0.298	0.649
Decision tree	Imbalance	128	78,377	6.74	0.148	0.845
Random Forest-15	Imbalance	128	78,377	5.73	0.232	0.891
Random Forest-50	Imbalance	128	78,377	19.11	0.448	0.919
Bootstrapping	Imbalance	128	78,377	42.06	0.403	0.864
Gradient boosting	Imbalance	128	78,377	51.58	0.248	0.566
Gradient boosting	Oversampled	128	166,989	114.42	0.649	0.735
XG-boosting	Imbalance	128	78,377	63.438	0.358	0.484
XG-boosting	Oversampled	128	166,989	127.90	0.801	0.717
Adaboost	Imbalance	128	78,377	25.98	0.819	0.443
Adaboost	Oversampled	128	166,989	50.68	1.72	0.614
KNN	Imbalance	92	78,377	3.08	42.65	0.871
KNN	Subsampled	102	41,023	1.49	5.437	0.831
KNN	Oversampled	114	166,989	10.18	83.91	0.964
KNN-Manhattan	Oversampled	118	166,989	9.90	108.37	0.970

NOF number of features, *NOE* number of samples

networks resulted in the new generation of power systems which is known as the smart grid. There are two major challenges for smart grids including a lack of energy resources and security issues.

These grids integrate Energy layer, ICT layer and Market layer which use a two-way communications that let them to digitally control the process of electricity supply procedure. It should also be noted that this smart approach provides more room for adversaries to attack the system.

Obviously traditional fuels are expensive and increase the greenhouse gases. As a clean and beneficial renewable energy, wind energy plays a significant role in power system. There are many technical issues for using wind energy due to the nature of wind which has not foresighted structure. Many factors like wind speed, temperature, pressure, wind direction, etc. must be considered to estimate the output energy of a turbine. So the traditional methods are not accurate and reliable for forecasting and it is needed to use intelligent approaches like ML methods.

In this chapter, application of machine learning to deal with two foremost challenges of smart grids have been studied. Two different methods have been experienced for wind power forecasting. These two procedures areselected due to the nature of wind sampling and the features. LSTM is one of the best techniques to deal with time series dataset and NN can handle many input and output considering the features and historical data.

The performance of LSTM in terms of accuracy is beter than NN while it is more complex and slow. Taking into account all aspects it seem using NN is more reasonable because there is not bif difference in accuracy of these two methods while NN is much faster than LSTM.

In the next part, we tried to leverage machine learning techniques to detect attacks. Many supervised learning methods have been compared based on F-score and runtime. As we know F-Score is a compound measure that can be used when we are facing a highly imbalanced dataset and the correctness of the decisions is crucial. Regarding the nature of smart grids and the dataset as well, F-Score is chosen to evaluate the algorithms in this project. Finally, among the considered methods, random forest performed a better task in terms of F-score and testing time.

References

1. H.M. Ruzbahani, H. Karimipour, Optimal incentive-based demand response management of smart households, in *Conference Record—Industrial and Commercial Power Systems Technical Conference*, vol. 2018 (2018), pp. 1–7
2. C.C. Sun, A. Hahn, C.C. Liu, Cyber security of a power grid: State-of-the-art. Int. J. Electr. Power Energ. Syst. **99**, 45–56 (2018)
3. X. Huang, Z. Qin, H. Liu, A survey on power grid cyber security: From component-wise vulnerability assessment to system-wide impact analysis. IEEE Access **6**, 69023–69035 (2018)
4. H.M. Ruzbahani, A. Rahimnejad, H. Karimipour, Smart households demand response management with micro grid, in *2019 IEEE Power and Energy Society Innovative Smart Grid Technologies Conference, ISGT 2019* (2019)
5. D. Koraki, K. Strunz, Wind and solar power integration in electricity markets and distribution networks through service-centric virtual power plants. IEEE Trans. Power Syst. **33**(1), 473–485 (2018)
6. H. Bakhtiari, R.A. Naghizadeh, Multi-criteria optimal sizing of hybrid renewable energy systems including wind, photovoltaic, battery, and hydrogen storage with ε-constraint method. IET Renew. Power Gen. **12**(8), 883–892 (2018)
7. G. Elbez, H.B. Keller, V. Hagenmeyer, A new classification of attacks against the cyber-physical security of smart grids, in *ACM International Conference Proceeding Series* (2018)
8. M. Esmalifalak, L. Liu, N. Nguyen, R. Zheng, Z. Han, Detecting stealthy false data injection using machine learning in smart grid. IEEE Syst. J. **11**(3), 1644–1652 (2017)
9. H.M. Rouzbahani, H. Karimipour, G. Srivastava, Big data application for renewable energy resource security, in *Handbook of Big Data and Privacy* (Springer, Cham, 2019)
10. J. Sakhnini, H. Karimipour, A. Dehghantanha, R.M. Parizi, G. Srivastava, Security aspects of Internet of Things aided smart grids: A bibliometric survey. Internet Things **9**, 100111 (2019)
11. M. Ozay, I. Esnaola, F.T. Yarman Vural, S.R. Kulkarni, H.V. Poor, Machine learning methods for attack detection in the smart grid. IEEE Trans. Neural Netw. Learn. Syst. **27**(8), 1773–1786 (Aug. 2016)
12. H. Wang et al., Deep learning-based interval state estimation of AC smart grids against sparse cyber attacks. IEEE Trans. Industr. Inform. **14**(11), 4766–4778 (2018)
13. Y. Zhang, L. Wang, W. Sun, R.C. Green, M. Alam, Distributed intrusion detection system in a multi-layer network architecture of smart grids. IEEE Trans. Smart Grid **2**(4), 796–808 (2011)

14. A.N. Jahromi, J. Sakhnini, H. Karimpour, A. Dehghantanha, A deep unsupervised representation learning approach for effective cyber-physical attack detection and identification on highly imbalanced data, in *Proceedings of the 29th Annual International Conference on Computer Science and Software Engineering* (IBM Corp., 2019), pp. 14–23
15. L.S. Adriaanse, C. Rensleigh, Comparing web of science, scopus and google scholar from an environmental sciences perspective. South Afr. J. Librar. Inform. Sci. **77**, 2 (2011)
16. S. Hochreiter, J. Schmidhuber, Long short-term memory. Neural Comput. **9**(8), 1735–1780 (1997)
17. MATLAB, The Mathworks., Inc. (Natick, Massachusetts, 2019)
18. F.L. van Rossum, Guido, and drake, in *Python 3* (Reference Manual. CreateSpace, Scotts Valley, CA, 2009)
19. K. Rahbar, C.C. Chai, R. Zhang, Energy cooperation optimization in microgrids with renewable energy integration. IEEE Trans. Smart Grid **9**(2), 1482–1493 (2018)
20. H. Quan, D. Srinivasan, A. Khosravi, Short-term load and wind power forecasting using neural network-based prediction intervals. IEEE Trans. Neural Netw. Learn. Syst. **25**(2), 303–315 (2014)
21. V. Yaramasu, B. Wu, P.C. Sen, S. Kouro, M. Narimani, High-power wind energy conversion systems: State-of-the-art and emerging technologies. Proc. IEEE **2015**, 740–788 (2015)
22. H. Karimipour, H. Leung, Relaxation-based anomaly detection in cyber-physical systems using ensemble kalman filter, *IET Cyber-Physical Systems: Theory & Applications* (2019)
23. L. da Xu, W. He, S. Li, Internet of things in industries: A survey. IEEE Trans. Industr. Inform **10**(4), 2233–2243 (2014)
24. H. HaddadPajouh, A. Dehghantanha, R.M. Parizi, M. Aledhari, H. Karimipour, A survey on internet of things security: Requirements, challenges, and solutions. Internet Things **1**, 100129 (2019)
25. A. Banerjee, K.K. Venkatasubramanian, T. Mukherjee, S.K.S. Gupta, Ensuring safety, security, and sustainability of mission-critical cyber-physical systems. Proc. IEEE **100**(1), 283–299 (2012)
26. M. Nassiri, H. HaddadPajouh, A. Dehghantanha, H. Karimipour, R.M. Parizi, G. Srivastava, Malware elimination impact on dynamic analysis: An experimental analysis on machine learning approach, in *Handbook of Big Data Privacy* (Springer, Cham, 2019), pp. 1–39
27. A. Namavar Jahromi et al., An improved two-hidden-layer extreme learning machine for malware hunting. Comput. Secur **89**, 101655 (2020)
28. H. Karimipour, A. Dehghantanha, R.M. Parizi, K.K.R. Choo, H. Leung, A deep and scalable unsupervised machine learning system for cyber-attack detection in large-scale smart grids. IEEE Access **7**, 80778–80788 (2019)
29. G. Kesavaraj, S. Sukumaran, A study on classification techniques in data mining, in *2013 4th International Conference on Computing, Communications and Networking Technologies, ICCCNT 2013* (2013)
30. L. Breiman, Random forests. Mach. Learn. **45**(1), 5–32 (Oct. 2001). https://doi.org/10.1023/A:1010933404324
31. S. Sun, Q. Chen, Hierarchical distance metric learning for large margin nearest neighbor classification. Int. J. Pattern Recognit. Artif. Intell. **25**(7), 1073–1087 (Nov. 2011)
32. M. Schmidt, H. Gish, Speaker identification via support vector classifiers, in *ICASSP, IEEE International Conference on Acoustics, Speech and Signal Processing—Proceedings,* vol. 1 (1996), pp. 105–108
33. K.W. Lau, Q.H. Wu, Online training of support vector classifier. Pattern Recogn. **36**(8), 1913–1920 (2003)
34. J.H. Friedman, Greedy function approximation: A gradient boosting machine. Ann. Stat. **29**(5), 1189–1232 (2001)
35. D.D. Margineantu, T.G. Dietterich, Pruning adaptive boosting. ICML-97 Final Draft

36. T. Chen, C. Guestrin, XGBoost: A scalable tree boosting system, in *Proceedings of the ACM SIGKDD International Conference on Knowledge Discovery and Data Mining*, 13–17 August 2016 (2016), pp. 785–794
37. A.G. Howard et al., MobileNets: Efficient convolutional neural networks for mobile vision applications (2017)
38. W. Xiang, P. Musau, A. A. Wild, D. M. Lopez, N. Hamilton, X. Yang, J. Rosenfeld, and T. T. Johnson, Verification for Machine Learning, Autonomy, and Neural Networks Survey, 2018. ArXiv:1810.01989
39. S. Xie, R. Girshick, P. Dollár, Z. Tu, K. He, Aggregated residual transformations for deep neural networks, in *Proceedings – 30th IEEE Conference on Computer Vision and Pattern Recognition, CVPR 2017*, vol. 2017, (Jan. 2017), pp. 5987–5995
40. V. Sze, Y.H. Chen, T.J. Yang, J.S. Emer, Efficient processing of deep neural networks: A tutorial and survey. Proc. IEEE **105**(12), 2295–2329 (2017)
41. F. Ghalavand, B. Alizade, H. Gaber, H. Karimipour, Microgrid islanding detection based on mathematical morphology. Energies **11**(10), 2696 (2018)
42. R.T.Q. Chen, Y. Rubanova, J. Bettencourt, D. Duvenaud, Neural ordinary differential equations. NIPs **109**, 31–60 (Jun. 2018)
43. Z. Zhang, Y. Shi, H. Toda, T. Akiduki, A study of a new wavelet neural network for deep learning, in *International Conference on Wavelet Analysis and Pattern Recognition* (vol. 1, 2017) pp. 127–131
44. H. Karimipour, V. Dinavahi, Robust massively parallel dynamic state estimation of power systems against cyber-attack. IEEE Access **6**, 2984–2995 (2017)
45. H. Karimipour, V. Dinavahi, Parallel relaxation-based joint dynamic state estimation of large-scale power systems. IET Gener. Transm. Distrib. **10**(2), 452–459 (2016)
46. H. Karimipour, V. Dinavahi, Extended Kalman filter-based parallel dynamic state estimation. IEEE Trans. Smart Grid **6**(3), 1539–1549 (2015)

Application of Machine Learning in State Estimation of Smart Cyber-Physical Grid

Shahrzad Hadayeghparast and Hadis Karimipour

1 Introduction

In recent years, we have witnessed an exponential growth in the development and deployment of various types of cyber-physical systems (CPSs). Increasing dependence on CPSs is growing in various applications such as energy, transportation, military, healthcare, and manufacturing. Many of such systems are deployed in the critical infrastructure (CI). Therefore, they are expected to be free of vulnerabilities and immune to all types of attacks, which, unfortunately, is practically impossible for all real-world systems. A very important and representative CPS is the supervisory control and data acquisition (SCADA) system, which is used in CI such as the smart grid [1].

A smart grid integrates distributed energy resources, includes a variety of interactions between customers and utility, and facilitates the participation of customers in the electricity market. The integration of renewable energy sources located in diverse locations and taking advantage of energy storage devices are made possible in the smart grid. The improvement in energy management as well as intelligent control of the distributed energy resources are advantages of moving from traditional power systems to the smart grid. The important acteristic of the smart grid is two way data and signal communication facilitating the participation of both customer and utility in load and energy management. An information network may connect the entire smart grid. Figure 1 illustrates the The Smart Cyber Physical Grid (SCPG), which consists of two parts: physical and cyber systems. The power grid

S. Hadayeghparast · H. Karimipour (✉)
School of Engineering, University of Guelph, Guelph, ON, Canada
e-mail: hkarimi@uoguelph.ca

H. Karimipour et al. (eds.), *Security of Cyber-Physical Systems*,
https://doi.org/10.1007/978-3-030-45541-5_9

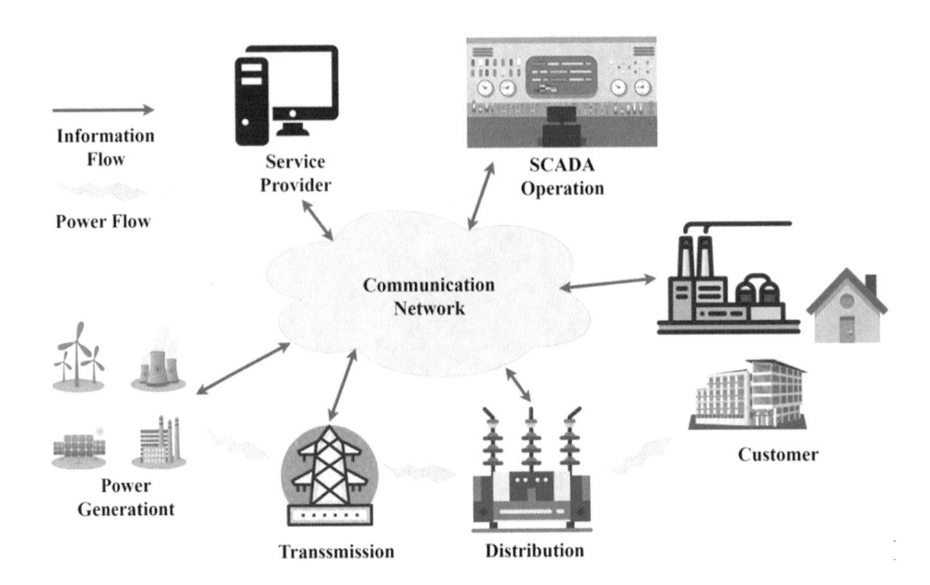

Fig. 1 Smart cyber physical grid

and field represent the physical system, while the cyber system includes the control center, electronic devices and communication architecture which are embedded throughout the physical system. The smart grid can be subject to either cyber-attacks into the information infrastructure or physical attacks by humans [2].

One of the critical issues in the smart grid is its increased vulnerability to cyber-threats [3]. a smart grid that enables two-way communications between consumer and utility, and thus more vulnerable to cyber-attacks [4]. Although the advancement of cyber technologies in sensing, communication and smart measurement devices significantly enhanced smart grid operation and reliability, its dependency on data communication makes it vulnerable to cyber-attacks [5]. The development of advanced information and communication technologies introduces lots of internet based entries, such as perimeter devices, poorly-configured firewalls and unsecure dialup connections. Undoubtedly, these entries produce potential risks and vulnerabilities from malicious cyber-attacks, threatening the economic health and security of the nations involved [6].

The cyber-attacks on the smart grid may obscure the control center into taking erroneous reasoning, lead to transmission congestion, or even cause catastrophic blackouts following cascading failures [6]. The negative impacts of cyber-attacks on the smart grid are enormous. Specifically, attackers can deliberately exploit the vulnerability of SCPGs to stealthily intrude into the communication channels and thus to contaminate confidential data, causing catastrophic blackouts [7] as well as environmental problems [8]. The failure of electric power gird will in return affect the bidirectional energy flow between power network, petroleum network and natural gas network, which is extremely harmful to maintain the reliability and security of SCPG [9]. Moreover, these cyber-attacks could greatly disturb the integration between emerging renewable energies and the traditional power grid, thereby endangering the power quality and energy supply [10]. Consequently, in order to mitigate the negative impacts of cyber-attacks on, it is urgent for us to develop an efficient detection mechanism against various malicious attacks [4]. Critical power system applications like contingency analysis and optimal power flow calculation rely on the power system state estimator. Hence the security of the state estimator is essential for the proper operation of the power system. In the future more applications are expected to rely on it, so that its importance will increase [11].

State estimation is the basic application of a power dispatching automation system. Its function is to obtain the best estimation of system state variables by a mathematical operation based on various real-time measurement data with errors obtained in the power system [12]. In modern smart grid architectures, information from smart meters is sent to the SCADA software system via PLCs (programmable logic controllers) or RTUs (remote terminal units). SCADA software can then process the received data and report the results to the operators for further decisions. One of the critical processes is state estimation, which provides relevant information to maintain the system in a stable and secure state. Within the monitoring procedure, there is a bad data detection routine that will try to identify bad measurements and remove them from the system before performing the state estimation. There are two widely used power flow models when considering state estimation: alternative current (AC) power flow model and direct current (DC) power flow model. An AC power flow model is formulated by nonlinear equations with two types of variables: bus voltage magnitudes and phase angles whereas a DC power flow model uses a linear model to approximate the AC power flow model [13].

The increasing size and complexity of modern power systems have made state estimation a slow and computationally expensive process [14]. Continued growth in demand followed by system development and complex interconnections within the new smart grid paradigm has led to significant operational and control problems. These problems necessitate the need for major changes in computational resources for real-time action by system operators in energy control centers which are hard to achieve using traditional measurements provided by SCADA [15]. Traditional power system state estimation using the bus voltage equations need to solve the nonlinear equations, resulting in a slow calculation by iteration. At the same time, the high-order terms are discarded in the process of solving the Jacobian matrix, which leads to the lowering of the calculation precision [12]. Traditional dynamic state estimation is also not scalable enough to process a large amount of data generated over the grid, and is prone to computational bottlenecks [16]. Moreover, in the process of state estimation, it is supposed to have bad data detection (BDD) procedure to identify problematic readings from smart meters to protect the system from erroneous readings or attacks of injected data. However, it is hard to use existing BDD mechanisms to distinguish bad data from normal since the bad data is always injected in carefully planned efforts [13]. In addition, the obvious question that arises from the enormous amount of data generation from the smart grid is efficient ways to analyze them for extracting valuable information. Without the extraction of useful information, the collected data holds little or no value. Machine learning appears as the tool required for the tall task of going through the massive amount of data generated. It fits in as the final piece of the smart grid system which is driven by data collection, analysis, and decision making. Machine learning techniques provide an efficient way to analyze, and then make appropriate decisions to run the grid; and thus enable the smart grid to function as it is intended to [17].

Machine learning techniques, which are the focus of this chapter, are being adopted to overcome the problem of slow computation caused by the traditional iterative calculation of state estimation, but also avoid the truncation error caused by iterative computation. Machine learning techniques have the characteristic of scaling to larger systems with a low computational cost; therefore, they are promising solutions to deal with state estimation in SCPGs, which are highly complex and large scale physical systems with a huge number of measurements. In other words, machine learning improves the speed and precision of state estimation and has great potential for engineering applications.

2 Security of State Estimation in Smart Cyber Physical Grid

2.1 State Estimation Security

Economic, environmental and technical challenges have given birth to the idea of transition from traditional power grids to a new paradigm of the smart grid [18]. The concept of the smart grid as a complex CPS is to insert adequate intelligence to augment control of the traditional electric power grid and make it more autonomous, fault-tolerant, reliable, and efficient due to the remarkable advancements in sensing, monitoring, control technologies, and also the tight integration with cyber infrastructure and advanced computing and communication technologies [18, 19].

It is extremely important to monitor the state of this complex system such that various control and planning tasks can be performed and the reliable operation of the power system is guaranteed [19]. Because bulk storage of the generated electric energy is not possible, generation and consumption should be closely equated; otherwise, there can be a deviation in the electrical quantities. Thus, a Power Control Center (PCC) needs to closely monitor the power network to make sure that the operation of the power system is safe and reliable. The state estimation is a well-organized method for online monitoring of states in power networks [18]. It is a vital part of smart grid operation and control which tries to roughly estimate voltage magnitude and phase angles in the network buses [20].

State estimation plays a very important role in the control center as depicted in Fig. 2. The state estimation is an on-line application that relies on redundant measurements and a physical model of the power system to periodically calculate an accurate estimate of the power system's state [21]. A BDD system is also included to

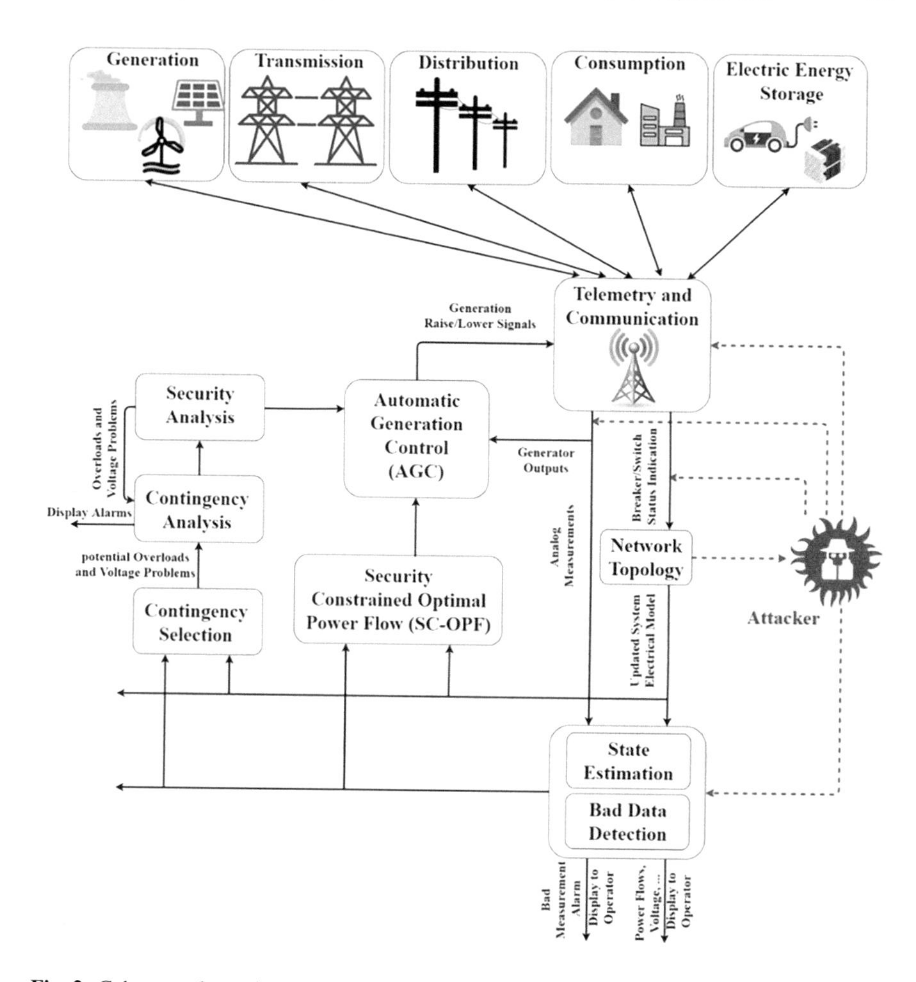

Fig. 2 Cyber-attacks against state estimation in smart cyber physical systems

detect faulty measurement data [11]. The measurements, including power injections into the buses and power flow in the branches, which are gathered from generation, transmission, distribution, consumption, and electric energy storage, are received by state estimator via communications links. These measurements are used to estimate the states, i.e., voltage magnitudes and angles, at buses. These state variables form the basis for correct decisions by the Energy Management System (EMS) about Automatic Generation Control (AGC) and Optimal Power Flow (OPF) to maintain the electric power systems in a safe operating zone [18]. Modern EMS provides information support in the control center for a variety of applications related to power network monitoring and control. One example is the optimal routing of power flows in the network, called OPF, which is to ensure cost-efficient operation. Another example is contingency analysis, which is an essential application to maintain the power system in a secure and stable state despite potential failures.

EMS is also expected to be integral components of future Smart Grid solutions, hence the secure and proper operation of the state estimation is of critical importance [11].

The communication network is the backbone of the scheme illustrated in Fig. 2. Due to the fact that communications infrastructure is prone to malicious cyber-attacks, relying on this network to transport system information introduces security risks. Some of the components require real-time data, and latency or data loss can have adverse effects on the electrical power grid [22]. Also, falsified state estimation results potentially mislead the operation and control functions of EMSs [23]. Consequently, a disruption to the state estimator can lead to loss of power or in extreme cases injury or loss of life [22]. Therefore, the security of state estimation in SCPGs has great importance.

2.2 *Vulnerabilities*

Security vulnerabilities increase by the transition from traditional power systems towards smart grids [24]. New security vulnerabilities will be introduced in the complex electrical power grid by Networking the different components together. Moreover, when all the components are networked together, the number of entry points which can be used for gaining access to the electrical power system will rise [22]. The most important vulnerabilities are listed below:

2.2.1 Customer Security

Massive amounts of data are collected and transferred to the service providers, consumers, and utility companies by smart meters. Private consumer information included in this data might be derived and misused [25].

2.2.2 Greater Number of Intelligent Devices

Many intelligent devices exist in the smart grid for managing both the electricity supply and network demand. Smart meters allow the participation of customers in energy management [2]. It is difficult to monitor and manage allof these devices due to the complexity of the smart grid; therefore, they may act as attack entry points into the network [25]. The field devices: IEDs, PLCs, Phasor Measurement Units (PMUs) RTUs, and smart meters have algorithms that can be manipulated by either cyber intruders or customers. The operators of the utilities can get misleading feedback by hampering and tampering with the data collection process in these devices, and as a result can cause interruptions which may ultimately result in a blackout [2].

2.2.3 SCADA System Security

A SCADA system can be put into danger by its previous employee or angry operator by putting and initiating a bug into the system. The grid could be damaged by intercepting or forging the access logs of this SCADA system. The state estimation data for the power system can tamper in the SCADA database. That may initiate a misleading operation of the smart grid. Energy management and efficient load forecasting are the primary functions of the smart grid. Thus, false load forecasting can misguide the decision of the distributed management system (DMS) [2].

2.2.4 Communication System

The intruders can exploit the communication network in the smart grid to damage different layers present in the grid. Eavesdropping may exist both in the wireless network and wired network in the smart grid. The network layer can be easily jammed by traffic injection. However, wireless and wired networks are vulnerable to traffic flooding and worm propagation attacks [2]

2.2.5 Physical Security

A smart grid includes many numerous critical equipment and field devices which are out of the utility's premises. They are used in remote locations to efficiently and effectively collect the data. Consequently, they might be vulnerable to physical access specialy in insecure physical locations [2, 25].

2.2.6 The Lifetime of Power Systems

Many outdated types of equipment are still in service in the physical system of SCPGs. This equipment might act as weak security points and might very well be incompatible with the current SCPG devices [25].

2.2.7 Implicit Trust Between Traditional Power Devices

Device-to-device communication in control systems is vulnerable to data spoofing where the state of one device affects the actions of another. For example, behavour of one device may be affected in an unwanted way by a false state from another device [25].

2.2.8 Different Team's Backgrounds

Another source of vulnerability is caused by bad decisions as a result of unorganized and inefficient communication between teams [25].

2.2.9 Using Internet Protocol (IP) and Commercial Off-the-Shelf Hardware and Software

The use of IP standards in smart grids is benefitial due to providing compatibility between the various components. However, devices using IP are inherently vulnerable to many IP-based network attacks such as IP spoofing, Denial of Service (DDoS), Tear Drop, and others [25].Also, some applications can be developed in the communication network that can intentionally change the MAC parameters that can lead to spoofing attacks and fake information passing. Distributed DDoS and malicious malwares on the internet also presents a huge threat to smart grid security [2].

2.2.10 More Stakeholders

A very dangerous attack called insider attacks may raise by having many stakeholders [25].

2.3 Smart Cyber Physical Grid Security Objectives

The security objectives of the SCPGs are different from most of the other industries. It is important that any security countermeasure implemented in the Smart Grid does not impede power availability. An example of this would be locking a system after too many failed password attempts. The power system always needs to be available, and locking the system during an emergency could cause safety issues. The security objectives being evaluated are confidentiality, integrity, and availability as shown in Fig. 3. In most industries confidentiality and integrity have higher precedence over availability. In the electrical power system, electricity must always be available, so this is the most important security objective. Integrity is the next important security objective followed by confidentiality [22].

2.3.1 Availability

Availability is the most important security objective. These systems continuously monitor the state of the electrical power grid, and a disruption in communications can cause a loss of power. The availability of the electrical power grid is its most

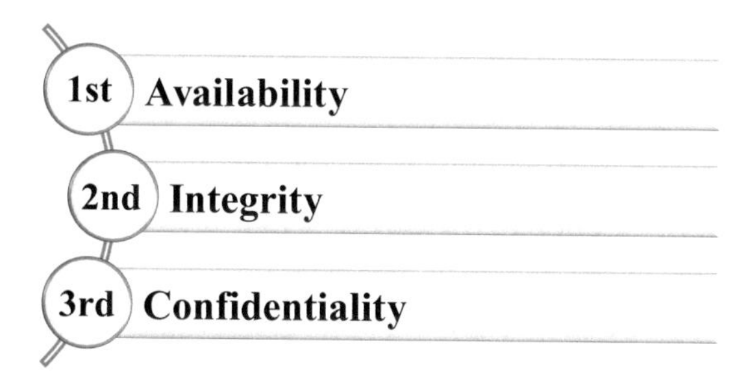

Fig. 3 Smart grid security objectives

important factor. By extension the most important security object of most of the smart grid components is also availability [22].

A typical cyber-attack targeting availability is Distributed Denial of Service (DDoS) Attack which targets AMI communication networks' data collector, preventing the normal communication between Wide Area Network (WAN) and Neighborhood Area Network (NAN) [26].

2.3.2 Integrity

Integrity is the next important security objective in the Smart Grid. The Smart Grid uses data collected by various sensors and agents. This data is used to monitor the current state of the electrical power system. The integrity of this data is very important. Unauthorized modification of the data or insertion of data from unknown sources can cause failures or damage in the electrical power system. The electricity in the smart grid not only needs to always be available, but it also has to have quality. The quality of the electrical power will be dependent on the quality of the current state estimation in the power system. The quality of the state estimation will rely on many factors, but the integrity of input data is very important [22].

FDIAs, which target integrity, introduce random and corrupted data within standard traffic activity in order to cause invalid measurements with the goal of disrupting the AMI network [26].

2.3.3 Confidentiality

The final security objective is confidentiality. The loss of data confidentiality in the Smart Grid has a lower risk than the loss of availability or integrity. There are certain areas in the Smart Grid where confidentiality is more important. The privacy of customer information, general corporation information, and electric market information are some examples [22].

Internet attacks and data confidentiality attacks are examples of cyber-attacks that target confidentiality. Data Confidentiality Attack attempts to compromise the information between electric utilities and end customers by targeting the hardware within the AMI communication network. Internet Attacks that compromise the software and systems in electric utilities [26].

2.4 *Cyber Kill Chain*

Recent trends in targeted cyber-attacks have increased the interest of research in the field of cyber security. The cyber kill chain is a model to describe cyber-attacks so as to develop incident response analysis capabilities. Cyber kill chain in simple terms is an attack chain, the path that an intruder takes to penetrate information systems over time to execute an attack on the target. The cyber Kill chain mainly consists of seven phases as shown in Fig. 4 [27].

2.4.1 Reconnaissance

Collecting information about the potential target is performed in the Reconnaissance stage. Reconnaissance itself can further be devided into target identification, selection and profiling. Information collected from reconnaissance is utilized in the next stages of the cyber kill chain for developing and delivering the payload [27].

2.4.2 Weaponize

Designing a backdoor and a penetration plan, using the information gathered from reconnaissance, to enable successful delivery of the backdoor is a task performed in the weaponize stage of the cyber kill chain [27].

2.4.3 Delivery

The critical part of the cyber kill chain is delivery which is responsible for an effective and efficient cyber-attack [27]. In this stage, the attacker tries to transfer the payload (from the preceding phase) to the target's environment and, in some cases, through another third party in order to exploit a trusted relationship between the third party and the target [28].

2.4.4 Exploitation

Once the cyber weapon is delivered, the target completes the required user interaction and weapon executes at the target side. triggering the exploit is the next step. Silently installing/executing the payload is the goal of an exploit [27].

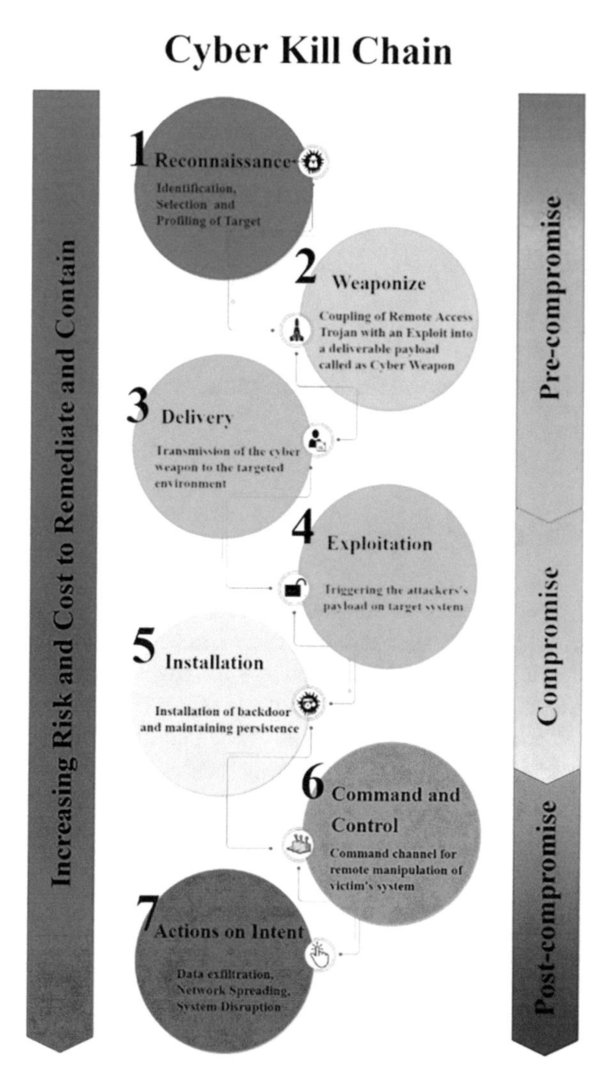

Fig. 4 Cyber kill chain

2.4.5 Installation

In this stage, the attackers will attempt to install access points, such as backdoors or other payloads, to gain persistent access to the target's system or network [28].

2.4.6 Command and Control

In this atage, the attacker establishes communication with the compromised host(s) and some Command and Control(C&C) server [28]. An important part of the remotely executed cyber-attacks is the C&C system, which is used to give remote covert instructions to compromised machines. Moreover, it acts as a place where all data can be exfiltrated [27].

2.4.7 Actions on Intent

The attacker takes action to attain his/her goals, which can be exfiltration or destruction [28].

2.5 *Impacts of Cyber Attacks*

By launching a cyber-attack, an attacker can either gain an economic benefit or disrupt the smart grid. Cyber-attacks could have an impact on the electricity market, power system operation, and distributed energy routing, respectively. This section reviews the impact of FDIAs on all three of these aspects [23].

The economic attack is a type of FDIA which can affect the operation of the deregulated electricity market. The electricity market consists of two markets: the day-ahead market and the real-time market. The specific operation model and settlement method differ according to countries. The independent system operator (ISO) in a wholesale electricity market collects data from various market participants. The locational marginal price (LMP) is an index that reflects the electricity price at each node. Based on state estimation obtained via the SCADA system, the ISO calculates the ex-post LMPs using the DC optimal power flow model [23].

The load redistribution (LR) attack is a type of FDIA which can affect the power grid operation by attacking the security-constrained economic dispatch (SCED). The objective of SCED is to minimize the total system operation cost (generation cost, load shedding cost, etc.) by re-dispatching the generation outputs. Once the estimated state is manipulated by an LR attack, the falsified SCED solution may drive the system to an uneconomic operating state. In a worst-case situation, this could lead to immediate load shedding or even wider load shedding in a delayed time without immediate corrective actions [23].

The main attacking processed is as follows. In the immediate attack, the attacker determines the injected vector and injects it into the targeted meters to maximize the operation cost of the system; the system operator then optimally reacts to the false state estimation that has been successfully manipulated by the injected vector. As a consequence of a successful immediate attack, the actual operation cost will increase no matter whether there is load shedding action or not [23].

In the delayed attack, the attacker determines the injected vector and injects it into the targeted meters; the control center performs the first round of SCED function based on the false state estimation, which would lead to the occurrence of the line overloading without being noticed; the control center then performs the second round of SCED after the tripping of the overload lines. A successful delayed attack results in line overloading undetected by the control center, which can lead to physical damage to the power system [23].

The energy deceiving attack is a type of FDIA which affects the distributed energy routing process [23].

3 Smart Grid State Estimation

This section presents the basics related to state estimation schemes and BDD operations in the smart grid. First, the network and measurement models will be described. Then it will be shown how the state estimator works. Next, the adversary model will be developed. After that, the way BDD identifies cyber-attacks will be discussed. Finally, different types of cyber-attacks will be introduced.

3.1 Network and Measurement Models

The network is represented by the power grid which is the physical system of SCPG. Measurement data received by the control center during system operation from sensors and meters fall into two categories namely digital and analogue data. The active and reactive parts of line flow measurements and bus injection vectors constitutes the analogue data $z = [z_1, z_2, \ldots, z_m]^T$. Also, the on and off states of line breakers and switches are digital network data demonstrated by $s = [s_1, s_2, \ldots, s_l]^T$. The system topology is denoted by $\mathcal{G} = \{\mathcal{V}, \mathcal{E}\}$, which can be observable by the control center with the help of s [29].

The system states are defined as the vector $x = [x_1, x_2, \ldots, x_n]^T$ which includes voltages phasors on all buses. In the case of no measurement noise and cyber-attacks, the measurement data z is related to the system state x and the system topology $\mathcal{G}$ in the AC power flow model as shown in Eq. (1) [13].

$$z = h(x, \mathcal{G}) + v \tag{1}$$

where h is a nonlinear function relating z to $(x, \mathcal{G})$ and $e = [e_1, e_2, \ldots, e_m]^T$. is the vector of measurement error.

The simplified DC model, where the nonlinear function h is linearized near the operating point is expressed in Eq. (2). It is noteworthy that H is dependent on the network topology $\mathcal{G}$ [30]:

$$z = Hx + v \tag{2}$$

where $H \in R^{m \times n}$ is the measurement matrix, $x \in R^n$ is the state vector including voltage phase angles at all buses except the slack bus, $z \in R^m$ includes the real parts of line flow and bus injection measurements, and $e \in R^m$ is the measurement noise.

3.2 State Estimator

A control center is used to monitor the devices in the power system. The control center will first observe the topology $\hat{g}$ as soon as it receives the network parameter p and network topology data s. When $\hat{g}$ is obtained, the system state can be estimated by the weighted least square (WLS) criterion, with the help of $\hat{g}$ and sensor measurements z. Then the estimated system state $\hat{x}$ will be sent to the bad data detector. If it is suspected to be manipulated with bad data, $\hat{x}$ will be forwarded to the bad data identification scheme for further processing; otherwise, the system is said to be free of bad data and will be sent out of the state estimation process for power distributions. The above process is called a generalized state estimation (GSE) [13].

For the state estimator using the AC power flow model, system states are commonly estimated by the iterative weighted least-square criterion (WLS).

$$\text{Min} \sum_{i=1}^{m} \sum\nolimits_{ii}^{-1} r_i^2 \; s.t. r_i = z_i - h_i\left(\hat{x}, \hat{\mathcal{G}}\right), \quad i = 1, \ldots, m \tag{3}$$

Considering the measurement functions are nonlinear, iterative algorithms are used. For every iteration, Δx can be obtained from $\Delta z = H\Delta x + e$, where H is the Jacobian matrix of $h\left(\hat{x}\right)$ and Δz is the residual vector. Then, the WLS estimator will be given by $\Delta \hat{x} = G^{-1} H^T R^{-1} \Delta z$ and and the estimated value of $\Delta \hat{z}$ [13]:

$$\Delta \hat{z} = H \Delta \hat{x} = K \Delta z \tag{4}$$

where K is called the hat matrix, for putting a hat on Δz. The diagonal elements of K are called the leverage scores, which can be collected in set $L = \{K_{ii} | \, i = 1, \ldots, m\}$.

3.3 Adversary Model

In order for an attack to be launched, entry points should be first exploited by an attacker. Then, in case of a successful entry, specific cyber-attacks can be delivered on the SCPG infrastructure such as meters, sensors and breakers [31].

Let $\overline{z}$ and $\overline{s}$ represent the observed measurements and breaker sensor measurements, which may contain malicious data, respectively. The adversary model can be described as follows [13]:

$$\begin{aligned} \overline{z} &= z + a \\ \overline{s} &= s + b \end{aligned} \tag{5}$$

3.4 Bad Data Detection and Identification

Bad measurements may be introduced due to minor physical errors or malicious attacks. The process of validating the topology and meter data is called bad data detection (BDD). The "residual principle" and χ^2-The test is widely used. After solving the WLS estimation problem, we can test whether

$$\sum_{i=1}^{m} \sum\nolimits_{ii}^{-1} r_i^2 \geq \chi^2_{(m-n),p} \tag{6}$$

where p is the value from the Chi-squares distribution table corresponding to confidence with probability p and $(m - n)$ degrees of freedom. If yes, the bad data alarm is triggered, and the identification can be accomplished by further processing residuals; otherwise, the system is said to be free of bad data.

Iterative Bad Data Identification and Removal (BDIR). If BDD detects the existence of at least one meter with bad data, BDIR is invoked to identify the bad data entries and remove them from the system [52]. A widely used criterion for determining a bad data entry is the Largest Normalized Residual $\left(r_N^{\max}\right)$Test. At the kth iteration, the state estimator uses $(z^{(k)})$as input. The measurement residual of the ith measurement can be normalized by

$$r_i^{N(k)} = \frac{\left|r_i^{(k)}\right|}{\sqrt{\sum_{ii}^{(k)}}} \tag{7}$$

The normalized residual vector follows a standard normal distribution with zero mean and unit covariance. Hence the meter with the largest $r_i^{(k)}$will be treated as a meter with bad data and then be removed from the system.

3.5 *Single-Period Attacks*

The research works done against power grid state estimation covered two major directions of attacks: single-period attacks and multi-period attacks. Single-period attacks are discussed in this section.

The single-period attack is a single attack or a series of attacks, which can be launched simultaneously. We can further divide single-period attacks into two kinds: state attacks and topology attacks. State attacks aim at injecting errors on meter measurements, and topology attacks aim at injecting errors on both meter measurements and breaker measurements. Single-period attacks work on injecting bias of system states by manipulating the meter readings (measurements).

3.5.1 State Attacks

This subsection reviews attack mechanisms for state attacks in the AC model. As explained in Sect. 1, the state attack assumes that one attacker can access a set of measurements and can control/manipulate them, whereas the breaker data is intact. In other words, the attacker can only manipulate measurements (z) and cannot modify the system topology (g). Examples of state attacks include false data injection attacks and data framing attacks [13].

False Data Injection Attacks (FDIAs)

FDIA can potentially inject errors on system states by only modifying the sensor data. Assuming that the meter data (z) to be sent to the control center is known and it can be altered to values specified by the attacker. Moreover, the attacker should obtain the topology of the system $h(\cdot)$. The data collected by the control center will be used to estimate the system states (x) and make operational decisions with possibly far-ranging consequences. The main idea of constructing an attack vector is to set a to a calculated sparse $m \times 1$ vector. If the breaker data is intact, the attack model will become,

$$\overline{z} = h\,(x + c) + a + v, \tag{8}$$

where c is the error injected after adding the attack a.

Data Framing Attack (DFA)

Data framing attack aims at misleading the control center about the source of a state attack. It was first proposed in the DC model by [32] and then be extended to the AC model by [33].

Unlike FDIA, DFA does not intend to bypass BDD. Instead it tries to confuse BDIR to remove "clean" measurements (those without malicious data) while keeping the malicious ones within the system, which finally will perturb the estimated system states. After i iterations of BDIR, $2i$ measurements will be removed out of the system. Specifically, in the kth process of BDIR, if P_j is identified as a bad measurement, the rows of P_j, Q_j and the rows of $h(\cdot)$ that correspond to P_j and Q_j will be removed and the updated measurement vector and measurement functions, denoted as $z^{(k+1)}$ and $h^{(k+1)}$, will be returned for the next iteration. Above all, the attack model can be represented as,

$$\overline{z}^{(k)} = h^{(k)}\,(x + c) + a^{(k)} + v^{(k)} \tag{9}$$

where $a^{(k)}$, $v^{(k)}$ are the $(m - 2k) \times 1$ vectors and $h^{(k)}(.)$ is a vector of $(m - 2k)$ nonlinear functions.

3.5.2 Topology Attacks

Different from state attacks, topology attack is to disorder the topology estimate (by perturbing breakers), as well as inject errors on the meter measurements. Thus, it will lead to the disorder of state estimation. In other words, the topology attack also considers the manipulations of power network topology data [13].

Leverage Point Attack (LPA)

The key feature of a leverage point, that is to say, the residue of the measurement corresponds with the leverage point is very small even it is injected with a very large error. Their attacking strategy is first to manipulate the network parameter p, which will increase K_{ii}. Then the premeditated error a_i added to this measurement z_i cannot be detected.

3.6 *Multi-Period Attacks*

A multi-period attack assumes that a sequence of attacks can be launched. Two kinds of attacks are reviewed in this section: sequential attacks and dynamic DoS attacks. A multi-period attack assumes that the attacks (either state attacks or topology attacks) can be injected sequentially according to a carefully designed time sequence. The adversary model can be represented as follows [13].

$$\begin{aligned} \overline{z}^{(t)} &= z^{(t)} + a^{(t)}, a^{(t)} \in A \\ \overline{s}^{(t)} &= s^{(t)} + b^{(t)}\,(\text{mod } 2)\,, b^{(t)} \in B \end{aligned} \tag{10}$$

where $a^{(t)}$ represents the meter data modification in the t th period by the adversary and $b^{(t)}$ represents the breaker sensor measurement modification in the t th period by the adversary. Above all, the multi-period attacks can be divided into two forms:

3.6.1 State Attack

State attacks in multi-periods: the sequence of attacks is only composed of state attacks A, where $A = \{a^{(1)}, a^{(2)}, \ldots, a^{(t)} | \, a \in A\}$;

The dynamic DoS attacks, which belong to the first form, aims at making the solution of real-time economic dispatch (RTED) an empty set. Since RTED can find the optimal generation adjustments to satisfy demands of the next time period, lack of RTED solution will lead to no future generation schedules. It forces us to use more high-price fast-ramping generators.

3.6.2 Topology Attack

Topology attacks in multi-periods: the sequence of attacks is only composed of topology attacks B, where $B = \{b^{(1)}, b^{(2)}, \ldots, b^{(t)} | \, b \in B\}$

The sequential attacks belong to the second form. Their attacking strategy is to build a sequential cascading failure simulator. Suppose there are l transmission lines, then we can have C_l^k choices of a k-link attack sequence ($k \leq l$) and implement these k attacks sequentially. When all attacks are launched, the simulator will quit, and the damage can be evaluated.

4 Application of Machine Learning Algorithms in State Estimation of SCPSs

In this section, first, studies carried out in the field of the application of machine learning algorithms in state estimation in smart grids are reviewed. These studies are divided into two categories n this chapter. Studies conducted in the first category use machine learning techniques for improving the performance of state estimation in terms of improving the accuracy of state estimation and reducing the calculation time. However, studies in the second category focus on the application of machine learning for cyber-attack detection against state estimation in smart grids. The studies in the first and second categories are described in Sects. 4.1 and 4.2 respectively. In addition, they are presented in Table 1. Finally, future works are discussed in Sect. 4.3.

Table 1 Notations

Notations	Descriptions
x	Vector of system state
z	Vector of meter measurements
s	Vector of breaker/switches states
n	Number of system states
m	Number of meter measurements
$\mathcal{V}$	Set of buses
$\mathcal{E}$	Set of transmission lines
$\mathcal{G}$	An undirected graph $\{\mathcal{V}, \mathcal{E}\}$ that indicate the topology of the system circuit
v	Vector of measurement errors
H	Measurement matrix
h	Nonlinear measurement functions
$\bar{s}$	A modified version of s by an adversary
$\bar{z}$	A modified version of z by an adversary
a	Attack vector injected into the meter measurements
b	Attack vector injected into the breaker measurements
r	Vector of measurement residuals
K	Hat matrix
c	Error bias/injected after the attack is implemented

4.1 Machine Learning for Performance Improvement of State Estimation

A fast method of state estimate based on Extreme Learning Machine (ELM) pseudo measurement is proposed in [12]. The method uses ELM to establish the pseudo-measurement model of voltages at each bus, and such a model contains the real part and the imaginary part of voltage. The active and reactive power injected into each bus are used as the input variables, and the real part and the imaginary part of the bus voltage vector are used as outputs in the ELM. The pseudo measurement and PMU measurements are used together to obtain the linear state estimation of the power system. The method proposed in this paper has better performance on fast state estimation in the system where PMU deployment can not cover all buses. Such a method not only can overcome the problem of slow computation caused by the traditional iterative calculation of state estimation, but also avoid the truncation error caused by iterative computation. In other words, it improves the precision of state estimation and has great potential for engineering applications.

Reference [34] dealt with real-time power system monitoring (estimation and forecasting) by building on data-driven Deep Neural Networks (DNN) advances. Prox-linear nets were developed for power system state estimation, that combines NNs with traditional physics-based optimization approaches. Deep Recurrent Neural Networks (RNNs) were also introduced for power system state forecasting from historical (estimated) voltages. Our model-specific prox-linear net based power system state estimation is easy-to-train, and computationally inexpensive. The proposed RNN-based forecasting accounts for the long-term nonlinear dependencies in the voltage time-series, enhance power system state estimation, and offers situational awareness ahead of time.

4.2 Machine Learning for Detection of Cyber Attacks Against State Estimation

A new defense mechanism based on an interval state predictor is proposed in [4] to effectively detect the malicious attacks. In this mechanism, the variation bounds of each state variable are formulated as a bi-level dual optimization problem. Any resultant state that falls outside the estimated bounds can be recognized as an anomaly, indicating a high possibility of data manipulating. In addition, a typical deep learning algorithm, termed as a deep belief network (DBN), is applied for electric load forecasting. DBN has a strong capability for nonlinear feature extraction, which will greatly improve the forecasting accuracy and thus narrow down the variation bounds of state variables, increasing the detection accuracy of the proposed defense mechanism. Two machine-learning-based techniques for stealthy attack detection are proposed in [35]. The first method utilizes supervised learning over labeled data and trains a distributed support vector machine (SVM). The second method requires no training data and detects the deviation in measurements. In both methods, principal component analysis is used to reduce the dimensionality of the data to be processed, which leads to lower computation complexities. In [36], machine learning algorithms are used to classify measurements as being either secure or attacked. An attack detection framework is provided to exploit any available prior knowledge about the system and surmount constraints arising from the sparse structure of the problem in the proposed approach. Well-known batch and online learning algorithms (supervised and semi-supervised) are employed with decision and feature level fusion to model the attack detection problem. The performance of three different classification techniques is tested with three heuristic FS techniques in [37]. The three machine learning algorithms used are SVM, K Nearest Neighbor (KNN) algorithm, and Artificial Neural Network (ANN). The three FS techniques are Binary Cuckoo Search (BCS), Binary Particle Swarm Optimization (BPSO), and Genetic Algorithm (GA). Machine learning and feature selection techniques are combined to take advantage of their strength and compensate for their weaknesses. These algorithms are compared based on their classification accuracy and computational efficiency. The results show that heuristic feature selection techniques are capable of selecting a subset of features that can obtain higher classification accuracy with a significantly lower number of features. Deep learning techniques are exploited in [30] to recognize the behavior features of FDIAs with the historical measurement data and employ the captured features to detect the FDI attacks in real-time. By doing so, the proposed detection mechanism effectively relaxes the assumptions on the potential attack scenarios and achieves high accuracy. In [38], a novel method is proposed to detect FDIAs based on a data-centric paradigm employing the margin setting algorithm (MSA). MSA is a relatively new machine learning algorithm that has been used in image processing fields. Two FDIA scenarios, playback attack and time attack, are investigated. Experimental results are compared with the SVM and ANN. Unsupervised anomaly detection for smart grids to recognize the behavior patterns of FDIAs using the historical measurement data is proposed in [39]. The goal is to capture dependencies between variables through associating of scalar energy to each variable, which serves as a measure of compatibility. A supervised machine learning–based scheme is proposed in [18] to detect a covert cyber deception assault in the state estimation–measurement feature data that are collected throughout a smart-grid communications network. The distinctive characteristic of the paper is that a genetic algorithm–based feature selection is used in our scheme to improve detection accuracy and reduce computational complexity. A combinatory anomaly detection approach considering the fact that any attack/fault in the smart-grid system is always reflected in the form of change in either voltage, current, or phase is proposed in [40]. The proposed method takes advantage of both wrapper and filter methods. For classification, least square support vector. machine (LSSVM) is selected. LSSVM is usually employed for large datasets [26]. In [41], two Euclidean distance-based anomaly detection schemes are proposed for covert cyber-assault detection in smart grid communications networks. The first scheme utilizes unsupervised-learning over unlabeled data to detect outliers or deviations in the measurements. The second scheme employs supervised-learning over labeled data to detect the deviations in the measurements. To improve detection accuracy and further reduce the computational complexity and the associated time delay, a genetic algorithm-based feature selection method is employed to choose the distinguishing optimal feature data subset as input to both of the proposed schemes. In [6], scenario based two-stage sparse cyber-attack models for smart grid with complete and incomplete network information are proposed. Then, in order to effectively detect the established cyber-attacks, an interval state estimation-based defense mechanism is developed innovatively. In this mechanism, the lower and upper bounds of each state variable are modeled as a dual optimization problem that aims to maximize the variation intervals of the system variable. At last, typical deep learning, i.e., stacked auto-encoder, is designed to properly extract the nonlinear and nonstationary features in electric load data. These features are then applied to improve the accuracy of electric load forecasting, resulting in a more narrow width of state variables. In [42], Deep autoencoder- based anomaly measurers are deployed throughout the power system to build a distributed Phasor measurement unit Data Manipulation Attacks (PDMA) detection framework. How to convert the historical PMU measurements into data samples for learning is also elaborated upon. Once trained, an anomaly measurer can assess the PDMA existence possibility of the new PMU measurements. By integrating the results of different anomaly measurers, the proposed distributed PDMA detection framework can detect PDMAs in the whole power system. A comparative study on the utilization of supervised learning classifiers to detect direct and stealth FDIAs in the smart grid is presented in [43]. Three widely used supervised learning based classifiers are chosen to design corresponding FDI detectors. An anomaly detection technique based on feature grouping combined with a linear correlation coefficient (FGLCC) algorithm is proposed in [40]. A decision tree is used as the classifier in the proposed method. In [44], unsupervised anomaly detection based on a statistical correlation between measurements is proposed. The goal is to design a scalable anomaly detection engine suitable for large-scale smart grids, which can differentiate an actual fault from a disturbance and an intelligent cyber-attack. The proposed method applies feature extraction utilizing symbolic dynamic filtering (SDF) to reduce computational burden while discovering causal interactions between the subsystems (Table 2).

4.3 Research Gaps and Future Works

While a number of efforts have been made on adopting machine learning algorithms in state estimation in smart grids, some problems are left for further studies as listed below:

- Although machine learning can improve the computational efficiency and accuracy of state estimation, only a small number of researches for performance improvement of state estimation based on machine learning techniques have been conducted. Consequently, there is a need for more studies in this field.
- Existing studies can defend one attack. However, no work can act against all existing cyber-attacks.
- Most of the studies carried out in this field have considered FDIAs based on the DC model, while machine learning algorithms can be employed for detecting more sophisticated FIDAs based on the AC model.
- Using machine learning techniques against more cyber-physical attacks such as topology attacks or multi-period attacks could be a promising research direction.
- Most of the studies are concerned with transmission systems while there is a need for more research in the application of machine learning for state estimation in distribution systems.
- Research can be expanded in the direction of using other advanced machine learning methods to detect cyber-attacks against state estimation in SCPG.
- Other appropriate feature selection techniques could be proposed for improving the computational efficiency of machine learning algorithms.
- A learning mechanism could be adopted to automatically update the machine learning algorithm with incoming test data to improve the detection accuracy

Table 2 Studies on the application of machine learning in state estimation

				Type of cyber attack			
				Single-period		Multi-period	
Ref. No.	Machine learning algorithm	AC/DC attack model	Name of cyber attack	State	Topology	State	Topology
Performance improvement							
[12]	ELM	–	–	–	–	–	–
[34]	DNN, RNN	–	–	–	–	–	–
Detection							
[4]	DBN	AC/DC	FDIA	✓	–	–	–
[6]	DL	DC	FDIA	✓	–	–	–
[42]	DL	DC	PDMA	✓	–	–	–
[30]	DL	DC	FDIA	✓	–	–	–
[35]	SVM, AD	DC	FDIA	✓	–	–	–
[41]	AD	DC	FDIA	✓	–	–	–
[38]	MSA	DC	FDIA	✓	–	–	–
[18]	SVM	DC	FDIA	✓	–	–	–
[37]	SVM, KNN, ANN	DC	FDIA	✓	–	–	–
[36]	SVM, KNN, perceptron	DC	FDIA	✓	–	–	–
[43]	ENN, SVM, KNN	DC	FDIA	✓	–	–	–
[39]	AD	DC	FDIA	✓	–	–	–
[40]	AD	DC	FDIA	✓	–	–	–
[44]	AD	DC	FDIA	✓	–	–	–

ELM extreme learning machine, *DNN* deep neural networks, *RNNs* recurrent neural networks, *DBN* deep belief network, *AN* anomaly detection, *DL* deep learning, *AD* anomaly detection, *SVM* support vector machine, *KNN* K nearest neighbor, *ANN* artificial neural network, *MSA* margin setting algorithm, *PDMA* phasor measurement unit data manipulation attacks, *ENN* extended nearest neighbor

5 Conclusion

In this chapter, the concept of state estimation security in SCPGs and the application of machine learning algorithms in state estimation were discussed. It was shown that the state estimation security is a critical issue, since it is the basis for correct decisions by EMS about AGC and OPF to maintain the electric power systems in a safe operating zone. Falsified state estimation results potentially mislead the operation and control functions of EMSs. Consequently, a disruption to the state estimator can lead to loss of power or in extreme cases injury or loss of life. Traditional BDD techniques are not efficient in coping with all types of cyber-attacks; therefore, machine learning algorithms are introduced as a promising solution. According to the studies carried out in this field, machine learning algorithms have proved to improve the performance of state estimation in terms of speed and precision. Also, they are capable of detecting various types of cyber-attacks, which are not detectable by transitional BDD techniques. However, much research is still needed in this field due to many existing research gaps. First, machine learning algorithms could be applied to various types of cyber-attacks such as multi-period attacks and topology attacks, which are not taken into consideration. Also, there is a need for more research on the performance improvement of state estimation using machine learning due to a limited number of works done in this field. Finally, more advanced machine learning algorithms along with appropriate feature selection techniques could be applied to get better results in state estimation in SCPGs.

References

1. A. Humayed, J. Lin, F. Li, B. Luo, Cyber-physical systems security—A survey. IEEE Internet Things J. **4**, 1802–1831 (2017)
2. M.A.H. Sadi, M.H. Ali, D. Dasgupta, R.K. Abercrombie, S. Kher, Co-simulation platform for characterizing cyber attacks in cyber physical systems, in *2015 IEEE Symposium Series on Computational Intelligence* (2015), pp. 1244–1251
3. J. Sakhnini, H. Karimipour, A. Dehghantanha, R.M. Parizi, G. Srivastava, Security aspects of Internet of Things aided smart grids: a bibliometric survey, in *Internet of Things* (Elsevier, Amsterdam, 2019), p. 100111
4. H. Wang, J. Ruan, Z. Ma, B. Zhou, X. Fu, G. Cao, Deep learning aided interval state prediction for improving cyber security in energy internet. Energy **174**, 1292–1304 (2019)
5. H. Karimipour and V. Dinavahi, On false data injection attack against dynamic state estimation on smart power grids, in *2017 IEEE International Conference on Smart Energy Grid Engineering (SEGE)* (2017), pp. 388–393
6. H. Wang, J. Ruan, G. Wang, B. Zhou, Y. Liu, X. Fu, et al., Deep learning-based interval state estimation of AC smart grids against sparse cyber attacks. IEEE Trans. Ind. Inf. **14**, 4766–4778 (2018)
7. Y. Xiang, L. Wang, Y. Zhang, Adequacy evaluation of electric power grids considering substation cyber vulnerabilities. Int. J. Electr. Power Energy Syst. **96**, 368–379 (2018)
8. G. Liang, S.R. Weller, J. Zhao, F. Luo, Z.Y. Dong, The 2015 ukraine blackout: implications for false data injection attacks. IEEE Trans. Power Syst. **32**, 3317–3318 (2016)

9. S. Hyysalo, J.K. Juntunen, M. Martiskainen, Energy Internet forums as acceleration phase transition intermediaries. Res. Policy **47**, 872–885 (2018)
10. S. Osorio, A. van Ackere, E.R. Larsen, Interdependencies in security of electricity supply. Energy **135**, 598–609 (2017)
11. O. Vukovic, K.C. Sou, G. Dan, H. Sandberg, Network-aware mitigation of data integrity attacks on power system state estimation. IEEE J. Sel. Areas Commun. **30**, 1108–1118 (2012)
12. Y. Wang, M. Xia, Q. Chen, F. Chen, X. Yang, F. Han, Fast state estimation of power system based on extreme learning machine pseudo-measurement modeling, in *2019 IEEE Innovative Smart Grid Technologies-Asia* (*ISGT Asia*) (2019), pp. 1236–1241
13. J. Wang, L.C. Hui, S.-M. Yiu, E.K. Wang, J. Fang, A survey on cyber attacks against nonlinear state estimation in power systems of ubiquitous cities. Pervasive Mob. Comput. **39**, 52–64 (2017)
14. H. Karimipour and V. Dinavahi, Accelerated parallel WLS state estimation for large-scale power systems on GPU, in *2013 North American Power Symposium* (*NAPS*) (2013), pp. 1–6
15. H. Karimipour, V. Dinavahi, Parallel relaxation-based joint dynamic state estimation of large-scale power systems. IET Gener. Transm. Distrib. **10**, 452–459 (2016)
16. H. Karimipour, V. Dinavahi, Extended Kalman filter-based parallel dynamic state estimation. IEEE Trans. Smart Grid **6**, 1539–1549 (2015)
17. E. Hossain, I. Khan, F. Un-Noor, S.S. Sikander, M.S.H. Sunny, Application of big data and machine learning in smart grid, and associated security concerns: A review. IEEE Access **7**, 13960–13988 (2019)
18. S. Ahmed, Y. Lee, S.-H. Hyun, I. Koo, Feature selection–based detection of covert cyber deception assaults in smart grid communications networks using machine learning. IEEE Access **6**, 27518–27529 (2018)
19. J. Hao, R.J. Piechocki, D. Kaleshi, W.H. Chin, Z. Fan, Sparse malicious false data injection attacks and defense mechanisms in smart grids. IEEE Trans. Ind. Inf. **11**, 1–12 (2015)
20. H. Karimipour, V. Dinavahi, Parallel domain decomposition based distributed state estimation for large-scale power systems, in *2015 IEEE/IAS 51st Industrial & Commercial Power Systems Technical Conference* (*I&CPS*) (2015), pp. 1–5
21. A. Abur, A.G. Exposito, *Power System State Estimation: Theory and Implementation* (CRC Press, Boca Raton, 2004)
22. T. Baumeister, Literature review on smart grid cyber security, collaborative software development laboratory at the University of Hawaii (2010)
23. G. Liang, J. Zhao, F. Luo, S.R. Weller, Z.Y. Dong, A review of false data injection attacks against modern power systems. IEEE Trans. Smart Grid **8**, 1630–1638 (2016)
24. H. Karimipour, V. Dinavahi, Robust massively parallel dynamic state estimation of power systems against cyber-attack. IEEE Access **6**, 2984–2995 (2017)
25. F. Aloul, A. Al-Ali, R. Al-Dalky, M. Al-Mardini, W. El-Hajj, Smart grid security: threats, vulnerabilities and solutions. Int. J. Smart Grid Clean Energy **1**, 1–6 (2012)
26. L. Wei, L. P. Rondon, A. Moghadasi, A. I. Sarwat, Review of cyber-physical attacks and counter defense mechanisms for advanced metering infrastructure in smart grid. in *2018 IEEE/PES Transmission and Distribution Conference and Exposition (T&D)* (2018), pp. 1–9
27. T. Yadav, A.M. Rao, Technical aspects of cyber kill chain. in *International Symposium on Security in Computing and Communication* (2015), pp. 438–452
28. P.N. Bahrami, A. Dehghantanha, T. Dargahi, R.M. Parizi, K.-K.R. Choo, H.H. Javadi, Cyber kill chain-based taxonomy of advanced persistent threat actors: analogy of tactics, techniques, and procedures. J. Inf. Process. Syst. **15**, 865–889 (2019)
29. J. Kim, L. Tong, On topology attack of a smart grid: undetectable attacks and countermeasures. IEEE J. Sel. Areas Commun. **31**, 1294–1305 (2013)
30. Y. He, G.J. Mendis, J. Wei, Real-time detection of false data injection attacks in smart grid: a deep learning-based intelligent mechanism. IEEE Trans. Smart Grid **8**, 2505–2516 (2017)

31. Y. Mo, T.H.-J. Kim, K. Brancik, D. Dickinson, H. Lee, A. Perrig, et al., Cyber–physical security of a smart grid infrastructure. Proc. IEEE **100**, 195–209 (2011)
32. J. Kim, L. Tong, R.J. Thomas, Data framing attack on state estimation. IEEE J. Sel. Areas Commun. **32**, 1460–1470 (2014)
33. J. Wang, L. C. Hui, and S.-M. Yiu, Data framing attacks against nonlinear state estimation in smart grid. in *2015 IEEE Globecom Workshops (GC Wkshps)* (2015), pp. 1–6
34. L. Zhang, G. Wang, G.B. Giannakis, Real-time power system state estimation and forecasting via deep unrolled neural networks. IEEE Trans. Signal Process. **67**, 4069–4077 (2019)
35. M. Esmalifalak, L. Liu, N. Nguyen, R. Zheng, Z. Han, Detecting stealthy false data injection using machine learning in smart grid. IEEE Syst. J. **11**, 1644–1652 (2014)
36. M. Ozay, I. Esnaola, F.T.Y. Vural, S.R. Kulkarni, H.V. Poor, Machine learning methods for attack detection in the smart grid. IEEE Trans. Neural Networks Learn. Syst. **27**, 1773–1786 (2015)
37. J. Sakhnini, H. Karimipour, A. Dehghantanha, Smart grid cyber attacks detection using supervised learning and heuristic feature selection. in *2019 IEEE 7th International Conference on Smart Energy Grid Engineering (SEGE)* (2019), pp. 108–112
38. Y. Wang, M.M. Amin, J. Fu, H.B. Moussa, A novel data analytical approach for false data injection cyber-physical attack mitigation in smart grids. IEEE Access **5**, 26022–26033 (2017)
39. H. Karimipour, S. Geris, A. Dehghantanha, H. Leung, Intelligent anomaly detection for large-scale smart grids. in *2019 IEEE Canadian Conference of Electrical and Computer Engineering (CCECE)* (2019), pp. 1–4
40. S. Geris and H. Karimipour, Joint state estimation and cyber-attack detection based on feature grouping, in *2019 IEEE 7th International Conference on Smart Energy Grid Engineering (SEGE)*, 2019, pp. 26–30
41. S. Ahmed, Y. Lee, S.-H. Hyun, I. Koo, Covert cyber assault detection in smart grid networks utilizing feature selection and euclidean distance-based machine learning. Appl. Sci. **8**, 772 (2018)
42. J. Wang, D. Shi, Y. Li, J. Chen, H. Ding, X. Duan, Distributed framework for detecting PMU data manipulation attacks with deep autoencoders. IEEE Trans. Smart Grid **10**(4), 4401–4410 (2018)
43. J. Yan, B. Tang, H. He, Detection of false data attacks in smart grid with supervised learning. in *2016 International Joint Conference on Neural Networks (IJCNN)* (2016), pp. 1395–1402
44. H. Karimipour, A. Dehghantanha, R.M. Parizi, K.-K.R. Choo, H. Leung, A deep and scalable unsupervised machine learning system for cyber-attack detection in large-scale smart grids. IEEE Access **7**, 80778–80788 (2019)

A Comparison Between Different Machine Learning Models for IoT Malware Detection

Sanaz Nakhodchi, Aaruni Upadhyay, and Ali Dehghantanha

1 Introduction

We live in a world of convergence where devices around us are increasingly becoming aware of their environment [1, 2]. These devices work to make our lives safer and more comfortable and become even more so effective when connected [3–6]. Such connected IoT (Internet of Things) devices are in constant dialogue with each other, sharing data and instructions over the local network and the internet [7, 8]. The pace at which these devices are being deployed today is staggering. According to McKinsey Global Institute [9], 127 new IoT devices are being connected every second and more than 64 Billion IoT devices are expected to be in use by 2025. The current growth in the IoT market is expected to remain strong too and is expected to double to $520 billion by 2021 [10]. IoT devices are now used across many industries and in critical infrastructure thereby making them an attractive target for attacks [11–13]. Authors in [14] present the current landscape of IoT security and discuss the areas that deserve more attention. With a 55%Year over Year surge [15] noticed in IoT based malware in 2019, the need for detection of malwares targeted at IoT devices has become even more urgent. Any attempt to secure IoT networks must start with protecting it against Malware based attacks [16]. Anti-viruses, which have been the traditional defense against malware, are unable to provide early detection due to the sheer volume of malwares engineered every day. Use of Machine Learning in malware analysis provides a promising

S. Nakhodchi (✉) · A. Upadhyay
School of Computer Science, University of Guelph, Guelph, ON, Canada
e-mail: nakhodcs@uoguelph.ca; aupadhyay@uoguelph.ca

A. Dehghantanha
University of Guelph, Guelph, ON, Canada
e-mail: adehghan@uoguelph.ca

H. Karimipour et al. (eds.), *Security of Cyber-Physical Systems*,
https://doi.org/10.1007/978-3-030-45541-5_10

alternative in malware detection due to its non-dependence on signatures creation for every malware family for detection.

Anti-virus software employs two primary methods [17] for malware detection: signature match and heuristic analysis. A new bot released in the wild may go undetected for considerable time before it is found and before vendors can react and create anti-virus signatures to counter it. The size of IoT networks and use of techniques to bypass heuristics-based detection makes it unfeasible for heuristic analysis to be performed on IoT networks [4]. Both the methods used above by the antivirus products make them less suitable for use in an IoT environment where quick detection is needed to stop the malware from compromising the whole network [18, 19].

Attacks on IoT networks seems to research topic with a high amount of interest among researchers [20–22]. For the authors in [23] discuss the various types of threats targeted towards IoT networks and the tool employed by the attackers to carry out the attack. Similarly, the issues faced by IoT devices today and various approaches for Malware analysis is discussed in [24, 25]. Several researchers have come up with their own novel approaches for IoT Malware detection. For example, in [26], authors present the use of CNN to identify IoT Malware by "converting program binaries to gray-scale images." Authors in [27] use the "Bayesian model updating method" for IoT network traffic analysis and Malware detection, they also provide a comparison with other machine learning classifiers such as KNN and SVM. Anomaly detection of power dissipation pattern of IoT devices using K-Means Clustering algorithm is used by authors in [28] for IoT malware detection.

In this paper we explore the application of five classification and clustering algorithms (SVM, Naive Bayes, KNN, K-Means and Spectral Clustering) along with a deep neural network (CNN) algorithm to classify malware using malware samples [29] from an IoT environment. We then compare our results for each of these algorithms and present our conclusions on the effectiveness of each with respect to our IoT dataset.

The rest of the paper is organized in the following sections. The Problem Statement we discuss our approach to the problem to IoT Malware detection. A survey of related work is presented in the Literature Review section. Next, we discuss our Methodology and present our results for all the algorithms used. This, lastly, is followed by the section where we present our conclusions and discuss the areas of possible future research.

2 Literature Review

Over the years, several researchers have focused on use of Machine Learning classifiers for detecting malwares in IoT devices [30, 31]. These research focus on classifying existing malware samples to train a ML model that can make (accurate) predictions on unseen malware samples. In [32], authors detail the two-step process used in ML malware detection namely: "feature extraction and classification/clustering". They go on to discuss the different methods of feature selection and classification algorithms like Artificial Neural Network (ANN), Decision Tree (DT), and Support Vector Machines (SVM). Moreover, [33] used SVM for detecting malware in android operating system and achieved 0.99 accuracy and precision for their generated dataset.

In [34], researchers use Naive Bayes classifier for malware detection in IoT devices based on Android platform. They achieved 98% accuracy when using the Naïve Bayes classification based on a decision tree. Moreover, [35] achieved 97% f-score on decision trees for malware detection approach. The researchers also implemented Naïve Bayes and Logistic Regression with f-score 51% and 94%, respectively. [36] Used KNN classifier with fingerprint feature for IoT malware detection on device layer and they achieved 98.2% accuracy.

Deep Eigenspace learning approach used in [21] and achieved 99.68% accuracy in IoT malware detection. In [37], authors survey three approaches for IoT malware detection: CNN on byte sequences, CNN on color images and CNN on assembly sequences. The authors used a training dataset created with 15,000 IoT malware and 1000 benign ware samples to prepare their training dataset. Their results showed that CNN on images and assembly sequences achieved higher accuracy than CNN on byte sequences.

In addition, [38] implemented ML technique (supervised, unsupervised and reinforcement learning) in IoT environment with malware detection, authentication and access control approaches. In [39], adversarial learning against attack used for IoT malware detection. "Off-the-shelf methods and Graph Embedding and Augmentation (GEA) methods" used which off-the-shelf gained 100% of misclassification and all malware classified as benign in GEA.

The previous work focused on implementing classifier or deep learning models on IoT malware datasets although in this research we selected three classifiers, two clustering and one deep learning model to compare the accuracy of detection on the IoT malware.

3 Methodology

This part considers the workflow of our research. Firstly, the IoT malware dataset was already collected and publicly available in CyberScienceLab [29]. In addition, the dataset consists of two folder which were Goodware and Malware with 268 and 244 Opcode files respectively. Secondly, the Natural Language Processing (NLP) technique used for preprocessing. Thirdly, SVM, Naive Bayes and KNN classifier, K-means, Spectral and CNN implemented on the creating dataset which are elaborating in the following section. Fig. 1 shows the implementation steps.

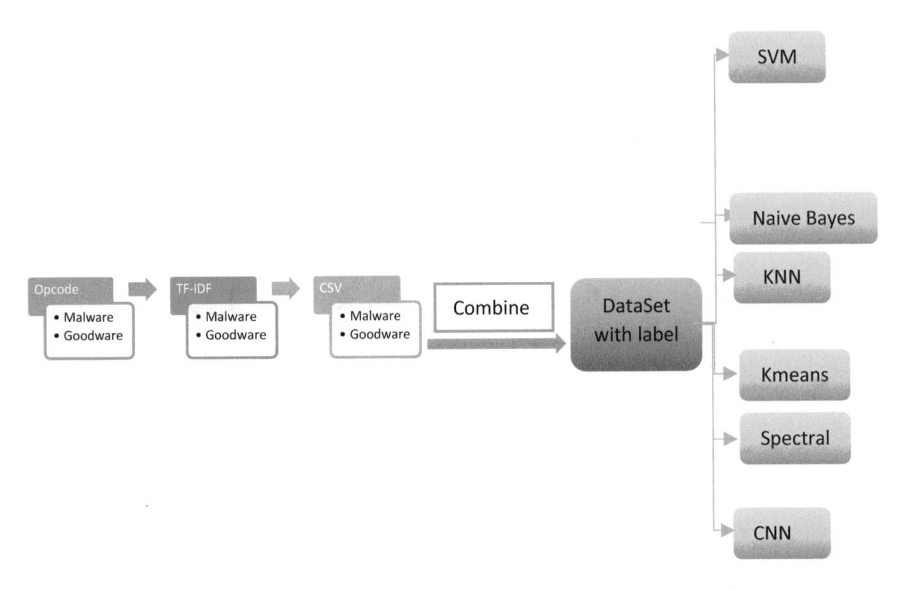

Fig. 1 Implementation process

3.1 Preprocessing

Preparing suitable dataset caused some challenges. The type of malware and goodware were Opcode which means that they were the text files. Thus, converting them to the number was essential parts and challenging one. We used Term Frequency- Inverse Document Frequency (TF-IDF) technique of NLP to facilitate this problem. It means that for each sample of malware and goodware, we got TF-IDF score and used it as the features. The results showed 252 and 362 features for malware and goodware. Then, we tried to put each feature in the header of a csv file and TF-TDF score of them as a row, which it means rotating and converting txt to csv. Consequently, we created two CSV files malware and goodware. In order to have a unit dataset, we combined csv files, kept similar features and created the dataset along with preprocessing our data. In the final stage, we had 236 features with 512 samples. Furthermore, we added an additional column "label" which was shows the type of samples: 0 indicated goodware and 1 indicated malware.

3.2 Implementing Machine Learning Models

We utilized tenfold Cross Validation (CV) and six different models were selected which were Support Vector Machin (SVM), Naive Bayes, KNN, K-means, Spectral and Convolutional Neural Network (CNN). We tried to test with different classifiers

(SVM, Naive Bayes and KNN), Clustering models (K-means and Spectral) and Deep learning model (CNN) as well.

SVM, Naïve Bayes classifiers implemented by Sklearn library with kernel = linear and the confusion matrix are generated. Moreover, KNN which is another classification algorithm based on feature similarity deployed with a value of k = 5. In addition, we used K-means and spectral which are clustering algorithms that attempts to find clusters within unlabeled datasets for comparing classification and clustering algorithms in selected dataset. Finally, deep learning model, Convolutional Neural Network (CNN), with three layers and utilizing Keras library implemented. In addition, we evaluated models with calculating accuracy, precision, recall and f1-score.

TP = True Positive, goodware classified as goodware

TN = True Negative, malware detected as a malware

FP = False Positive, malware detected as a goodware

FN = False Negative, goodware classified as malware

Accuracy = $\frac{TP+TN}{TP+TN+FP+FN}$, how methods could detect malware and goodware accurately

Precision = $\frac{TP}{TP+FP}$, ratio of malware samples that are correctly predicted

Recall = $\frac{TP}{TP+FN}$, ratio of predicting malware that are correctly labeled as malware

F1-score = $2 * \frac{Precision*Recall}{Precision+Recall}$, is the harmonic of precision and recall

4 Finding

In this section the results of each models consider. In general, Precision shows the rate for relevant results. In addition, Recall is another measurement for sensitivity of the relevant results and F1-Score is the value for estimating the performance of the system.

As can be seen in Table 1, CNN gave best accuracy rather than the other models. In addition, KNN located in second rank of detecting malware in the IoT dataset although K-means and spectral which are clustering models are in the next stage. SVM was the next classifier with 98.03 accuracy and Naïve Bayes is the least which achieved 96.07% accuracy in malware detection. Consequently, deep learning model (CNN), obtained best accuracy and classifiers and clustering models are in the follow are in the second and third location of detecting malware in the selected IoT dataset.

We improved the accuracy of malware detection in IoT with different models. For example, [36] achieved 98% and we increased about 1–99% in KNN. Moreover, we boost accuracy in using CNN of deep learning from 94 [30] to 100.

Table 1 Comparing results

	SVM	Naïve bayes	KNN	K-means	Spectral	CNN
Accuracy	98.03	96.07	99.02	99	98	100
Precision	98	97	99	98	87	100
Recall	98	96	99	98	84	99
F1-score	98	96	99	98	93	99

5 Conclusion and Future Work

IoT technology and being attractive target for attackers are becoming more prominent [40, 41]. Thus, providing the system for detecting malware of IoT systems can have a significant impact to create a safe environment. In this paper, we considered some classifiers, clustering and deep learning models for achieving best rate of malware detection. CNN obtained best accuracy in malware detection in the specific dataset. The other deep learning models can test for the future works. In addition, CNN model can test with more data and real dataset with verity of features.

References

1. M. Pruthvi, S. Karthika, N. Bhalaji, 'Smart college'-study of social network and IoT convergence, in *Proceedings of the International Conference on I-SMAC (IoT in Social, Mobile, Analytics and Cloud), I-SMAC 2018* (IEEE, 2019), pp. 100–103. https://doi.org/10.1109/I-SMAC.2018.8653787.
2. M. Conti, A. Dehghantanha, K. Franke, S. Watson, Internet of things security and forensics: challenges and opportunities. Future Gener. Comput. Syst. **78**, 544–546 (2018). https://doi.org/10.1016/j.future.2017.07.060
3. M. Jerabandi, M.M. Kodabagi, A review on home automation system, in *Proceedings of the 2017 International Conference On Smart Technology for Smart Nation, SmartTechCon 2017* (IEEE, 2018), pp. 1411–1415. https://doi.org/10.1109/SmartTechCon.2017.8358597.
4. H. HaddadPajouh, A. Dehghantanha, R.M. Parizi, M. Aledhari, H. Karimipour, A survey on internet of things security: requirements, challenges, and solutions. Internet Things, 100129 (2019). https://doi.org/10.1016/j.iot.2019.100129
5. G. Srivastava, R.M. Parizi, A. Dehghantanha, K.-K.R. Choo, Data sharing and privacy for patient IoT devices using blockchain, in *International Conference on Smart City and Informatization* (Springer, Singapore, 2019), pp. 334–348
6. S. Mohammadi, V. Desai, H. Karimipour, Multivariate mutual information-based feature selection for cyber intrusion detection, in *2018 IEEE Electrical Power and Energy Conference (EPEC)* (IEEE, 2018), pp. 1–6. https://doi.org/10.1109/EPEC.2018.8598326.
7. S. Yousefi, F. Derakhshan, H. Karimipour, H.S. Aghdasi, An efficient route planning model for mobile agents on the internet of things using Markov decision process. Ad Hoc Netw. **98**, 102053 (2020). https://doi.org/10.1016/j.adhoc.2019.102053
8. S. Geris, H. Karimipour, Joint state estimation and cyber-attack detection based on feature grouping, in *2019 IEEE 7th International Conference on Smart Energy Grid Engineering (SEGE)* (IEEE, 2019), pp. 26–30. https://doi.org/10.1109/SEGE.2019.8859926.
9. C. Petrov, Internet Of things statistics from 2019 to justify the rise of IoT (2019), https://techjury.net/stats-about/internet-of-things-statistics/. Accessed 25 Oct 2019

10. L. Columbus, IoT market predicted to double by 2021, reaching $520B (2018), https://www.forbes.com/sites/louiscolumbus/2018/08/16/iot-market-predicted-to-double-by-2021-reaching-520b/#768bbd9d1f94. Accessed 13 Dec 2019
11. A. Namavar Jahromi et al., An improved two-hidden-layer extreme learning machine for malware hunting. Comput. Secur. **89**, 101655 (2020). https://doi.org/10.1016/j.cose.2019.101655
12. J. Sakhnini, H. Karimipour, A. Dehghantanha, R.M. Parizi, G. Srivastava, Security aspects of internet of things aided smart grids: a bibliometric survey. Internet Things, 100111 (2019). https://doi.org/10.1016/j.iot.2019.100111
13. M.R. Begli, F. Derakhshan, H. Karimipour, A Layered intrusion detection system for critical infrastructure using machine learning, in *IEEE Int. Conf. on Smart Energy Grid Engineering (SEGE)* (IEEE, 2019), pp. 1–5
14. M. Binti Mohamad Noor, W.H. Hassan, Current research on internet of things (IoT) security: a survey. Comput. Netw. **148**, 283–294 (2019). https://doi.org/10.1016/j.comnet.2018.11.025
15. G. Blaine, Mid-year update: 2019 sonicwall cyber threat report (SocinWall, 2019)
16. E.M. Dovom, A. Azmoodeh, A. Dehghantanha, D.E. Newton, R.M. Parizi, H. Karimipour, Fuzzy pattern tree for edge malware detection and categorization in IoT. J. Syst. Archit. **97**, 1–7 (2019). https://doi.org/10.1016/j.sysarc.2019.01.017
17. M. Al-Asli, T.A. Ghaleb, Review of signature-based techniques in antivirus products, in *2019 International Conference on Computer and Information Sciences (ICCIS)* (IEEE, 2019), pp. 1–6. https://doi.org/10.1109/ICCISci.2019.8716381.
18. H.H. Pajouh, R. Javidan, R. Khayami, A. Dehghantanha, K.K.R. Choo, A two-layer dimension reduction and two-tier classification model for anomaly-based intrusion detection in iot backbone networks. IEEE Trans. Emerg. Top. Comput. **7**(2), 314–323 (2019). https://doi.org/10.1109/TETC.2016.2633228
19. S. Mohammadi, H. Mirvaziri, M. Ghazizadeh-Ahsaee, H. Karimipour, Cyber intrusion detection by combined feature selection algorithm. J. Inf. Secur. Appl. **44**, 80–88 (2019). https://doi.org/10.1016/j.jisa.2018.11.007
20. A. Azmoodeh, A. Dehghantanha, M. Conti, K.K.R. Choo, Detecting crypto-ransomware in IoT networks based on energy consumption footprint. J. Ambient Intell. Humaniz. Comput. **9**(4), 1141–1152 (2018). https://doi.org/10.1007/s12652-017-0558-5
21. A. Azmoodeh, A. Dehghantanha, K.-K.R. Choo, Robust malware detection for internet of (battlefield) things devices using deep eigenspace learning. IEEE Trans. Sustain. Comput. **4**(1), 88–95 (2018). https://doi.org/10.1109/tsusc.2018.2809665
22. H. Karimipour, A. Dehghantanha, R.M. Parizi, K.K.R. Choo, H. Leung, A deep and scalable unsupervised machine learning system for cyber-attack detection in large-scale smart grids. IEEE Access **7**, 80778–80788 (2019). https://doi.org/10.1109/ACCESS.2019.2920326
23. A. Lohachab, B. Karambir, L.A. Lohachab, Critical analysis of DDoS-an emerging security threat over IoT networks. J. Commun. Inf. Netw. **3**(3), 57–78 (2018). https://doi.org/10.1007/s41650-018-0022-5
24. S.W. Soliman, M.A. Sobh, A.M. Bahaa-Eldin, Taxonomy of malware analysis in the IoT, in *Proceedings of ICCES 2017 12th International Conference on Computer Engineering and Systems* (IEEE, 2018), pp. 519–529. https://doi.org/10.1109/ICCES.2017.8275362.
25. S. Sharmeen, S. Huda, J.H. Abawajy, W. Nagy Ismail, M.M. Hassan, Malware threats and detection for industrial mobile-IoT networks. IEEE Access **6**, 15941–15957 (2018)
26. J. Su, V. Danilo Vasconcellos, S. Prasad, S. Daniele, Y. Feng, K. Sakurai, Lightweight classification of IoT malware based on image recognition, in *2018 IEEE 42nd Annual Computer Software and Applications Conference (COMPSAC)*, vol. 2 (IEEE, 2018), pp. 664–669. https://doi.org/10.1109/COMPSAC.2018.10315.
27. F. Wu, L. Xiao, J. Zhu, Bayesian model updating method based android malware detection for IoT services, in *2019 15th International Wireless Communications and Mobile Computing Conference, IWCMC 2019* (IEEE, 2019), pp. 61–66. https://doi.org/10.1109/IWCMC.2019.8766754.
28. S. Papafotikas, A. Kakarountas, A machine-learning clustering approach for intrusion detection to IoT devices, in *2019 4th South-East Europe Design Automation, Computer Engineering,*

Computer Networks and Social Media Conference (SEEDA-CECNSM) (IEEE, 2019), pp. 1–6. https://doi.org/10.1109/SEEDA-CECNSM.2019.8908520.
29. CyberScienceLab, IoT malware detection dataset - Cyber Science Lab (2019), https://cyberscience lab.org/iot-malware-detection-dataset/. Accessed 25 Oct 2019
30. H. HaddadPajouh, A. Dehghantanha, R. Khayami, K.K.R. Choo, A deep recurrent neural network based approach for internet of things malware threat hunting. Futur. Gener. Comput. Syst. **85**, 88–96 (2018). https://doi.org/10.1016/j.future.2018.03.007
31. A. Kumar, T.J. Lim, EDIMA: early detection of IoT malware network activity using machine learning techniques, in *2019 IEEE 5th World Forum on Internet of Things (WF-IoT)* (IEEE, 2019), pp. 289–294. https://doi.org/10.1109/wf-iot.2019.8767194.
32. Y. Ye, T. Li, D. Adjeroh, S.S. Iyengar, A survey on malware detection using data mining techniques. ACM Comput. Surv. **50**(3), 1–40 (2017). https://doi.org/10.1145/3073559
33. H.S. Ham, H.H. Kim, M.S. Kim, M.J. Choi, Linear SVM-based android malware detection for reliable IoT services. J. Appl. Math., 2014, 594501 (2014). https://doi.org/10.1155/2014/594501
34. R. Kumar, X. Zhang, W. Wang, R.U. Khan, J. Kumar, A. Sharif, A multimodal malware detection technique for android IoT devices using various features. IEEE Access **7**, 64411–64430 (2019). https://doi.org/10.1109/ACCESS.2019.2916886
35. Z. Markel, M. Bilzor, Building a machine learning classifier for malware detection, in *WATeR 2014 - Proceedings of the 2014 2nd Workshop on Anti-Malware Testing Research* (IEEE, 2015). https://doi.org/10.1109/WATeR.2014.7015757.
36. T. Duc Nguyen, S. Marchal, A.-R. Sadeghi, DÏoT: a self-learning system for detecting compromised IoT devices, in *Proc. 39th IEEE Int. Conf. Distrib. Comput. Syst.* (IEEE, 2019)
37. K.D.T. Nguyen, T.M. Tuan, S.H. Le, A.P. Viet, M. Ogawa, N. Le Minh, Comparison of three deep learning-based approaches for IoT malware detection, in *Proceedings of 2018 10th International Conference on Knowledge and Systems Engineering, KSE 2018* (IEEE, 2018), pp. 382–388. https://doi.org/10.1109/KSE.2018.8573374.
38. L. Xiao, X. Wan, X. Lu, Y. Zhang, D. Wu, IoT security techniques based on machine learning: how do IoT devices use AI to enhance security? IEEE Signal Process. Mag. **35**(5), 41–49 (2018). https://doi.org/10.1109/MSP.2018.2825478
39. A. Abusnaina, A. Khormali, H. Alasmary, J. Park, A. Anwar, A. Mohaisen, Adversarial learning attacks on graph-based IoT malware detection systems, in *2019 IEEE 39th International Conference on Distributed Computing Systems (ICDCS)* (IEEE, 2019), pp. 1296–1305. https://doi.org/10.1109/ICDCS.2019.00130.
40. H. Karimipour, S. Geris, A. Dehghantanha, H. Leung, Intelligent anomaly detection for large-scale smart grids, in *2019 IEEE Canadian Conference of Electrical and Computer Engineering (CCECE)* (IEEE, 2019), pp. 1–4
41. J. Sakhnini, H. Karimipour, A. Dehghantanha, Smart grid cyber attacks detection using supervised learning and heuristic feature selection, in *Proceedings of 2019 the 7th International Conference on Smart Energy Grid Engineering, SEGE 2019* (IEEE, 2019), pp. 108–112. https://doi.org/10.1109/SEGE.2019.8859946.

A Bibliometric Analysis on the Application of Deep Learning in Cybersecurity

Sanaz Nakhodchi and Ali Dehghantanha

1 Introduction

Upon using widespread Internet, miscellaneous threats increase among users. Variety of threats are growing as well. For instance, Multi-vectored and multi-staged are the new generation of threats, which an attacker by using multiple meaning of propagation and by infiltrating networks can exfiltrate valuable data [1, 2]. In addition, the significance of threat impact cannot overlook which can be destruction, corruption and disclosure of information, theft and denial of services, elevation of privilege and illegal usage [3, 4]. For example, Facebook claim that hackers by exploiting a vulnerability of theirs networks could get access to 30 million of user accounts. Moreover, in 2018 Ander Armour said that MyFitnessPal application was the target of attackers and 150 million users affected [5]. Moreover, cyberattack on critical infrastructure can have catastrophic impact on industry and society. Ukraine power grid cyberattack in 2015 was one of the successful cyberattack which 30 substations were shut down and 230,000 people were affected. Thus, several companies in the glob are trying to design new technologies in order to protect computer systems and their networks from malicious attacks and actors [6].

It is difficult to detect the new generation of cyber-attacks by traditional cybersecurity procedures [7–9]. In recent years, Artificial Intelligence methods have collected success in a different application domains [10]. Pattern recognition, natural language processing, speech recognition along with Deep Learning (DL) algorithms have a crucial role in finding solution for complicated problems [6, 11].

S. Nakhodchi (✉)
School of Computer Science, University of Guelph, Guelph, ON, Canada
e-mail: nakhodcs@uoguelph.ca

A. Dehghantanha
University of Guelph, Guelph, ON, Canada
e-mail: adehghan@uoguelph.ca

H. Karimipour et al. (eds.), *Security of Cyber-Physical Systems*,
https://doi.org/10.1007/978-3-030-45541-5_11

Upon using DL methods for attack detection, it can be a flexible approach due to capability of high-level feature extraction from complex large-scale data [12–14]. DL also helps companies to update their intrusion detection systems with low cost and time [6, 15–17]. Moreover, using machine learning and DL for attack detection in the large scale data such as smart grid and IoT are becoming interesting topics recently [18, 19].

However popular and enormous companies such as Salesforce Facebook, Google and Microsoft have already implemented DL in their products, "the cybersecurity industry is still playing catch up" [20]. It is worth to mention that the two areas that DL has improved significantly are "malware detection and network intrusion detection". Noor et al. [21] create a framework that based on attack patterns of cyber threat intelligence reports can facilitate cyber threat attribution. It also used natural language processing and DL as a classifier for higher accuracy. In addition, because of the problems of current NIDS methods, Shone et al. [22] proposed new NDAE method with unsupervised approach. Dl can detect other anomality as well. For example, Kobojek and Saeed [23] by using keystroke dynamics, it able to evaluate a user with recurrent neural networks (LSTM), and monitoring framework based on SDN and forensic approach with deep learning models along with evaluating deep packet inspection considered in [24]. In addition, dealing with high imbalanced data in cyber-physical attack detection proposed [25]. The researchers used unsupervised deep learning models for improve accuracy.

The aforementioned examples elaborate that research activity in this topic is crucial. Although there are many articles published for supporting research activity, any bibliometric that focus on research impact and trends of publication cannot find up to 2018. Bibliometric helps researchers to better understand research patterns and structures of a topic. It is "a field that uses mathematical and statistical techniques, from counting to calculus, to study publishing and communication patterns in the distribution of information [26]". The advantages of bibliography studies rely on authors, institutions and researchers which can demonstrate, evaluate and predict the previous and next research's domains.

In order to consider the increase of using DL in cybersecurity approach, this paper aims to provide an investigation of the comprehensive evaluation of DL in cybersecurity research activities published in WoS from 2010 to 2018. The approach includes the evaluation of using DL in cybersecurity research, publication patterns, research topics and assessment. To address this research, the following questions are considered: (a) What is the trend of publications in DL for cybersecurity study; (b) how does this trend help to identify the future direction of using DL algorithm in cybersecurity study?

Using deep learning AND cybersecurity as the main keywords, it is identified 69 articles. The search also expanded to use more key words based on [6, 27, 28] such as intrusion detection, malware detection, cyber analytics and convolutional neural network. The query of all keywords which is used for retrieving articles is "(deep learning OR CNN OR convolutional neural network) AND (forensic OR Cyber-security OR Cyber-security OR cyber-attack OR defense OR cyber analytics OR threat intelligence OR threat hunting OR malware detection OR intrusion detection OR malware classification)" with 739 papers. The other keywords such as incident and anomaly detection are considered. Due to irrelevant result, these keywords are removed from search. All data exploited from Web of Science Core Collection. The analysis is based on relationship between the abstract, title, publication, citation, research area, geographical location and the keywords. Moreover, in this paper, the classification of using DL with different aspects of cybersecurity is discussed by considering the frequency of words used in keywords and title.

The rest of this paper is organized as follows. Section 2 describes the methodology. Section 3 demonstrate findings and information of DL studies in cybersecurity, and Sect. 4 is the discussion and conclusion of the study.

2 Methodology

Bibliometrics is a way for "measurement of science and technology" [26]. It means that it is a method for evaluating and ranking institution, people, research topic and countries. In the other word, it is oldest research method in science and library. This paper used bibliometric based on [29], bibliometric is used for evaluation research with two perspectives: general instructions and publication analysis. General instruction is about search engine and searching articles in order to give a suitable result. However, publication analysis relates to the assessment of publication with considering impact factor, publisher and citation. This method is used in several researches such as [30] comparing and correlation occupation therapy journals in English in 2015, [31] considering the scientific development trend in food chemistry from 1976 to 2016 (40 years), [32] studying 60 years researches focused on the benchmarking procedure, [33] analyzing about 12,000 articles in renewable energies.

In this paper, for retrieving information, several strategies were used. The first step is choosing the keywords which is important because it relates to research's information along with direction and interest of publications [34]. In this study, manual and automatic methods are applied for retrieving articles. The main key words are "Deep learning AND cybersecurity" and 69 articles are covered from 2010 to 2018. Upon applying other keywords, 760 articles are found which by applying some filters such selecting English language, time span from 2010 to 2018, excluding "Arts & Humanities Citation Index (A&HCI)", "Social Science Citation Index (SSCI)" and "Conference Proceedings Citation Index-Social Science & Humanities (CPCI-SSH)" from Core Collection, 739 articles remained for analysis. The analysis was considered based on the, (a) impact journals, (b) highly cited articles, (c) research areas, (d) productivity, (e) keyword frequency, (f) institutions and (g) authors. For visualizing the result VOSviewer tool is used which is the powerful free application for analysis bibliometric.

2.1 Web of Science

Web of Science (WoS), Google Scholar, Science Direct, IEEE Explore, Elsevier's Scopus are the popular databases used for indexing journals. The three main bibliometric sources are WoS, Google scholar and Elsevier's Scopus [35–37] which are used for ranking journals especially about their productivity and citations. In this paper, WoS database selected based on firstly, it is the first tool in bibliometric analysis before emersion of Scopus and Google Scholar, secondly WoS has the highest impact factor journals from Scopus (94%). Moreover, Scopus and Google scholar are excluded because of avoiding overlap and low data quality respectively [37].

In addition, WoS is selected because of being powerful in visualizing analysis bibliometric in science [38]. Furthermore, the literature in WoS is related to since 1900 although Scopus has literature publications from 1996 [36]. The other advantage of using WoS is that it has all types of articles with their components including citations, organizations, authors along with bibliometric references for all of these [37].

In this paper, WoS is as a search engine which has collected "over 20,000 peer-reviewed and high-quality journals published" on the globe. In addition, more than "250 science, social science and humanities disciplines topics are covered in WoS [39]. This search engine has variety options for searching such as filtering based on publication dates, recently added, number of times cited, relevance, author name along with excluding by years, institutions and countries. These facilities made this study more conductive.

3 Finding

This section is about how to find related articles in deep learning and cybersecurity and is divided into seven sub-topics: productivity, research area, institution, authors, impact journals, high-cited articles and keyword frequency. The importance of this finding is about showing publishing rates with bibliometric data. Furthermore, it can demonstrate high-quality research which is useful for creating novel knowledge and making sure that pursuit into deep learning in security studies is more in-depth. Figure 1 is provided to show the number of publications in 2010–2018. It shows six categories of publications including proceedings paper, article, review, editorial material, meeting abstract and book chapter. The proceedings paper's category has the highest proportion with a total 71.31%. This amount is followed by the Article which is 28.01% while Book chapter and Editorial Material have the lowest percentage of publications, 0.13%.

Based on Fig. 1 in 2018, the number of proceeding and articles have increased significantly. This is possibly because of taking time for publishing articles in the journals which is affected on the number of publications.

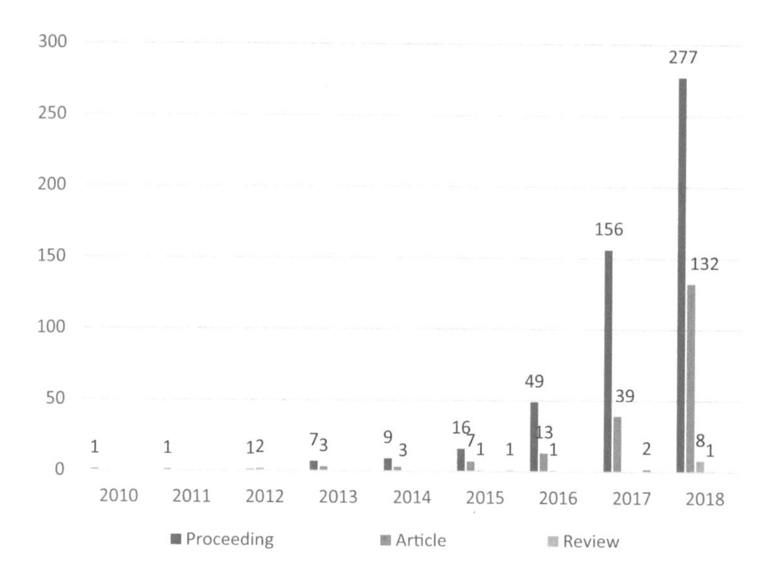

Fig. 1 Number of publications

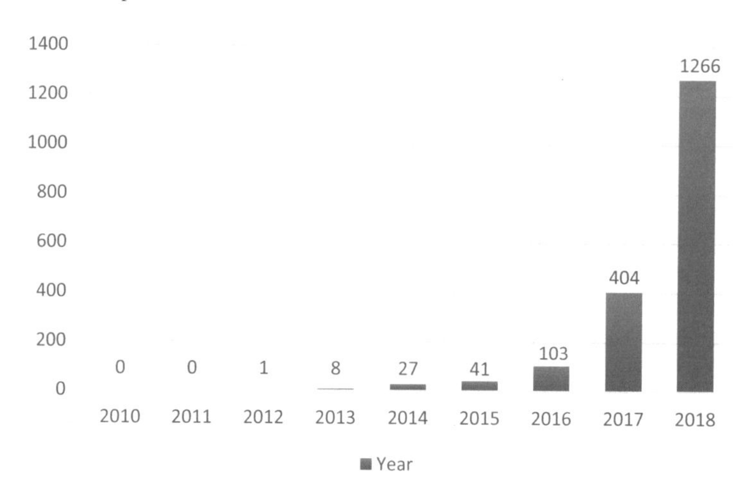

Fig. 2 Citation distributions

Upon previous discussion, citation analysis is useful for assessing the frequency of the journals built by extracting data from the citation index. This is useful for evaluating researchers' performance in academic. It also demonstrates the information about researchers using shared resources along with providing a comprehensive sight of the subject researched. Figure 2 illustrates the citation distributions.

Figure 2 shows the number of citations of publications between 2010 and 2018. It can figure out that the number of publications affected the number of citations which it means if an article published earlier it has more citations.

The total number of citations received by publication was 1850 and the annually average was 372 during 2010–2018.

3.1 Productivity

This section considers productivity in the continents. Productivity in publications rely on the frequency of publications. It is able to be used as a tool for calculating the number of articles published among continents. Considering articles based on growing of productivity has a significant impact on the researcher due to focusing and expanding the research components along with analysis on deep learning in cybersecurity. The crucial aspect of concentrating in the analysis of productivity is that it is useful for increasing and improving the productivity of the publications. Moreover, it helps to contribute in the new technology or methods in research. It also uses for evaluating of countries and continents that have published more papers.

Table 1 shows the number of publications with respect to the time during 2010–2018 according to continents. Table 1 also illustrates that continents of Asia is the major contribution in publishing articles with the China being the most significant. This is followed by the continent of North America and Europe. Australia, South America and Africa comparatively conducted lesser research about deep learning in cybersecurity topic. The table shows that the United States with 32.20% contribution is located after China with 26.25%. This is followed by India, South Korea, Italy and England.

Based on the above information, the continent of Asia and North America are the most productive in publishing articles. Asia seems to be significantly in front of the continent of North America, and this is probably due to the lesser amount of research funding provided by the respective countries in conducting research in similar research areas. In any research area, the funding is crucial due to enabling researchers to perform new studies which can impact on the number of publications. For instance, $176.8 billion has been spent on R&D in the United Stated in 2018 while the Chines government has provided $291 billion fund for research in 2018 [40, 41].

3.2 Research Areas

This part considers a number of publications in specific research areas. Research areas include two sections: single and multiple disciplinary with goal of developing scientific aspects of a certain research area. Moreover, Research areas are important factors that helps others to measure and evaluate the performance of a research based on citation and publication rates. The trend of the publication during

Table 1 Productivity

List of continents	Number of articles	Number of articles (%)
Asia	*329*	
China	194	26.25
India	45	6.08
South Korea	44	5.95
Japan	19	2.57
Taiwan	13	1.75
Turkey	11	1.48
Vietnam	10	1.35
Pakistan	10	1.35
Saudi Arabia	10	1.35
Malaysia	9	1.21
Singapore	9	1.21
Iran	8	1.08
United Arab Emirates	7	0.94
Israel	4	0.54
Indonesia	3	0.40
Qatar	3	0.40
Egypt	2	0.27
Thailand	2	0.27
Bahrain	1	0.13
Bangladesh	1	0.13
Philippines	1	0.13
North America	*270*	
United States	238	32.20
Canada	23	3.11
Russia	7	0.94
Mexico	2	0.27
South America	*13*	
Brazil	9	1.21
Colombia	2	0.27
Argentina	1	0.13
Chile	1	0.13
Peru	1	0.13
Europe	*104*	
Italy	36	4.87
England	28	3.78
Germany	20	2.70
Spain	15	2.03
France	8	1.08
Switzerland	6	0.81
Austria	5	0.67
Netherlands	5	0.67

(continued)

Table 1 (continued)

List of continents	Number of articles	Number of articles (%)
Portugal	4	0.54
Belgium	4	0.54
Finland	4	0.54
Denmark	4	0.54
Poland	4	0.54
Greece	3	0.40
Ireland	3	0.40
Romania	3	0.40
North Ireland	3	0.40
Scotland	3	0.40
Ireland	3	0.40
Romania	3	0.40
Lithuania	2	0.27
Norway	2	0.27
Bulgaria	2	0.27
Czech Republic	2	0.27
Wales	2	0.27
Estonia	1	0.13
Sweden	1	0.13
Hungary	1	0.13
Jordan	1	0.13
Malta	1	0.13
Australia	*29*	
Australia	29	3.92
Africa	*6*	
Morocco	4	0.54
Algeria	1	0.13
Uganda	1	0.13

times can be shown by performance of any research areas. Database of Web of Science has an index of different disciplines with 150 scientific research areas such as Computer Science, Engineering, Telecommunications, Imaging science photographic technology and Optics, etc.

Table 2 has 15 research areas of publications. As can be seen, the majority of papers are published in Computer science and Engineering domains. New technologies and methods are come from these two areas which can be included datamining, computer architecture, artificial engineering, machine learning and security, etc.

Table 2 Research areas

Research areas	Number of publications
Computer science	491
Engineering	334
Telecommunications	133
Imaging science photographic technology	30
Automation control systems	29
Optics	21
Acoustics	13
Mathematical computational biology	12
Science technology other topics	12
Mathematics	9
Operations research management science	8
Physics	8
Chemistry	7
Education educational research	7
Materials science	7

3.3 Institutions

This section considers the number of publications based on institutions. The aim of this section is to identify which institutions are active and also to measure their quality by comparing institutions according to publications. Table 3 shows a part of list of institutions which have publications in deep learning with cybersecurity approach.

It can be seen that institutions from Asia have the highest number of papers. North America and Europe are in the followed. It becomes visible that China Academic of Science from China and State University System of Florida from United States have the highest rank in the number of publications. In addition, it seems that speed rate of publications in China is faster than the other Asian countries due to having more prestigious universities in China. In addition, Table 3 shows that University of Chinese Academy of Sciences Cas from China, Amrita Vishwa Vidyapeetham University from India and Korea Advanced Institute of Science Technology Kaist are the same place with 14 publications each. Based on evidences, there is an intense competition among institutions of Asia and North America in terms of number of publications.

3.4 Authors

This section considers the number of publications based on authors under countries. The goal is to identify who is more active in terms of authorship in countries. Table 4

Table 3 List of institutions

Institutions	Publications	Country
Chinese academic of science	26	China
State University System of Florida	17	United States
Amrita Vishwa Vidyapeetham University	14	India
Korea Advanced Institute of Science Technology Kaist	14	Korea
University of Chinese Academy of Sciences Cas	14	China
University of California System	13	United States
Shanghai Jiao Tong University	12	China
Polytechnic University of Milan	11	Italy
University of Texas System	11	United States
Beijing University of Posts Telecommunications	10	China
Shenzhen University	10	China
Sun Yat Sen University	10	China
National University of Defense Technology China	9	China
United States Department of Defense	9	United States
United States Department of Energy Doe	9	United States
Drexel University	8	United States
Purdue University	7	United States

Table 4 List of authors

Authors	Publications	Country
K. P Soman	12	India
Rahul K. Vigneswaran	12	India
Prabaharan Poornachandran	11	India
Paolo Bestagini	9	Italy
Stefano Tubaro	9	Italy
Wei Wang	8	China
Luca Bondi	7	Italy
Ye Liu	7	China
Matthew C Stamm	7	United states
Xiangui Kang	6	China
Jonghyun Kim	6	South Korea
Heung-Kyu Lee	6	South Korea
Bo Li	6	China
Khaled Alrawashdeh	5	United states
Belhassen Bayar	5	United states

provides the information of authors and their countries. It can be seen that most of the authors are from Asian countries such as India and China. American authors are active in this area as well. Moreover, Italy is another country that has activities in terms of publishing papers in cybersecurity with deep learning approach.

Table 4 shows that K. P Soman with Rahul K. Vigneswaran from India has the most publications (12) and Prabaharan Poornachandran with one difference (11) has the second rank. With investigating, it identified that K. P Soman and Prabaharan

Poornachandran were college in the same university and worked together and Rahul K. Vigneswaran who was PhD student and researcher at the university with mainly interested in cyber threat and machine learning.

By closer view, it is indicated that Dr. K. P Soman is a head and a professor of department Computational Engineering and Networking at Amrita Vishwa Vidyapeetham University. In addition, he has two labs under names Cybersecurity Lab and Computational Thinking Lab. Moreover, Dr. Poornachandran is assistant professor along with working in the Center for Cybersecurity Systems and Networks at the university.

3.5 *Impact Journals*

This section considers the impact journals based on the computer science area which can help researchers to improve their activities by publishing in high quality journals.

Table 5 shows the journals title with the greatest number of publications. As can be seen, Lecture Notes in Computer Science has located in top with 32 publications followed by Proceeding of SPIE and IEEE Access in Computer Science. Lecture Notes in Computer Science is a book series with the greatest number of publications. Since it has lots of publishing services in different domains such as information technology and computer science [42]. Proceeding of SPIE also is the record of conference which is stand for Society of Photo-Optical Instrument Engineering. It is indexed in many databases such as Index to Scientific and Technical Proceedings, Astrophysics Data System and Chemical Abstracts. Moreover, it has more than 480,000 papers with 1800 as average increasing [43].

Table 5 demonstrates that IEEE Access was one of the journals that had lots of publications (21). IEEE Access is multidisciplinary journal with 100% open access which all IEEE fields are covered. The other benefits of Publishing in the IEEE Access are short time reviewing and accepting (4–6 weeks), integrating multimedia contents, satisfying "open access (OA) publishing requirements" and "benefits from practical innovation" which means the authors may work in industry and be government leaders in R&D who don't have any publication in IEEE although they would like to have "an outlet for their "innovation [44]. In addition, Special Section is one of the useful parts in IEEE Access that allow researchers to publish their methods if they work in specific area. Artificial Intelligence Technologies for Electric Power Systems and Emerging Approaches to Cyber Security are two Special Sections that recently accepting article.

Moreover, Table 6 shows that IEEE Access has received 187 citations over years followed by Lecture Notes in Computer Science with 47 citations. Furthermore, 35 citation belongs to IEEE Transactions on Information Forensics and Security. It has articles about forensics, security, biometrics, surveillance and systems applications [45].

Table 5 List of journals

Journals title with the greatest number of publications	IF	Q	P
Lecture Notes in Computer Science	0.6	Q2	32
Proceedings of SPIE	–	–	23
IEEE Access	4.098	Q1	21
Communications in Computer and Information Science			8
IEEE Transactions on Information Forensics and Security	6.21	Q1	7
Lecture Notes in Artificial Intelligence			7
Signal Processing Image Communication	2.81	Q2	7
IEEE Signal Processing Letters	3.26	Q1	6
Journal of Intelligent Fuzzy Systems	1.63	Q2	6
Computers Security	3.06	Q1	5
Digital Investigation	1.66	Q2	5
Security and Communication Networks	0.72	Q2	5
Journal of Visual Communication and Image Representation	2.25	Q2	4
Pattern Recognition Letters	2.81	Q2	4
Advances in Neural Information Processing Systems	7.79	Q3	3

Table 6 Highly cited journals

Most cited journals title	IF	Q	C
IEEE Access	4.098	Q1	187
Lecture Notes in Computer Science	0.6	Q2	47
IEEE Transactions on Information Forensics and Security	6.21	Q1	35
Multimedia Tools and Applications	2.10	Q2	33
Sensors	3.03	Q2	31

To sum up, the impact factor of journals can absorb researchers to publish their activities due to widely read by other researchers with increasing their citations.

3.6 *Highly-Cited Articles*

This section considers the journals citations which has reflected on research quality. Table 7 shows the top five cited articles with the information of the number of citations, published journals, year of publications and research areas. The research areas which contribute to articles were Engineering and Computer Science.

The majority of citations belong to "The Limitations of Deep Learning in Adversarial Settings". The method of this paper was about exploring the adversary's behaviors in deep neural networks (DNNs). The application could generate samples and classified them correctly with 97% adversarial success rate.

Table 7 List of top five highly-cited articles

Titles	Times cited	Published journal	Year	Research area
The Limitations of Deep Learning in Adversarial Settings	137	1st IEEE European Symposium on Security and Privacy	2016	Computer Science; Engineering
Distillation as a Defense to Adversarial Perturbations against Deep Neural Networks	111	2016 IEEE Symposium on Security and Privacy (Sp)	2016	Computer Science; Engineering
Spiking Deep Convolutional Neural Networks for Energy-Efficient Object Recognition	102	International Journal of Computer Vision	2015	Computer Science
Median Filtering Forensics Based on Convolutional Neural Networks	93	IEEE Signal Processing Letters	2015	Engineering
Structural Design of Convolutional Neural Networks for Steganalysis	83	IEEE Signal Processing Letters	2016	Engineering

3.7 *Keywords Frequency*

This section considers the keywords that most of researchers used in their works. The importance of this section is related to detecting articles in the recent and past research issues. The web of Science has provided author keywords and an article theme descriptions since 1990 [34] which is helpful for analyzing research trends and identifying research gaps. Table 8 shows the keywords and title occurrences between 2010 and 2018.

Table 8 shows that intrusion detection and feature are the most relevant in titles and keywords. For example, "Intrusion Detection System Using Deep Neural Network for In-Vehicle Network Security" and "Deep Neural Network Based Malware Detection Using Two Dimensional Binary Program Features" titles have the two terms "detection" and "feature". Figure 3 shows further information for deep analysis.

The clusters of Fig. 3 shows that research related to "deep learning" and "cybersecurity" have increased. It demonstrates two main clusters are the algorithms of deep learning (green) and security topics (red) which are used. For example, "malware", "detection method", "ransomware" and "ids" are the terms related to security and "convolutional neural network" and "cnn" are terms which has the relation with deep learning. Figure 4 is provided just for more visibility of some important nodes. Moreover, the blue cluster shows some of terms that make a relation with the main keywords.

Table 8 Relation between Deep learning, cybersecurity and keywords

Titles	Frequency	Keywords	Frequency
Intrusion detection	34	Feature	487
Identification	32	Data	465
Malware detection	31	Attack	412
Deep neural network	24	Detection	332
Intrusion detection system	22	Technique	324
Network intrusion detection	19	Algorithm	286
Defense	18	CNN	258
Deep learning approach	18	Accuracy	235
Anomaly detection	15	Malware	213
Evaluation	13	Deep learning	210
Malware classification	11	Classifier	192
Convolutional neural networks	10	Convolutional neural networks	174
Deep convolutional neural networks	10	Architecture	163
Recurrent neural networks	10	Layer	156
Camera model identification	10	Security	111
Android malware detection	9	Deep neural networks	107
Deep neural networks	9	Neural networks	105
Internet	9	Solution	105
Machine learning	6	Training	98
Artificial intelligence	6	IDS	95
Network intrusion detection system	6	Intrusion detection system	94
Anomaly	6	Challenge	93
Long short term memory	5	Machine learning	88
Cyber security	5	Intrusion detection	84

4 Concluding Remarks

One of the "critical national policy issue" is information security which it means that a computer system has availability, integrity and confidentiality on its data [46]. On the one hand, by connecting the system to the Internet leads to increase cyber-attacks which can cause data breaches and data theft. Moreover, in the large-scale companies, it can have a major impact on their products and costs. Consequently, there is a requirement to detect cyber-attacks and even in critical infrastructure predict them before happened. On the other hand, artificial intelligence has powerful methods and algorithm such as machine learning and deep learning that they can have a significant impact on intrusion detection along with attack prediction on a computer system.

Deep learning models, due to being successful in various type of big data, absorb the interests in cybersecurity fields. For instance, [47] used generative adversarial network (GAN) to detect zero-day malware attacks, [48] implemented Convolutional Neural Network (CNN) to develop the detection of malware variants,

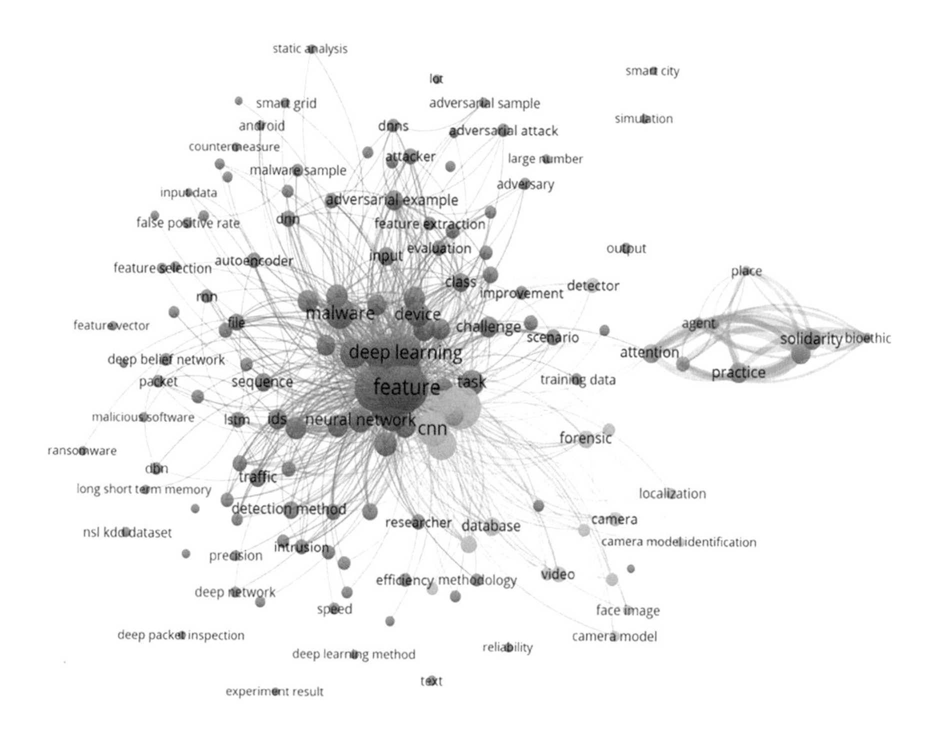

Fig. 3 Keyword clusters

and [49] developed Recurrent Neural Network (RNN) with restricted Boltzmann machines (RBM) to identify anomaly in a network and so on. Since, there are lots of deep learning algorithms and variety of cyber-attacks in different infrastructures, the bibliometrics model can give a wide picture of these trends and future works.

In this paper, bibliometric analysis was used for creating a comprehensive research about the trend of activities in deep learning with cybersecurity approach between 2010 and 2018. Seven criteria were considered: productivity, research area, institution, author, highly cited journal, impact journal and keywords frequency. In the last 8 years, it was shown that the number of publications and citations about this topic had increased.

In this study, we collected and analyzed the publications from 2010 to 2018. First, these were considered based on continents. It was mentioned that Asia has greatest productions of publications. Second, Computer Science was popular in terms of research area. Third, it was noted that Chinese academic of science from China had the most contribution. Forth, K. P Soman with Rahul K. Vigneswaran from India were highlighted as the active authors. Fifth and Sixth were considered the impact journals and most cited articles which were Lecture Notes in Computer Science and The Limitations of Deep Learning in Adversarial Settings, respectively. Last but not least, in terms of keywords frequency, a map analysis has also been used to show variety of keyword used in deep learning and cybersecurity domains.

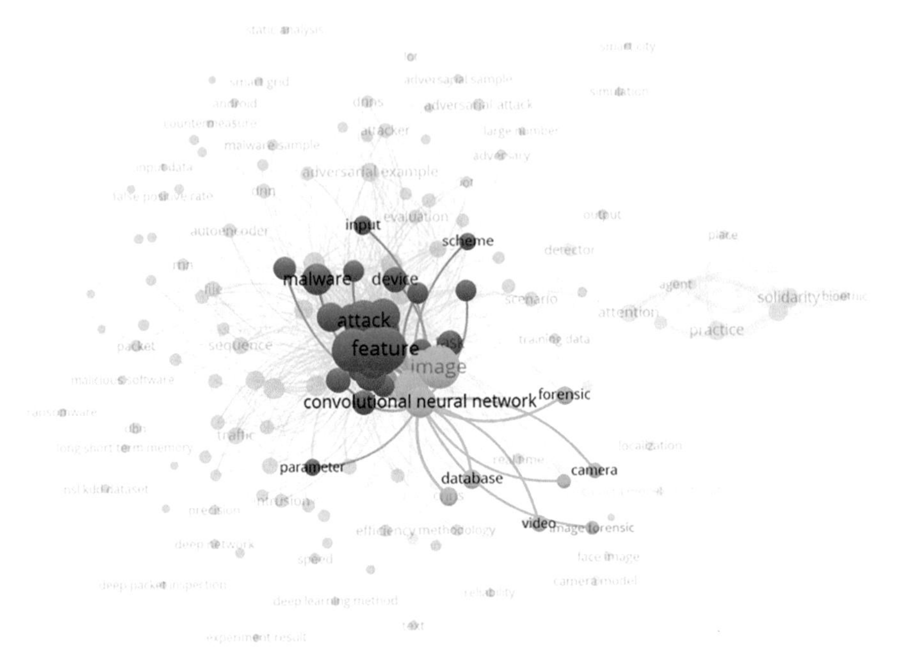

Fig. 4 More visibility in important nods

To sum up, the result of this study can help researchers to find out which fields or topics in using deep learning and cybersecurity still require search and investigation.

References

1. W. Tounsi, H. Rais, A survey on technical threat intelligence in the age of sophisticated cyber attacks. Comput. Secur. **72**, 212–233 (2018). https://doi.org/10.1016/j.cose.2017.09.001
2. S. Yousefi, F. Derakhshan, H. Karimipour, H.S. Aghdasi, An efficient route planning model for mobile agents on the internet of things using Markov decision process. Ad Hoc Netw. **98**, 102053 (2020). https://doi.org/10.1016/j.adhoc.2019.102053
3. M. Jouini, L.B.A. Rabai, A. Ben Aissa, Classification of security threats in information systems. Procedia Comput. Sci. **32**, 489–496 (2014). https://doi.org/10.1016/j.procs.2014.05.452
4. S. Mohammadi, V. Desai, H. Karimipour, Multivariate mutual information-based feature selection for cyber intrusion detection, in *2018 IEEE Electrical Power and Energy Conference, EPEC 2018*, (IEEE, Piscataway, 2018). https://doi.org/10.1109/EPEC.2018.8598326
5. D. Rafter, Cyberthreat trends: 2019 cybersecurity threat review. Symantec (2019). https://us.norton.com/internetsecurity-emerging-threats-cyberthreat-trends-cybersecurity-threat-review.html. Accessed 25 June 2019
6. S. Mahdavifar, A.A. Ghorbani, Application of deep learning to cybersecurity: a survey. Neurocomputing **347**, 149–176 (2019). https://doi.org/10.1016/j.neucom.2019.02.056

7. B. Geluvaraj, P.M. Satwik, T.A. Ashok Kumar, *The Future of Cybersecurity: Major Role of Artificial Intelligence, maChine Learning, and Deep Learning in Cyberspace* (Springer, Singapore, 2019), pp. 739–747
8. H. HaddadPajouh, A. Dehghantanha, R.M. Parizi, M. Aledhari, H. Karimipour, A survey on internet of things security: requirements, challenges, and solutions, in *Internet of Things*, (2019), p. 100129. https://doi.org/10.1016/j.iot.2019.100129
9. A. Namavar Jahromi et al., An improved two-hidden-layer extreme learning machine for malware hunting. Comput. Secur. **89**, 101655 (2020). https://doi.org/10.1016/j.cose.2019.101655
10. M.Z. Alom et al., A state-of-the-art survey on deep learning theory and architectures. Electronics **8**(3), 292 (2019). https://doi.org/10.3390/electronics8030292
11. E.M. Dovom, A. Azmoodeh, A. Dehghantanha, D.E. Newton, R.M. Parizi, H. Karimipour, Fuzzy pattern tree for edge malware detection and categorization in IoT. J. Syst. Archit. **97**, 1–7 (2019). https://doi.org/10.1016/j.sysarc.2019.01.017
12. A.A. Diro, N. Chilamkurti, Distributed attack detection scheme using deep learning approach for Internet of Things. Futur. Gener. Comput. Syst. **82**, 761–768 (2018). https://doi.org/10.1016/J.FUTURE.2017.08.043
13. J. Sakhnini, H. Karimipour, A. Dehghantanha, Smart grid cyber attacks detection using supervised learning and heuristic feature selection, in *Proceedings of 2019 the 7th International Conference on Smart Energy Grid Engineering, SEGE 2019*, (2019), pp. 108–112. https://doi.org/10.1109/SEGE.2019.8859946
14. DHS & FBI, GRIZZLY STEPPE – Russian Malicious Cyber Activity Summary (2016)
15. S. Mohammadi, H. Mirvaziri, M. Ghazizadeh-Ahsaee, H. Karimipour, Cyber intrusion detection by combined feature selection algorithm. J. Inf. Secur. Appl. **44**, 80–88 (2019). https://doi.org/10.1016/j.jisa.2018.11.007
16. M. Begli, F. Derakhshan, H. Karimipour, A layered intrusion detection system for critical infrastructure using machine learning, in *Proceedings of 2019 the 7th International Conference on Smart Energy Grid Engineering, SEGE 2019*, (2019), pp. 120–124. https://doi.org/10.1109/SEGE.2019.8859950
17. S. Geris, H. Karimipour, Joint state estimation and cyber-attack detection based on feature grouping, in *Proceedings of 2019 the 7th International Conference on Smart Energy Grid Engineering, SEGE 2019*, (2019), pp. 26–30. https://doi.org/10.1109/SEGE.2019.8859926
18. H. Karimipour, V. Dinavahi, On false data injection attack against dynamic state estimation on smart power grids, in *2017 5th IEEE International Conference on Smart Energy Grid Engineering, SEGE 2017*, (2017), pp. 388–393. https://doi.org/10.1109/SEGE.2017.8052831
19. H. Karimipour, S. Geris, A. Dehghantanha, H. Leung, Intelligent anomaly detection for large-scale smart grids, in *2019 IEEE Canadian Conference of Electrical and Computer Engineering, CCECE 2019*, (2019). https://doi.org/10.1109/CCECE.2019.8861995
20. S. Singh and Balamurali A. R, Using the power of deep learning for cyber security (part 1) (2018). https://www.analyticsvidhya.com/blog/2018/07/using-power-deep-learning-cyber-security/. Accessed 27 June 2019
21. U. Noor, Z. Anwar, T. Amjad, K.-K.R. Choo, A machine learning-based FinTech cyber threat attribution framework using high-level indicators of compromise. Futur. Gener. Comput. Syst. **96**, 227–242 (Jul. 2019). https://doi.org/10.1016/j.future.2019.02.013
22. N. Shone, T.N. Ngoc, V.D. Phai, Q. Shi, A deep learning approach to network intrusion detection. IEEE Trans. Emerg. Top. Comput. Intell. **2**(1), 41–50 (2018). https://doi.org/10.1109/TETCI.2017.2772792
23. P. Kobojek, K. Saeed, Application of recurrent neural networks for user verification based on keystroke dynamics. Telecommun. Inf. Technol **3**, 80–90 (2016)
24. G. De La Torre Parra, P. Rad, K.-K.R. Choo, Implementation of deep packet inspection in smart grids and industrial Internet of Things: challenges and opportunities. J. Netw. Comput. Appl. **135**, 32–46 (2019). https://doi.org/10.1016/J.JNCA.2019.02.022
25. A. Namvar Jahromi, J. Sakhnini, H. Karimpour, A. Dehghantanha, A deep unsupervised representation learning approach for effective cyber-physical attack detection and identification on highly imbalanced data, in *Proceedings of the 29th Annual International Conference on Computer Science and Software Engineering*, (2019)

26. Y. Gingras, *What is bibliometrics?* (MIT Press, Cambridge, 2016)
27. Y. Xin et al., Machine learning and deep learning methods for cybersecurity. IEEE Access **6**, 35365–35381 (2018). https://doi.org/10.1109/ACCESS.2018.2836950
28. D.S. Berman, A.L. Buczak, J.S. Chavis, C.L. Corbett, A survey of deep learning methods for cyber security. Information (Switzerland) **10**(4). MDPI AG (2019). https://doi.org/10.3390/info10040122
29. J. Koskinen et al., How to use bibliometric methods in evaluation of scientific research? An example from Finnish schizophrenia research. Nord. J. Psychiatry **62**(2), 136–143 (2008). https://doi.org/10.1080/08039480801961667
30. T. Brown, S.A. Gutman, Impact factor, eigenfactor, article influence, scopus SNIP, and SCImage journal rank of occupational therapy journals. Scand. J. Occup. Ther. **26**(7), 475–483 (2019). https://doi.org/10.1080/11038128.2018.1473489
31. J.P. Kamdem et al., Research trends in food chemistry: a bibliometric review of its 40 years anniversary (1976–2016). Food Chem. **294**, 448–457 (2019). https://doi.org/10.1016/j.foodchem.2019.05.021
32. G. La Scalia, R. Micale, M. Enea, Facility layout problem: bibliometric and benchmarking analysis. Int. J. Ind. Eng. Comput. **10**, 453–472 (2019). https://doi.org/10.5267/j.ijiec.2019.5.001
33. J.L. Aleixandre-Tudó, L. Castelló-Cogollos, J.L. Aleixandre, R. Aleixandre-Benavent, Renewable energies: Worldwide trends in research, funding and international collaboration. Renew. Energy **139**, 268–278 (2019). https://doi.org/10.1016/j.renene.2019.02.079
34. J. Sun, M.-H. Wang, Y.-S. Ho, A historical review and bibliometric analysis of research on estuary pollution. Mar. Pollut. Bull. **64**(1), 13–21 (2012). https://doi.org/10.1016/J.MARPOLBUL.2011.10.034
35. A. Abrizah, A.N. Zainab, K. Kiran, R.G. Raj, LIS journals scientific impact and subject categorization: a comparison between Web of Science and Scopus. Scientometrics **94**(2), 721–740 (2013). https://doi.org/10.1007/s11192-012-0813-7
36. J. Mingers, L. Leydesdorff, A review of theory and practice in scientometrics. Eur. J. Oper. Res. **246**(1), 1–19 (2015). https://doi.org/10.1016/J.EJOR.2015.04.002
37. P. Mongeon, A. Paul-Hus, The journal coverage of Web of Science and Scopus: a comparative analysis. Scientometrics **106**(1), 213–228 (2016). https://doi.org/10.1007/s11192-015-1765-5
38. J. Zhang, Q. Yu, F. Zheng, C. Long, Z. Lu, Z. Duan, Comparing keywords plus of WOS and author keywords: a case study of patient adherence research. J. Assoc. Inf. Sci. Technol. **67**(4), 967–972 (2016). https://doi.org/10.1002/asi.23437
39. M. Ruccolo, Web of science core collection: introduction (2019)
40. Science News Staff group, Updated: U.S. spending deal contains largest research spending increase in a decade. Science (2018). https://doi.org/10.1126/science.aat6620
41. T. Ng, J. Cai, China's funding for science and research to reach 2.5 per cent of GDP in 2019 | South China Morning Post. South China Morning Post (2019). https://www.scmp.com/news/china/science/article/2189427/chinas-funding-science-and-research-reach-25-cent-gdp-2019. Accessed 16 Jul 2019
42. Springer, *Lecture notes in computer science* (Springer, Berlin, 2019)
43. SPIE, SPIE, The internation Society for optics and photonics (2019)
44. IEEE Access Team, Benefits of publishing in IEEE Access - IEEE AccessIEEE Access. IEEE Access (2019). https://ieeeaccess.ieee.org/benefits-of-publishing-in-ieee-access/. Accessed 19 Jul 2019
45. IEEE Groups, IEEE Transactions on Information Forensics and Security, IEEE, Piscataway (2019)
46. B. Cashell, W.D. Jackson, M. Jickling, B. Webel, *CRS Report for CONGRESS the economic Impact of Cyber-Attacks the Economic Impact of Cyber-Attacks* (CRS, Washington DC, 2004)
47. J.Y. Kim, S.J. Bu, S.B. Cho, Zero-day malware detection using transferred generative adversarial networks based on deep autoencoders. Inf. Sci. (2018). https://doi.org/10.1016/j.ins.2018.04.092

48. Z. Cui, F. Xue, X. Cai, Y. Cao, G.G. Wang, J. Chen, Detection of malicious code variants based on deep learning. IEEE Trans. Ind. Inform. (2018). https://doi.org/10.1109/TII.2018.2822680
49. C. Li, J. Wang, X. Ye, Using a recurrent neural network and restricted boltzmann machines for malicious traffic detection. NeuroQuantology **16**(5), 823–831 (2018)

Dynamical Analysis of Cyber-Related Contingencies Initiated from Substations

Koji Yamashita, Chee-Wooi Ten, and Lingfeng Wang

1 Cyber-Physical Systems (CPS) Attacks Upon IP-Based Substations

An electronic remote access to unmanned IP-based substations[1] is enabled due to the regular maintenance for high voltage substation to deal with ongoing data analytics that is instrumented from the remote terminal units (RTUs), intelligent electronic devices (IEDs), and phasor measurement units (PMUs). At the same time, it enables remote access from "uninvited guests" to the networks [1–3]. While boundary technologies may restrict the remote access from certain IP addresses, it does not perform deep packet inspection on the contents of control and data flowing between the boundary of two networks. As such, extensive prevention of studies based on the base cases of power grid topology and statuses could help system planner to understand the rare, extreme scenarios in preparing the

[1]The terms *substations* and *buses* are interchangeable and are used throughout the chapter. The "bus" has been used in literature. Both terms indicate a node of power transmission networks. The hypothesized nodal removal is introduced due to the possibilities of remote access to those substations, implied to the possibilities of "intruders" through electronic means. By electrically disconnecting the compromised substation(s) through switching attack, it may introduce grid congestion, leading to various sorts of instability issues from the subsequent events.

K. Yamashita (✉) · C.-W. Ten
Department of Electrical and Computer Engineering, Michigan Technological University, Houghton, MI, USA
e-mail: kyamashi@mtu.edu; ten@mtu.edu

L. Wang
Department of Electrical Engineering, University of Wisconsin–Milwaukee, Milwaukee, WI, USA
e-mail: wang289@uwm.edu

H. Karimipour et al. (eds.), *Security of Cyber-Physical Systems*,
https://doi.org/10.1007/978-3-030-45541-5_12

worst cases of switching attacks through compromised IP-based substations [4, 5]. Cyberattacks are presumed either from the combinations of insiders and outsiders through coordination [6].

A CPS attack through intrusion to a substation network can be possibly executed through two possible ways within the network, i.e., through local computer that has complete access to all breakers and controlled switchgears in the substation, or through a digital relay that has partial switching control to the circuit breakers associated with the relay in that substation. The notation of S and R are referred to as total substations and total relays in a power system. The corresponding of $k = 1, 2, \cdots$ till the upper limit of k is determined. Elaboration of combinations is discussed in the next subsections. However, all switching attacks are studied using steady-state approaches and there is no association of time with respect to consequential events and local relay models are implemented.

1.1 Disruptive Switching in Substations

Plotting for a CPS attack upon unmanned substations can be one of the growing concerns in cybersecurity point of view. Especially after the Ukraine substation switching attacks in 2015 and 2016 [7–11], researchers and practitioners from around the world realize that the cyberattacks in substations are no longer a science fiction [12]. In addition, such events made us aware of the fact that the ubiquitous deployment of advanced sensors in managing the cyber-physical world. The cybersecurity [13] is specifically referenced as the intrusion paths to the deployed substation automation networks where individuals plot to maximize the damage by extracting information from the local console of control and data acquisition in order to help attackers understand the implications of electronic "sabotage."

The term "security" in power system has been referred in design emphasizing on reliability [14, 15] and contingency [16–18] triggered by an event such as system fault, such as, lightning strokes with a sudden disconnection of components that can be either *generator tripping* and *load shedding*. Because almost all large blackouts were initiated by the system event, power system engineers have paid attention to the possible future threat that can cause the violation of the power system security, i.e., large blackouts. The dependence on electricity has increased and uninterrupted electricity is vital for the current society. Such cyberattack to substation can even threaten human life. Therefore, the substation switching attack is highly likely to be one of the future threats not only for the power companies but the entire nation.

1.2 S-k Contingencies

This is a classical study of cyber-based contingency under steady-state analysis. The presumed outages of components associated with each node of transmission networks (referred to substation outages) are removed from the base cases. However, such studies do not include sequential events in "time" to demonstrate the cascading [19] and it can be challenging to capture blackout/brownout in steady-state simulation. The "divergence" of power flow solutions is the only indicator to show that system reaches its limit and such simulation study can be pre-screening. Investigation of dynamic study may exhibit the scenarios otherwise, depending on the relay models that capture the cascades. This approach is improved computationally by eliminating the combinations in the early stage of k so that a larger number of combinations would not carry forward to the higher number of k. Such an approach can significantly reduce total combination when incorporated the cascading effect using power flow methods (steady-state). The advantage of this approach can stimulate the combinatorial cases in a shorter time using practically-sized transmission networks that can be very useful for simulating a studied network for actuarial science that can be quantitatively utilized to estimate risks.

1.3 R-k Contingencies

Each substation is gradually replaced with digital relays for modernization of IP-based substations. Depending on the relay types, each relay is connected to a different part of the circuit breakers in a substation. While this may prove the usefulness of details in terms of impacts, the types of deployed relays in a power system can be complex to mimic the real scenarios. This sort of contingency can be structured between S-k and N-k/N-1 [20]. Without systematic elimination techniques, it can be computationally challenging for each base case, especially when the size of a power system is larger than the 1000-bus system. This framework is under development.

1.4 Capturing Spatiotemporal Tripping Sequences

This book chapter is to establish the cyber-related contingency analysis of S-k under the dynamical simulation environment in which the relay and control models are provided in the hypothesized scenarios. This section elaborates on the characteristics of bad actors (insiders and outsiders), key factors that can affect the dynamical behaviors of a grid, and modeling of dynamics under switching attacks.

1.4.1 Attack Modeling

Initial Plot

A substation switching attack is defined by an electrical disconnection of a large number of power equipment such as transformers, transmission lines, and buses that are connected to the substations through compromised local SCADA networks.

For worst, switching attacks can initiate cascading tripping that may lead to brownouts or blackouts. However, these may not always be detrimental to grids. For example, when a large capacity substation is isolated from the main grid and a large sub-transmission system that is connected to the disconnected substation, such a sub-transmission system can maintain the stability and continue to operate the system with the aid of emergency control systems, such as, special protection system (SPS) or system integrity protection scheme (SIPS) [21]. From the power system operator's perspectives, many of the disconnections of power equipment are not categorized as *N-k* contingency except the generator tripping. In this chapter, the dynamic behavior initiated by switching attacks in substations elaborates on the sequential events of tripping in the numerical case studies.

Sequential Events

The blackout occurs when all generators are electrically disconnected from a power grid. After the initiation of substation switching attacks, protective relays at different locations are operated successively during the cascaded stage until all generators are disconnected due to relay operations. The protective relays that are operated during cascaded events depend on how angle differences change, how bus voltages change, and how the power imbalance changes. The increasing angles difference causes rotor angle stability, while the increasing power imbalance causes frequency stability. Because multiple stability issues can occur during cascaded events and emerging stability issues vary depending on substation switching attacks, sequential events also vary, depending on the switching attacks.

1.4.2 Key Factors of Dynamical Behaviors

Voltage Threshold

The voltage level of substations with 200 kV or higher can be vulnerable to switching attacks. Therefore, the transmission network can become the major target and analysis of the base case and how the bottleneck of a system can be identified and later implemented with security protection technologies. Although removing a lower voltage substation may not initialize a cascade of the blackout, the combinatorial analyses should be included in a study.

Types of System Stability

Grid's well-being is observed by its abnormal phenomena that are generalized into three types: (1) transient stability, (2) frequency stability, (3) voltage stability, and overload [22]. Any of the stability issues below can occur triggered by the initial events of sequential switching attacks in substations.

- **Transient stability**: Transient stability is not generally violated without a significant voltage drop. Therefore, the possibility of the violation of transient stability due to substation switching attacks is relatively low. However, poorly damped power swing oscillation or negative damping oscillation can happen, when the dominant frequency component of the power swing oscillation totally changes, or when steady-state stability is violated after the substation switching attack. In addition, the angle difference between synchronous generators and synchronous condensers can be expanded during the cascaded events, which results in the out-of-step status in the grid (see a case study in Sect. 1.5.4).
- **Frequency stability**: Losing one or more substations due to switching attacks often create a abrupt mismatch in the power balance between generation and loads, which results in a change in the system frequency. In general, as k of "S-k" becomes larger, the frequency stability issue is more likely to happen (but that is not always the case). Both significant frequency drop or rise can occur. However, the frequency drop is more likely to be observed because the frequency rise leads to the generator tripping by overfrequency protections. This is due to the frequency starts to decrease after the generator tripping.
- **Voltage stability**: Undervoltage load shedding can happen once the significant voltage drop mainly due to system faults or deficiency of the reactive power support. However, some loads are disconnected due to their local protection in the case of low voltage before activation of undervoltage load shedding. In addition, system faults do not occur during substation switching attacks. Therefore, the voltage collapse caused by voltage stability is unlikely to happen. On the other hand, the significant voltage rise can also happen especially when loads with a large number of reactive consumption are lost all of a sudden (see a case study in Sect. 1.5.3). That condition is likely to happen especially when a large amount of loads and/or transformers are disconnected due to substation switching attacks. As k of "S-k" becomes larger, the violation of voltage instead of the voltage stability issue, is more likely to happen.
- **Overload**: Overloading conditions can also be a major abnormal phenomenon as a result of substation switching attacks. Once a heavily loaded line (or tie-line) between a large power station and load centers is disconnected, the power flow over the disconnected line is rerouted to the remaining available lines and those lines can be overloaded.

The stability limit may not necessarily be restricted by the thermal limit and transient stability. The limit for frequency stability can restrain the power flow over the transmission line. In this case, such a transmission line has a sufficient margin for the overloading before the switching attacks, and the possibility of the overloading due to substation switching attacks is unlikely. Transformers show time-inverse characteristics for the overloading and some transformers have short-term and long-term limits. In sum, the combination of instability introduced can implicate the overall control for destabilizing the grid.

Protective Relaying

Major protections are classified into: (1) line protection, (2) transformer protection, (3) bus protection, and (4) generator protection. It should be emphasized that the protections in transmission networks are normally designed to detect the system faults with a significant change in the voltage, current, and frequency to minimize the faulted section and to prevent power outages. Only when such a significant change in electric quantities is observed/detected, the protections can initiate their actions. However, substation switching attacks do not always cause such a change in electric quantities. Therefore, protective relays that are designed for clearing the system faults are unlikely to operate in the case of substation switching attacks.

- **Line protection**: Unlikely to operate except overloading protection
- **Transformer protection**: May operate over- or under-voltage protection
- **Bus protection**: Unlikely to operate
- **Generator protection**: Likely to operate

Wide-Area Control and Protection

The local protection is designed to arm the grid avoiding equipment damage. On the contrary, the scheme of wide-area control and protection involves multiple substation coordination. Such schemes are categorized as:

- Protective relays to protect a component
- Protective relays to prevent the blackout.

The first type of relay operates at a component level, while the second type of relay operates at a system level. An advanced relay system to prevent blackout is also called an event-based special protection system/scheme [23]. Such an advanced relay system is normally initiated after the targeted circuit breaker operation. However, in order to avoid the malfunction (referred to unwanted operation) of the relay system, a change in electric quantity, such as voltage, is usually used as an "AND" function. Because the disruptive switching attack will unlikely to cause the same level of behavior as system faults, the advanced relay system to prevent blackouts is highly unlikely to operate for this type of substation switching attack [23]. Therefore, the event-based special protection system is out of scope and is not further discussed.

Grid Size

Statistically, an exhaustive enumeration of total substation outage combinations grows exponentially with the size of a power system [24]. The correlation is system-dependent where the same size of the system under different topological setup would result in different combinations of diverged power flow solutions.

The total number of substations in each power system could have different impact based on the elimination methods at the lower k values, which is the number of substations with no electricity. Efficient algorithms to include cascading outage and without outage have shown a significant reduction of combination at the lower number of k; model with cascading outage would help to avoid large enumeration of evaluation with steady-state analysis [25, 26]. This chapter would adapt these algorithms to study combinations validating through dynamic simulation methods. On the other hand, complete enumeration without power system analysis tools can be computationally intensive [27] due to each combination would have to be validated to know if a scenario may lead to cascading blackout or brownout. An accurate assessment with time-domain analysis is needed in determining the significant impacts of the grid.

1.4.3 Modeling the System Dynamics Under Switching Attacks

Cascaded events triggered by switching attacks from compromised substations hypothetically can be simulated in detail using time-domain simulation tools [24]. If loads are connected to the compromised substation, those loads are manually disconnected by hackers and brownout can happen, the loss of electricity (LOE) is not zero. As k of S-k increases, the electricity loss caused by the manually disconnected power equipment is likely to increase because more substations are disconnected from the main grid. On the other hand, as k of S-k increases, the risk of additional disconnection caused by protective relays, i.e., cascaded events, also increases. The risks of cascaded events can lead to the major brownout or blackout and a large number of protective relays operate before the power outage. Therefore, protective relays are modeled to capture the sequential dynamics in the simulation where it captures cascade of grid instability.

However, protective relays for clearing short circuit faults or ground faults are unlikely to operate because fault currents are not created during substation switching attacks; meaning, protective relays that operate without fault currents are the pre-conditions of the grid before further analysis. In addition, the protective relays that operate *first* for the same power system dynamics (i.e. protective relays whose operating conditions are more close to the normal operating condition), need to be properly modeled in the more realistic studies.

Frequency Relay

Frequency relay models are considered for synchronous generators, including synchronous condensers, and loads. The relay settings of overfrequency relays and underfrequency relays generally consist of a frequency level, a timer that imposes delays and the undervoltage blocking function. The undervoltage blocking function disables tripping power equipment when the measured voltage is low, because the frequency relay fails to calculate the frequency accurately.

Because the frequency relay is often operated in multiple steps/stages, the multiple relay settings are often employed with the common undervoltage blocking function. In order to avoid the mechanical damage of the power plant, over speed protective relays are activated when the frequency significantly increases. However, over-speed relays are unlikely to operate before the overfrequency relay operation because the frequency threshold level of over-speed relays is higher than that of overfrequency relays. Therefore, over-speed relays may not kick in the wake of the attack events.

Overvoltage Relay

Overvoltage relay models should be implemented for transformers and synchronous condensers, if any. The relay settings of overvoltage relays generally consist of a voltage level and a timer. Overvoltage relays may have a higher voltage level with the shorter timer, which is known as the instantaneous overvoltage relay.

Out-of-Step (OOS) Relay

Out-of-Step (OOS) relays [28] should be used for transmission lines as well as generators (including synchronous condensers, if any). OOS relay is designed using either the impedance or the voltage angle difference [29]. Although OOS relays are not always placed in all of transmission lines in many countries, it is recommended to implement OOS relays to all of the lines for substation switching attack case studies from the numerical stability point of view (Otherwise, the time-domain simulation can unexpectedly be terminated due to the numerical instability during cascaded events.) as well as in the stability point of view.

Voltage/Frequency (V/F) Relay

The volts per hertz relay is generally implemented in generators. This relay can operate especially when the system frequency significantly goes down or when the terminal voltage significantly goes up. This protection may be skipped if the overvoltage relay and underfrequency relay are modeled. Over-excitation limiters (OEL) [28] may be used on behalf of overvoltage relays because over-excitation limiters play a role in reducing the terminal voltage of generators. On the contrary, the OEL should be used if it is deployed in a generating unit.

Undervoltage Relay

Undervoltage relays should be used for loads. However, the typical relay settings of undervoltage relays are the threshold voltage level and its timer, and the typical relay setting of the timer is 0.5 s or longer. On the other hand, the loads have their own disconnection characteristics in response to the voltage level and this self-disconnection characteristic reacts nearly instantaneously, i.e., significantly faster than the operation of the undervoltage relay. Therefore, the modeling of the load self-disconnection characteristics is critical.

Automatic Voltage Controller

In order to represent the cascaded event more precisely, not only proper protective relay models, but also accurate generator models with controller models are vital. A representative exciter model should be implemented in generator models including synchronous condenser models. For the selection of the exciter model, the AC exciter is a general choice. In the case of the bulk power system with large capacity generators and poorly damped inter-area oscillations, the thyristor-based exciter with the power system stabilizer (PSS) may be selected for a large capacity generator, such as 1 GVA class generators [30].

The aforementioned exciters must include an over-exciter limiter (OEL) in modeling. When the voltage imbalance significantly changes the filed voltage of generators and exceeds its upper limit, the OEL starts to decrease the field voltage, which results in lower terminal voltage of the generator and lower reactive power support from the generator. Although the response speed of the OEL is much slower compared to that of the AVR, the dynamic response of the OEL can shorten the time to blackouts especially when it takes a longer time to result in blackouts. The generic model with its parameters is provided in references [31, 32].

Frequency Controller

The primary frequency model should be implemented into generator models excluding synchronous condenser models. Because most countries use the thermal power unit as a major source, the simplified steam turbine governor model may be used as the generic power plant type. The generic model with its parameters is provided in references [32, 33].

The secondary and tertiary frequency controllers may be skipped for substation switching attack case studies mainly because the response time is not quite fast compared to the cascaded events after substation switching attacks. However, when the long-term stability study is required, the secondary frequency control, also known as automatic generation control (AGC) [28], should be considered in the modeling.

The power plant controller model [28, 33] that can precisely represent the dynamics of boiler and fuel controllers is suggested to be included especially when advanced gasification combined power plants (AGCCs) [33] need to be modeled. The exhausted gas control reacts relatively fast and exhausted gas control can be prioritized depending on the ambient temperature and the active power output level.

Generally speaking, the power plant controller model that represents modes of operations [28], either turbine-following mode or boiler-following mode, is critical and can significantly change the system frequency. However, there is currently no available generic power plant controller model with its parameters. Therefore, only when the real-life case study is performed and that information is available, the power plant controller can be included in the modeling.

1.5 Simulated Studies

Simulation cases were performed using a commercially available time-domain simulation tool that is currently used by power companies [34]. The power flow snapshot was generated referring to [35]. The dynamic simulation models with their generic/standard parameters were provided from publicly available references.

1.5.1 Modeling Specifics

The system diagram of the IEEE 14-bus system model is shown in Fig. 1. Ten substation locations are also shown using figures in a star box in the same figure. Fig. 2 shows the power flow solution based on [35]. Because this system model is represented as of a part of American Electric Power System in 1962 [30, 35],

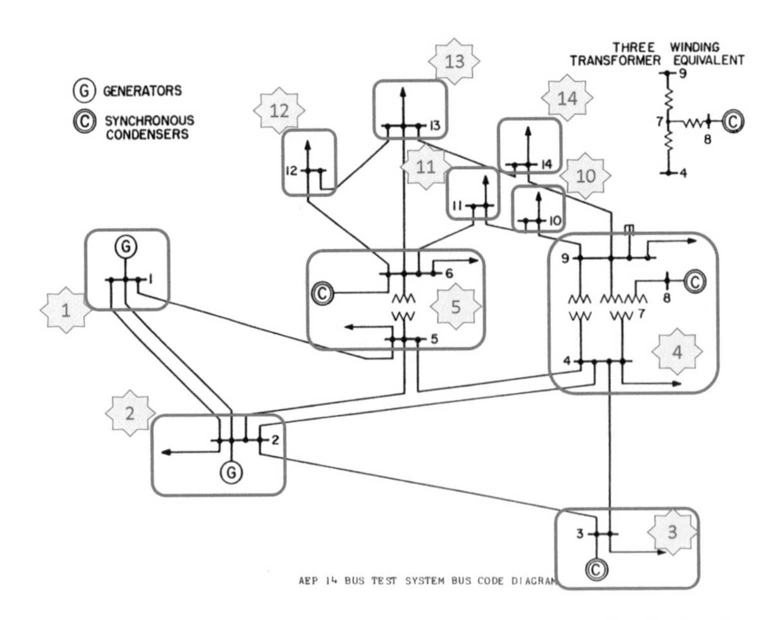

Fig. 1 Single diagram of the IEEE 14 bus system model and assumed substation locations

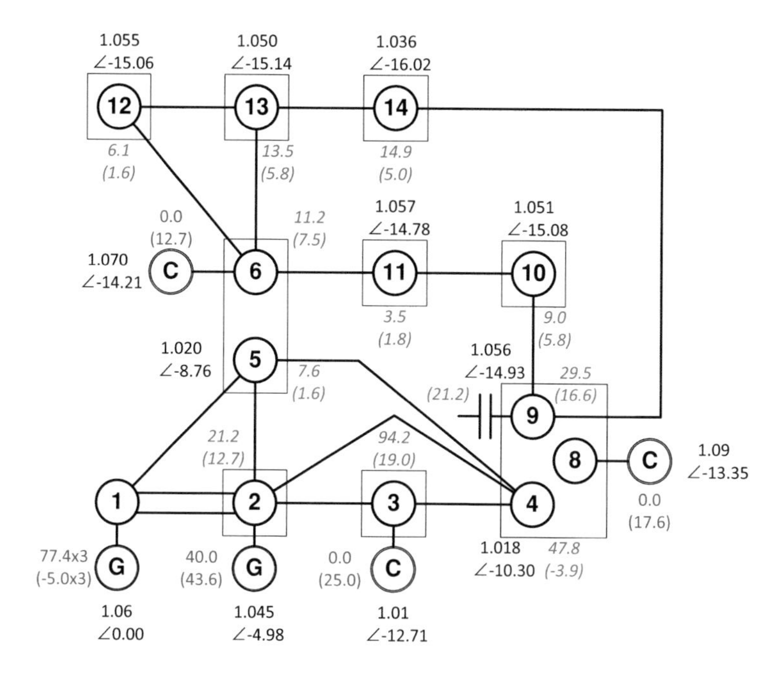

Fig. 2 A power flow solution using IEEE 14-bus system model. Blue: Active power output of generator [MW] (Reactive power output [MVAr]). Pink: Active power load consumption [MW] (Reactive power load consumption [MVAr]). Black: Magnitude of bus voltage [p.u.] with angle of bus voltage [degrees]

the DC exciter model with the OEL is used as the exciter (see Fig. 3) and the primary frequency controller of steam turbines are used for both generators as the frequency controller of generators (see Fig. 4). Both units are assumed to be operated in regulated machines with the turbine output upper limit of 105% and the power plant controller and automatic generation control (AGC), i.e., secondary frequency control are not implemented in the model. It is noted that parameters shown in Figs. 3 and 4 are example parameters.

Because the rated capacity of a single thermal power unit in 1960s was less than 500 MVA, it is assumed that the Substation 1 includes three 100 MVA units with an output of 77.5 MW each. The assigned generator constants are shown in Table 1.

Frequency Relay Model

The following settings of overfrequency and underfrequency relays for generators (including synchronous condensers) are as follows: [36, 37].

- 61.8 Hz (overfrequency relay)
- 58.0 Hz (underfrequency relay)

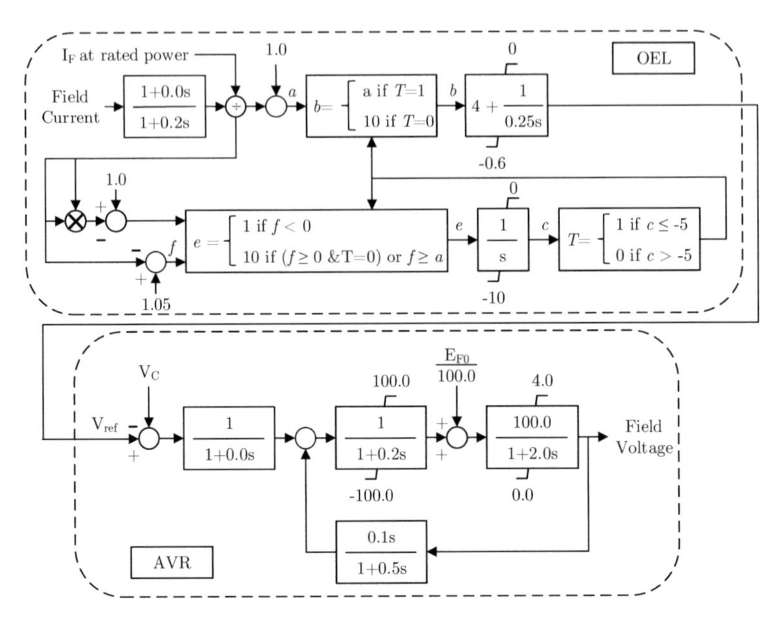

Fig. 3 AVR model with OEL model [32]

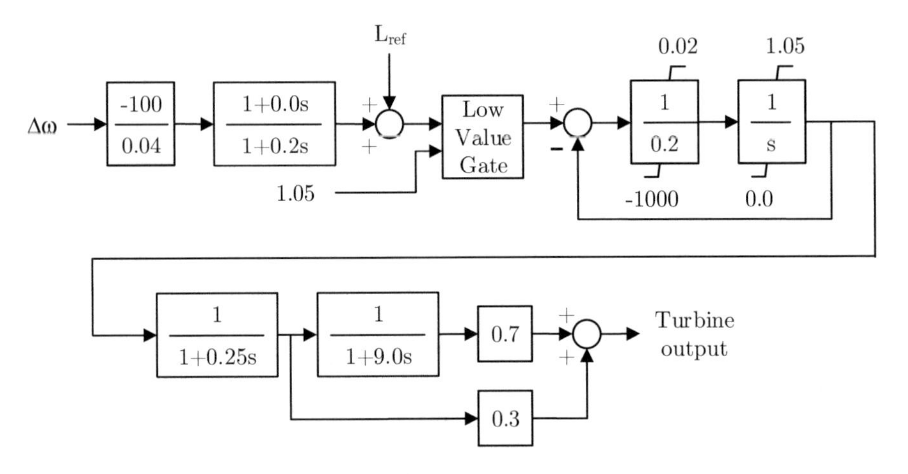

Fig. 4 Primary frequency control model [32]

Example relay settings of underfrequency relays for loads are in the following [36, 37].

- 59.3 Hz (5% reduction with relative to the total load)
- 58.9 Hz (additional 10% reduction with relative to the total load)
- 58.5 Hz (additional 10% reduction with relative to the total load)

Table 1 Generator constants

Specification	G1			G2
	G1-1	G1-2	G1-3	G2
Rated capacity [MVA]	100	100	100	60
Rated power [MW]	85	85	85	48
Inertia constant, M [s]	6	6	6	6
D-axis reactance [p.u.]	1.7	1.7	1.7	1.7
D-axis transient reactance [p.u.]	0.35	0.35	0.35	0.35
D-axis sub-transient reactance [p.u.]	0.25	0.25	0.25	0.25
Q-axis reactance [p.u.]	1.7	1.7	1.7	1.7
Q-axis sub-transient reactance [p.u.]	0.25	0.25	0.25	0.25
D-axis transient time constant [s]	1	1	1	1
D-axis sub-transient time constant [s]	0.03	0.03	0.03	0.03
Q-axis open circuit time constant [s]	0.206	0.206	0.206	0.206
Q-axis sub-transient time constant [s]	0.03	0.03	0.03	0.03
Armature leakage reactance [p.u.]	0.225	0.225	0.225	0.225
Armature time constant [s]	0.4	0.4	0.4	0.4
Zero phase reactance [p.u.]	1.7	1.7	1.7	1.7
Negative phase reactance [p.u.]	0.25	0.25	0.25	0.25

Note that the maximum load shedding amount is of 25% for the total loads. Because the undervoltage locking level is often set for 0.4 p.u., the value is then used, which means the frequency relay does not trip the loads when the voltage level is under 40%. The other parameters include timer is set to zero in the case studies, while the circuit breaker operation time is set as 70 ms.

Overvoltage Relay Model

The following parameters of overvoltage relays are implemented in [38, 39]:

- 115% (Overvoltage relay with time = 1 s)
- 125% (instantaneous overvoltage relay with time = 30 ms)

In addition to the timer imposed, the circuit breaker operation time is set as 70 ms.

Out-of-Step Relay Model

The voltage angle difference between the terminal voltage and the internal induced voltage is used for the OOS relay for synchronous generators, while the voltage angle difference between transmission lines is used for the OOS relay for networks. When the voltage angle difference is used for OOS relays, the relay setting is set as 180° between the two buses according to the definition of the OOS condition.

Although the timer is set as zero, the circuit breaker operation time is set as 30 ms. A typical setup of a relaying combination for each substation is illustrated in Fig. 6. This setup is used for simulation of switching attack simulation.

Load Model

Load characteristics are represented using the exponential load model with the voltage and frequency dependent as shown in Eq. (1) [28, 40]. Load voltage characteristics indices are based on widely used parameters for power system analysis in utilities and system operators [41]. The coefficients of Load frequency characteristics are set based on the International Council on Large Electric Systems (CIGRE) working group which goes through substantial studies on the utility's grid systems from around the world [42].

$$P = P_0 \left(\frac{V}{V_0}\right)^1 \left(1 + \frac{3.33}{100}\Delta f\right) = P_0 \left(\frac{V}{V_0}\right) \left(1 + \frac{3.33}{100}\Delta f\right) \quad (1)$$

$$Q = Q_0 \left(\frac{V}{V_0}\right)^2 \left(1 + \frac{0}{100}\Delta f\right) = Q_0 \left(\frac{V}{V_0}\right)^2 \quad (2)$$

where,

P_0: Initial active power consumption of loads
Q_0: Initial active power consumption of loads
V_0: Initial load bus voltage
P: Active power consumption of loads
Q: Reactive power consumption of loads
V: Load bus voltage

The load defined here is an aggregated electricity consumption of thousands of consumers at a high voltage transmission network. The aggregated load in distribution networks [43] represents a fraction of high voltage loads. The modeling of loads is an empirical study that accumulates the statistical correlation between voltage and current magnitudes from the current and potential transformers (CT/PT).

Undervoltage Load Shedding

The load self-disconnection characteristics are summarized in the following [44]:

Load Self-disconnection Voltage	Below 80%
Load Self-disconnection Starting Voltage	80%
Load Self-disconnection Saturation Voltage	40%
Maximum load Self-disconnection Amount	25%

This characteristic illustrated in Fig. 5 are employed by all loads (both active power loads and reactive power loads) and superimposed on the load voltage and

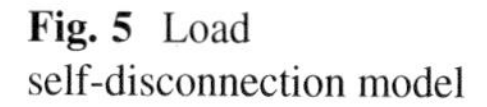

Fig. 5 Load self-disconnection model

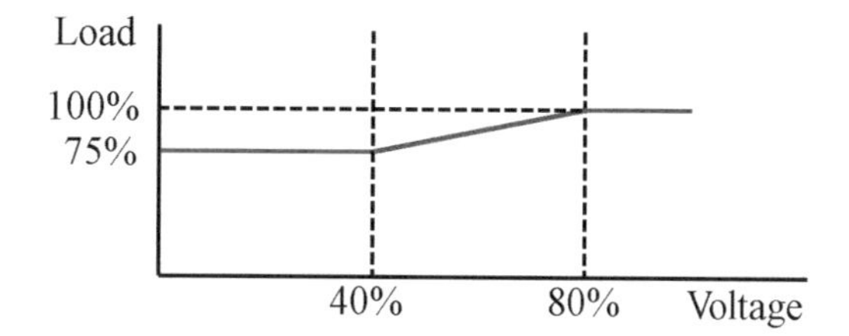

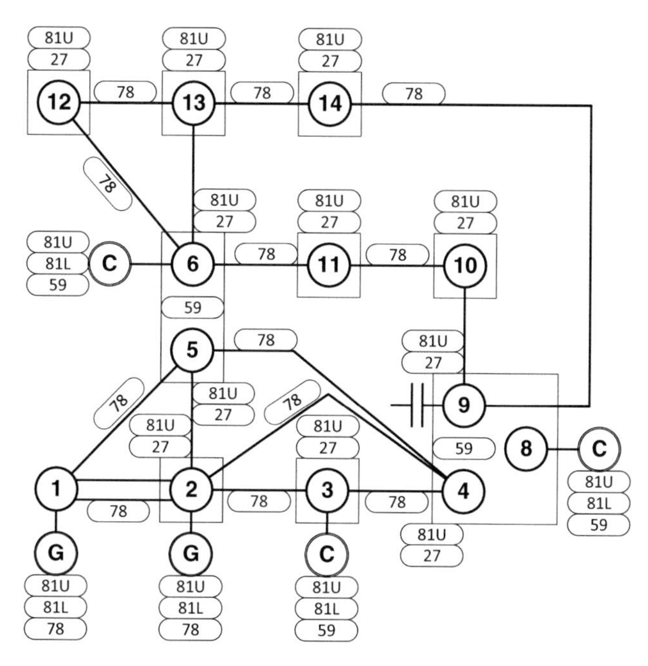

Fig. 6 Deployment of relay models in IEEE14-bus system model. 81U: Overfrequency relay; 81L: Underfrequency relay; 78: Out-of-step relay; 59: Overvoltage relay; 27: Underfrequency relay emulated by load self-disconnection characteristics. The assigned numbers are based on [45]

frequency characteristics shown in Eq. (1). Because disconnected loads do not normally recover quickly and it takes over a minute for many loads to recover, load recovery characteristics are not considered in the model. For example, once the load bus voltage level is below 40% under the rated frequency, 25% of the loads are disconnected, and the remaining 75% loads that stay connected reduce the active and reactive power consumption to 40% and 16% (=0.4^2), individually, due to the load voltage characteristics. Deployed relay models in IEEE 14-bus model system are illustrated in Fig. 6.

1.5.2 Case 1: Sequence of Relay Operation

The sequential relays are initiated by the switching attack on substations 2 and 12 in Fig. 1 where it is referred to as one combination of S-2 cases. This serves as the initiating events where assumptions made here is that attackers successfully hack into the substation through the substation firewall and have compromised the substation console that is accessed to all substation breakers.

As shown in Fig. 7, protective relays at different locations are operated sequentially during the cascaded stage. As shown in Fig. 8, the underfrequency relay operations for load at 0.06 s through 0.36 s are aggregated. In this example substation switching attack, one of the generators is isolated due to the attack and tripped 0.3 s after the switching attack due to the overfrequency relay operation. The protective relay operation for the isolated component is not included in Fig. 8.

In this study case, the synchronous generator at Bus 1 is accelerated after the switching attack, while the synchronous condenser is decelerated after the switching attack. This discrepancy comes from the opposite magnitude relation between the mechanical output and electrical output of those synchronous machines.

The output of synchronous condensers is zero at the steady-state. However, it immediately increases using its rotating energy, i.e., the inertia shown during the occurrence of attack at substations 2 and 12. Because the electrical output is larger than the mechanical output for that time period, the rotating speed of synchronous condensers decreases, which is indicated by the measured frequency decrease.

On the other hand, the electrical output of the synchronous generator at Bus 2 decreases when the attack at substations 2 and 12 occurs because the deficiency of the active power output is compensated by synchronous condensers as well as the synchronous generator at Bus 2. Therefore, the electrical output is smaller than the

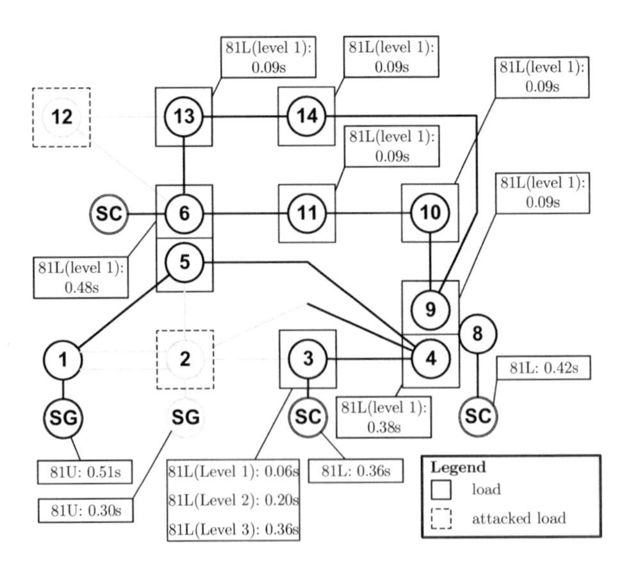

Fig. 7 Relay operation time for substation switching attack on Substations 2 and 12

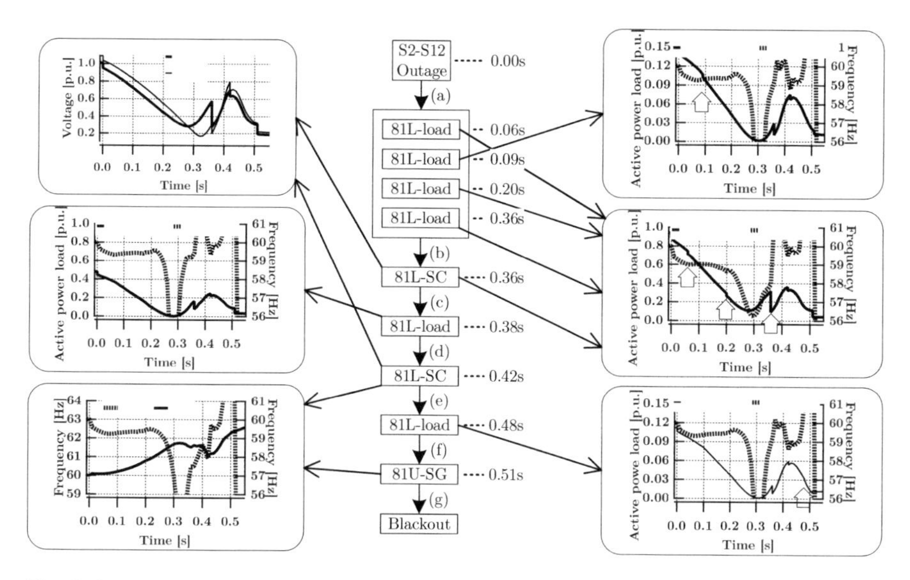

Fig. 8 Sequence of relay operation with corresponding waveform

mechanical output following the substation switching attack, the rotating speed of the synchronous generator at Bus 1 increases, i.e., the measured frequency increases.

Therefore, the underfrequency relays for the loads and synchronous condensers operate, while the overfrequnecy relay for the synchronous generator at Bus 1 operates at 0.51 s, which results in the blackout.

1.5.3 Case 2: Sequence of Relay Operation

The sequential relays are initiated by the switching attack on substations 1, 3, 10, 11, 12, 13, and 14 in Fig. 9 where it is referred to as combinations of S-7 cases. Protective relays at different locations including overvoltage relays are operated sequentially during the cascaded stage. As shown in Fig. 10, the underfrequency relay operations for load at 0.05 s through 0.15 s are aggregated. In this example substation switching attack, one of the generators is isolated due to the switching attack and tripped 0.24 s after the switching attack due to the overfrequency relay operation. The protective relay operation for the isolated component due to the switching attack is not shown in Figs. 9 and 10.

In this study case, the synchronous generator at Bus 2 is decelerated after the switching attack, while the disconnected synchronous generator at Bus 1 is accelerated after that. As shown in Fig. 11, the electric power output at Bus 2 increased immediately right after the switching attack. Because the electric output exceeds the mechanical input of the generator at Bus 2, the rotating speed of the generator starts to decrease, which results in a decrease in the system frequency.

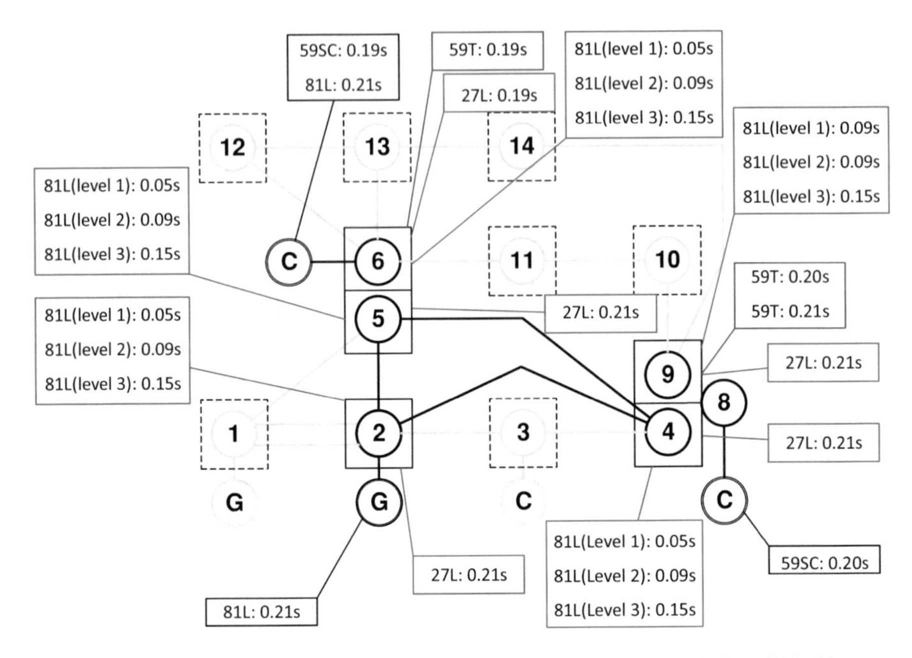

Fig. 9 Relay operation time for substation switching attack on Substations 1, 3, and 10–14

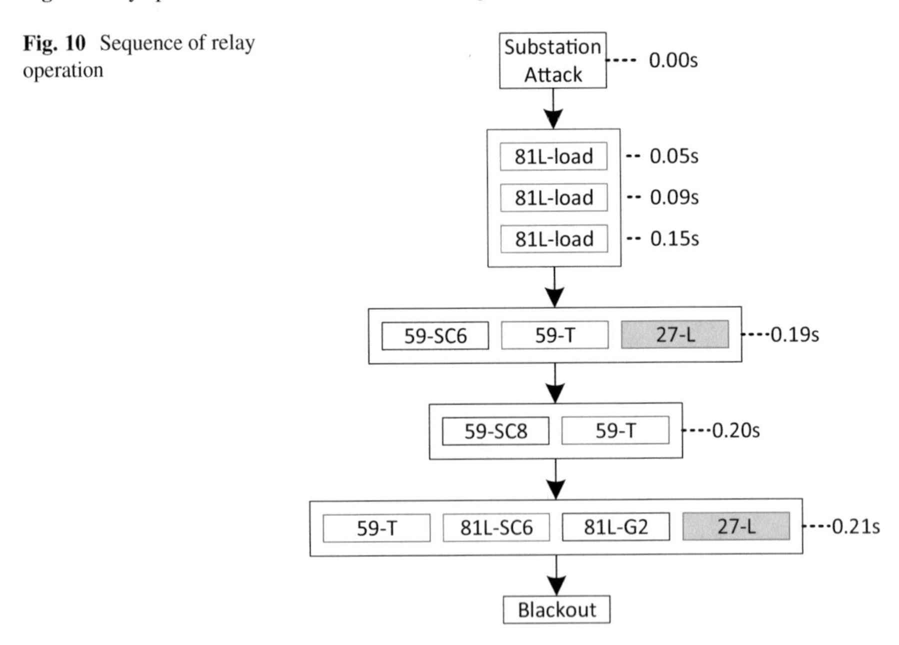

Fig. 10 Sequence of relay operation

On the other hand, the terminal voltage of synchronous condensers at buses 6 and 8 increased by over 10% and reach to over 115% right after the switching attack. Synchronous condensers generally contribute to increasing the system voltage

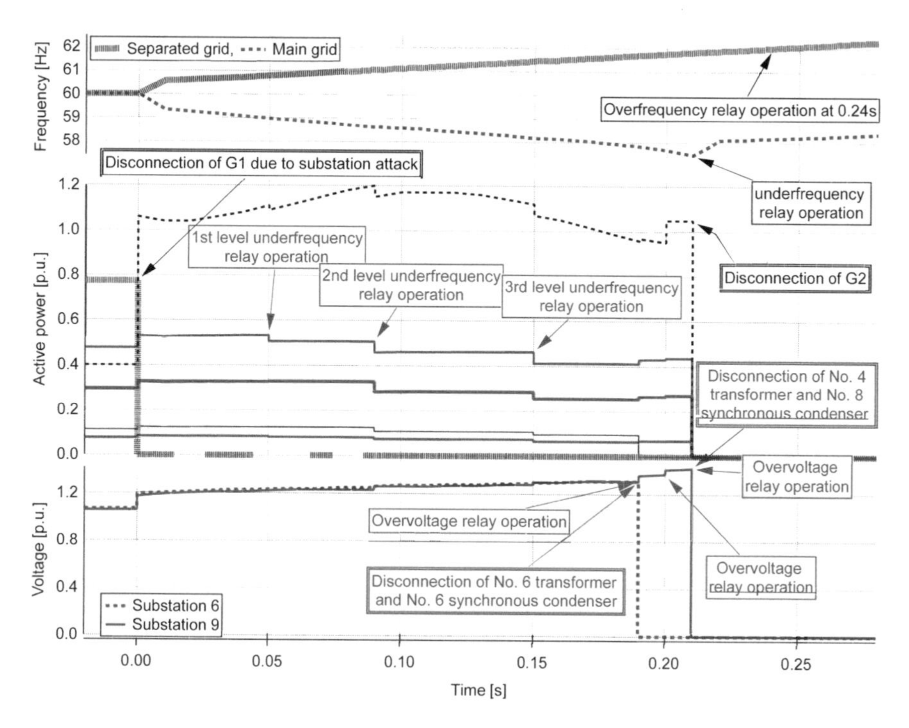

Fig. 11 Phenomenon of system dynamics for the IEEE 14-bus system [35] initiated by a substation switching attack upon substation 7

providing reactive power at heavy load conditions. Before the substation switching attack, the synchronous condensers at buses 6 and 8 are connected to the medium point of the grid between power stations and loads. However, after the substation switching attack, those condensers are eventually connected to the end of the grid as shown in Fig. 11, losing the supporting loads, and their terminal voltages are raised.

As the system frequency decreases, the connected loads are gradually disconnected due to underfrequency relays, which leads to lighter loading condition, and the system voltage further increases and reaches to over 125%. Therefore, instantaneous overvoltage relays of one synchronous condenser at bus 6 and transformers near bus 6 at 0.19 s. It is noted that undervoltage relay operation at bus 6 in Figs. 10 and 11 is the results of the disconnection of the transformer at bus 6, not the reason for the substation switching attack. After disconnecting them, the system voltage additionally increases and the rest synchronous condenser and the transformer are disconnected due to the overvoltage relay at 0.20 s, which is a typical cascading event.

The immediate increase in the system voltage at 0.20 s contributes to increasing the total demand due to the load voltage characteristics, which accelerates the decrease in the system frequency and trip the remaining generator at bus 2 due to the underfrequency relay at 0.21 s. It is noted that undervoltage relay operations at buses 2, 4, and 9 in Figs. 10 and 11 are the results of the blackout, not the reason for the blackout. Therefore, the underfrequency relays for the loads and synchronous condensers operate, while the overfrequnecy relay for the synchronous generator at Bus 1 operates at 0.51 s, which results in the blackout.

1.5.4 Case 3: Sequence of Relay Operation

The sequential relays are initiated by the switching attack on substations 2, 10, and 11 in Fig. 12 where it is referred to as one combination of S-3 cases. Protective relays including OOS protections at different locations are operated sequentially during the cascaded stage. Figure 13 illustrates the underfrequency relay operations for load between 0.06 and 0.24 s may be aggregated. In this example substation switching attack, one of the generators at bus 2 is isolated due to the switching attack and tripped 0.30 s after the switching attack due to the overfrequency relay operation.

Referred to Sect. 1.5.2, the connecting synchronous generator at bus 1 is accelerated while synchronous condensers decelerate. In this case, as the discrepancy of the rotor speed between the synchronous generator and the synchronous condensers increases, (the twist of) the angle difference between them also increases, which results in the out-of-step conditions at two locations (over the line between buses 4 and 5 and the line between buses 6 and 12) at 0.33 and 0.34 s. After the loss of two lines, the electrical distance between the source and loads is expanded, and the remaining generator at bus 1 is tripped by overfrequency relay.

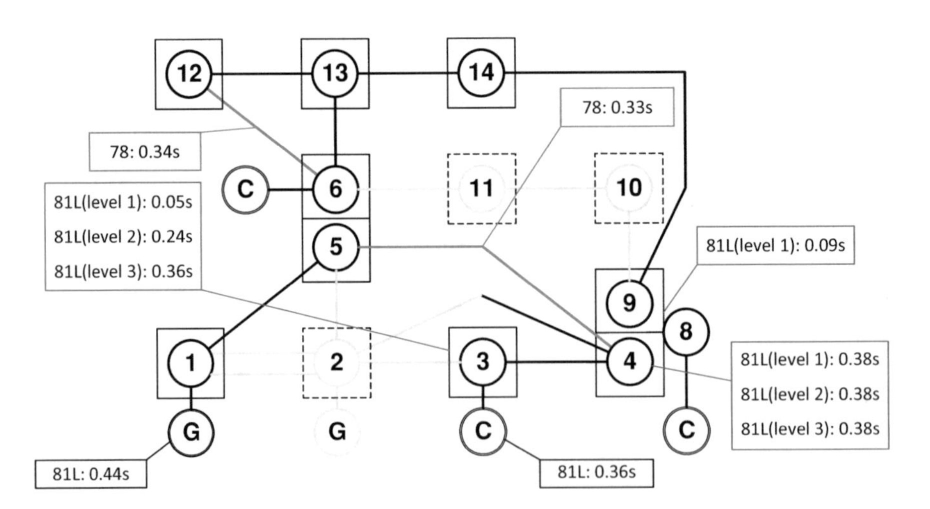

Fig. 12 Relay operation time for substation switching attack on Substations 2, 10, and 11

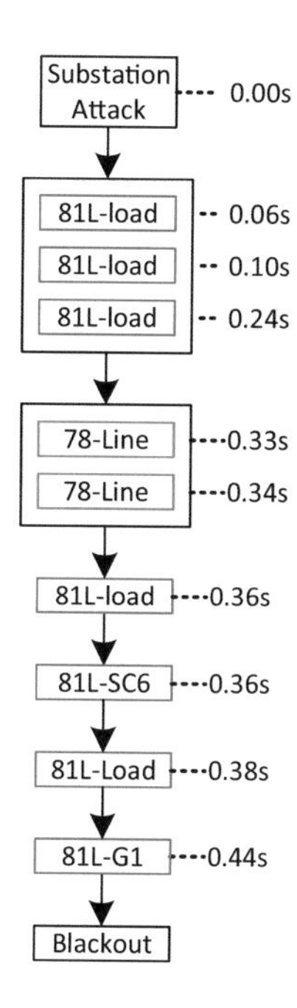

Fig. 13 Sequence of relay operation initiated by tripping substations

1.5.5 Loss of Electricity

The loss of electricity (LOE) is often considered the precursor of grid instability. Statistically, the following simulated statistics of LOE for the case study in IEEE 14-bus system is shown below:

0 ~9.99%: 21 cases
10 ~19.99%: 30 cases
20 ~29.99%: 12 cases
30 ~39.99%: 30 cases
40 ~49.99%: 2 cases
50 ~59.99%: 0 cases
60 ~69.99%: 0 cases
70 ~79.99%: 0 cases
80 ~89.99%: 8 cases
90 ~99.99%: 0 cases
100%: 950 cases

Under this study of simulation, the percentage of an evident number of blackouts for k from 1 to 10 of S-k has reached 92.3%. The simulation results have shown that the original hypothesis of assumption, which is the total combinations of blackout cases, decreases as the size of the grid is larger for $k \leq 10$. The constant 10 is arbitrarily assigned based on the knowledge of simulation study that optimally provides the intuition of study that shows the cutoff of large combinations for the first 10 orders.

1.6 Summary

The new cyber-based contingency analysis has introduced a new direction of operational planning challenges in cybersecurity with elimination algorithms to identify and confirm critical combinations of cascade problems. This chapter introduces how initial events of substation switching attack, by disconnecting the system node off the grid, result in cascading blackouts via time-domain simulation studies. The stability that is relevant to power system phenomena caused by substation switching attacks is briefly discussed. The protective relays that are likely to operate during the cascaded events initiated by the switching attacks in IP-based substation. An IEEE 14-bus system model is selected as the system model and the power system dynamics models including protection models are implemented into the system model. A commercially available time-domain simulation tool is used to simulate the cascaded events. Three phenomena representing unique scenarios are introduced: (1) frequency event only, (2) frequency event with overvoltage, (3) frequency event with the out-of-step phenomenon. The loss of electricity for all *S-k* contingencies in the IEEE 14-bus system is also demonstrated. This chapter facilitates future research in attack combinatorics based on hypothetical scenarios and cascades on transmission grids. The other possibilities of this work include permutations of switching and confirmation of system-wide cascades. Such studies can be incorporated as part of the risk assessment, which has a potential contributing to actuarial science frameworks.

References

1. C.-W. Ten, C.-C. Liu, M. Govindarasu, Vulnerability assessment of cybersecurity for SCADA system. IEEE Trans. Power Syst. **23**(4), 1836–1846 (2008)
2. R. Bulbul, P. Sapkota, C.-W. Ten, L. Wang, A. Ginter, Intrusion evaluation of communication network architectures for power substations. IEEE Trans. Power Syst. **30**(30), 1372–1382 (2015)
3. D. Leversage, E. James, Estimating a system's mean time-to-compromise. IEEE Secur. Priv. **6**(1), 52–60 (2008)
4. J. Hong, C.-C. Liu, Intelligent electronic devices with collaborative intrusion detection systems. IEEE Trans. Smart Grid **10**(1), 271–281 (2019)
5. C. Moya, J. Wang, Developing correlation indices to identify coordinated cyber-attacks on power grids. IET Cyber-Phys. Syst. Theory Appl. **3**(4), 178–186 (2018)
6. R.P. Stanley, Topics in algebraic combinatorics, course notes for mathematics 192. Harvard University (2013). http://www-math.mit.edu/~rstan/algcomb/algcomb.pdf
7. Electricity Information Sharing and Anlaysis Center (E-ISAC), Analysis of the cyber attack on the Ukrainian power grid, 18 Mar 2016. https://www.nerc.com/pa/CI/ESISAC/Documents/E-ISAC_SANS_Ukraine_DUC_18Mar2016.pdf
8. Industrial Control Systems Cyber Emergency Response Team (ICSCERT), Cyber-attack against Ukrainian critical infrastructure, Feb 2016. https://ics-cert.us-cert.gov/alerts/IR-ALERT-H-16-056-01
9. E. Perez, First on CNN: U.S. investigators find proof of cyberattack on Ukraine power grid, 3 Feb 2016. http://www.cnn.com/2016/02/03/politics/cyberattack-ukraine-power-grid/

10. K. Zetter, Everything we know about Ukraine's power plant hack, 20 Jan 2016. http://www.wired.com/2016/01/everything-we-know-about-ukraines-power-plant-hack/
11. ICS-CERT Alert (IR-ALERT-H-16-056-01), Cyber-attack against Ukrainian critical infrastructure, 25 Feb 2016. https://ics-cert.us-cert.gov/alerts/IR-ALERT-H-16-056-01
12. L. Pietre-Cambacedes, M. Tritschler, G.N. Ericsson, Cybersecurity myths on power control systems: 21 misconceptions and false beliefs. IEEE Trans. Power Del. **26**(1), 161–172 (2011)
13. S. Zonouz, K.M. Rogers, R. Berthier, R.B. Bobba, W.H. Sanders, T.J. Overbye, SCPSE: security-oriented cyber-physical state estimation for power grid critical infrastructures. IEEE Trans. Smart Grid **3**(4), 1790–1799 (2012)
14. Y. Zhang, L. Wang, Y. Xiang, C.-W. Ten, Inclusion of SCADA cyber vulnerability in power system reliability assessment considering optimal resources allocation. IEEE Trans. Power Syst. **31**(6), 4379–4394 (2016)
15. Y. Zhang, L. Wang, Y. Xiang, Power system reliability analysis with intrusion tolerance in SCADA systems. IEEE Trans. Smart Grid **7**(2), 669–683 (2016)
16. M.A.C. Camargo, A.J. Rivera, R.R. Pena, Impact assessment of substation contingencies in power systems, in *Transmission Distribution Conference and Exposition - Latin America (PES T D-LA), 2014 IEEE PES*, Medellin, Sept 2014, pp. 1–6
17. S.H. Song, S.H. Lee, T.K. Oh, J. Lee, Risk-based contingency analysis for transmission and substation planning, in *Proc. IEEE Trans. Dist. Conf. Exposition: Asia and Pacific*, Seoul, Oct. 2009, pp. 1–4
18. M. Paramasivam, S. Dasgupta, V. Ajjarapu, U. Vaidya, Contingency analysis and identification of dynamic voltage control areas. IEEE Trans. Power Syst. **30**(6), 2974 – 2983 (2015)
19. M.J. Eppstein, P.D. Hines, A "random chemistry" algorithm for identifying collections of multiple contingencies that initiate cascading failure. IEEE Trans. Power Syst. **27**(3), 1698–1705 (2012)
20. Q. Chen, J. McCalley, Identifying high risk N-k contingencies for on-line security assessment. IEEE Trans. Power Syst. **20**(2), 823–834 (2005)
21. V. Madani, Global industry experiences with system integrity protection schemes (SIPS). IEEE Power System Relaying Committee, Tech. Rep. Working Group C4 2009 Report, Oct 2009
22. P. Kundur, J. Paserba, V. Ajjarapu, G. Andersson, A. Bose, C. Canizares, N. Hatziargyriou, D. Hill, A. Stankovic, C. Taylor, T.V. Cutsem, V. Vittal, Definition and classification of power system stability IEEE/CIGRE joint task force on stability terms and definitions. IEEE Trans. Power Syst. **19**(3), 1387–1401 (2004)
23. W. G. C2/C4.37, A proposed framework for coordinated power system stability control. CIGRE, Paris, Tech. Rep. TB742, Sept 2018
24. C.-W. Ten, K. Yamashita, Z. Yang, A. Vasilakos, A. Ginter, Impact assessment of hypothesized cyberattacks on interconnected bulk power systems. IEEE Trans. Smart Grid **9**(5), 4405–4425 (2018)
25. C.-W. Ten, A. Ginter, R. Bulbul, Cyber-based contingency analysis. IEEE Trans. Power Syst. **31**(4), 3040–3050 (2016)
26. Z. Yang, C.-W. Ten, A. Ginter, Extended enumeration of hypothesized substations outages incorporating overload implication. IEEE Trans. Smart Grid **9**(6), 6929–6938 (2018)
27. Z. Huang, Y. Chen, J. Nieplocha, Massive contingency analysis with high performance computing, in *Power Energy Society General Meeting, 2009. PES '09. IEEE*, Calgary, AB, July 2009, pp. 1–8
28. V. Vittal, J.D. McCalley, P. Anderson, A. Fouad, *Power System Control and Stability*, 3rd edn. (Wiley-IEEE Press, Hoboken, 2019)
29. CIGRE Working Group B5.19, Protection Relay Coordination. Technical Brochure 432, pp. 1–179, Oct 2010
30. P. Kundar, *Power System Stability and Control*, 1st edn. (McGraw-Hill, New York, 1994)
31. IEEE Recommended Practice for Excitation System Models for Power System Stability Studies. IEEE Std. 421.5, 2016
32. H. Taniguchi, *Power System Analysis – Modeling and Simulation (Japanese)*, 1st edn. (IEEJ and Ohmsha, Tokyo, 2009)

33. P. Pourbeik, Dynamic models for turbine-governors in power system studies. IEEE PES, Tech. Rep. PES-TR1, Jan 2013
34. Power System Stability Study Group, Integrated Analysis Software for Bulk Power System Stability. CRIEPI Report: ET90002, July 1991
35. R.D. Christie, Power systems test case archive, Aug 1999. https://labs.ece.uw.edu/pstca/
36. ERCOT, ERCOT nodal operating guides – section 2: system operations and control requirements, May 2019
37. PJM Manual 36: System Restoration Revision: 25, PJM System Operations Division, June 2018. https://www.pjm.com/~/media/documents/manuals/m36.ashx
38. IEEE SCC21 Standards Coordinating Committee on Fuel Cells, Photovoltaics, Dispersed Generation, and Energy Storage, in *IEEE Standard for Interconnection and Interoperability of Distributed Energy Resources with Associated Electric Power Systems Interfaces*, April 2018 (IEEE, Piscataway, 2018)
39. J.L. Blackburn, T.J. Domin, *Protective Relaying: Principles and Applications*, 4th edn. (CRC Press, Boca Raton, 2014)
40. W.G. C4.605, Modeling and aggregation of loads in flexible power networks. CIGRE, Paris, Tech. Rep. TB566, Feb 2014
41. J.V. Milanovic, K. Yamashita, S.M. Villanueva, S. Djokic, L.M. Korunovic, International industry practice on power system load modeling. IEEE Trans. Power Syst. **28**(3), 3038–12 (2013)
42. S.M. Villanueva, K. Yamashita, L.M. Korunovic, S.Z. Djokic, J. Matevosyan, A. Borghetti, Z.Y. Dong, J.V. Milanovic, CIGRE WG C4.605 – modeling and aggregation of loads in flexible power networks – scope and status of the work by June 2012, in *Proc. of CIGRE C4 Colloquium*, Hakodate (2012), pp. 109–114
43. Y. Tang, S. Zhao, C. Ten, K. Zhang, L. Thillainathan, Establishment of enhanced load modeling by correlating with occupancy information. IEEE Trans. Smart Grid **11**, 1–12 (2019)
44. K. Yamashita, S.Z. Djokic, F. Villella, J.V. Milanovic, Self-disconnection and self-recovery of loads due to voltage sags and short interruptions, in *Proc. of CIGRE Symposium*, Lisbon, April 2013, pp. 22–24
45. IEEE Standard for Electrical Power System Device Function Numbers, Acronyms, and Contact Designations, IEEE Std. C37.2, 2008

Distributed Attack and Mitigation Strategies for Active Power Distribution Networks

Jingyuan Liu and Pirathayini Srikantha

1 Introduction

The advent of the Internet of Things (IOT) has equipped a large number of traditionally passive devices in the DN with unprecedented communication and decision-making capabilities. This digital upgrade of grid systems has led to increased efficiency in power system operation as well as improved accommodation capability for diverse load and generation elements in the power network. However, this improvement in grid information structure is also associated with increased risk of system vulnerabilities, which can be exploited by adversarial groups to inflict costly and sometimes irreversible system-wide disruptions.

In this chapter, we investigate the role of cyber-capable load and generation grid elements in DN attack and mitigation strategies. Based on our reviews of the existing state-of-the-art in power system security, we argue that the ability to coordinate various cyber-capable load and generation elements in the smart grid is crucial to the attacker and defender alike. Here, we introduce an attack strategy that aims to cause widespread disruptions of DN operation through incremental actuation of compromised distributed load devices such as Electric Vehicles (EVs), as well as a countering strategy that utilizes utility-controlled active power injection devices to effectively suppress the adverse effects of the attack.

Existing literature on improving DN resiliency can be divided into three major categories. Works in the first category focus on the detection of intrusion attempts on smart grid communication protocols and robustifying the system against jamming

J. Liu
York University, Toronto, ON, Canada
e-mail: jliu2325@eecs.yorku.ca

P. Srikantha (✉)
Lassonde School of Engineering, York University, Toronto, ON, Canada
e-mail: psrikan@yorku.ca

H. Karimipour et al. (eds.), *Security of Cyber-Physical Systems*,
https://doi.org/10.1007/978-3-030-45541-5_13

and denial-of-service attacks [6, 17, 22]. Works in the second category emphasize the detection and mitigation of the adverse effects of malicious system data tempering [10, 12]. Proposals in the final category focus on the maintenance of system limits in the face of the increasing proliferation of unregulated distributed loads [13, 23]. Nevertheless, few among the existing state of the art address the risks posed by compromised load devices in the power system, and to the best of our knowledge no mitigation strategy for addressing the threat posed by a large number of compromised sources acting in coordination has been proposed.

For the attack strategy, we exploit a significant flaw in the Zigbee wireless communication protocol prevalent in power networks to devise a DN attack scheme. As showcased by reference [19], this flaw allows potential adversaries to bypass standard security mechanisms in the smart grid and to remotely monitor as well as actuate cyber-enabled grid components. After establishing secure communication links with the compromised devices, the attacker leverages solution concepts in Stakelberg games outlined in reference [5] to systematically target specific buses in DN that are most susceptible to adversarial stress. Next, the attacker proceeds to deploy its resources in these buses according to a stealthy actuation scheme that seeks to maximize the damage to the DN without triggering the protection mechanisms. The actuation scheme derived from the principles of population game theory then accomplishes its objective through distributed coordination of compromised loads.

Since it is not economically feasible to equip every bus in the DN with enough generation/storage capacity to compensate for heavy power draws caused by concentrated attacks from adversarial agents, the mitigation scheme we formulated seeks to offset the attacks through effective coordination among defense mechanisms such as utility-controlled storage devices and power injection elements. Finally, we derive a guarantee on the convergence performance of the mitigation scheme from the principles of dual update in convex optimization. The performance of both the proposed attack strategy and its mitigation scheme are then verified through simulations conducted on practical systems such as the IEEE-33 bus DN, IEEE-69 bus DN and the Brazilian 136 bus DN.

2 System Model

2.1 System Settings

Accurate modeling of the cyber-enabled smart grid systems is instrumental in the development of attack strategies and their countermeasures. To reflect the realities in a cyber-capable smart DN that hosts both EPU-managed power injecting components and a large number of distributed smart loads, we make the following assumptions in our attack-mitigation model:

1. The adversary will be able to monitor the power load/generation on each bus in the DN.

2. The adversary can compromise the cyber-capable distributed load devices in the DN through the communication channel.
3. The adversary will be able to establish secure communication links with compromised devices and appliances.
4. The adversary has no information about the state of operation of EPU-controlled resources after the commencement of the attack.
5. The EPU can infer the onset of an attack based on the irregularities in the DN load profile.
6. The EPU will be able to provide accurate measurements of load profile deviations in real-time.
7. The EPU will be able to securely communicate with the resources under its control.
8. The resources controlled by both the adversary and the EPU are real load based.

Assumptions 1 and 2 stem from existing vulnerabilities in communication protocols and software systems in power grid. As described by reference [4, 16, 18, 24], these vulnerabilities can be exploited by the potential attackers to compromise cyber-enabled equipment and devices ranging from distribution control centers to smart appliances. Since the adversary will be able to have complete control over the compromised appliances, Assumption 3 is also feasible. As the EPU will drastically increase the level of security for system monitoring after the onset of an attack has been identified [9], the adversary will not be able to receive information pertaining to the operating state of EPU-controlled devices after the attack has been launched. Therefore, Assumption 4 is realistic. Assumption 5 is supported by the proliferation of Advanced Metering Infrastructure (AMI) in power grids, which allows modern EPUs to access real-time power consumption/generation measurements from every bus in the DN. Since attack techniques based on compromising consumer loads and IoT devices inevitably leads to excessive, irregular stress in the DN load profile, the EPU will be able to detect such attacks through revision of load profile measurements. Assumption 6 is based on the recent advancements in countermeasures against false data injection attacks [10] as well as the measurement capabilities of the EPUs. As the communication protocols used by the power networks guarantee secure communication between the EPU and its ancillary devices [18], Assumption 7 is also practical. Finally, since the type of power network discussed in this work is lower voltage DNs with resistive lines, both reactive loads and devices that maintain power system stability through the regulation of reactive power (e.g. synchronous condensers and static var compensators) operate at vastly reduced efficiency. Therefore, the last assumption is also valid.

2.2 Cyber-Physical Model for the Smart DN

Next, we introduce the model for the cyber-enabled smart DN adopted from reference [14]. In this model, the set of all buses in the DN is denoted by $\mathcal{B}$ and the number of buses is denoted by n (i.e. $|\mathcal{B}| = n$). The voltage on bus b_i is denoted by V_i. Moreover, the voltage magnitude $|V_i|$ for every $b_i \in \mathcal{B}$ is subjected to the upper limit $\overline{V}_i$ and lower limit $\underline{V}_i$ (i.e. $\underline{V}_i \leq |V_i| \leq \overline{V}_i$) to ensure the safe operation of the DN. The total power consumption/generation is denoted by $\mathcal{E}^i$. If there exists a line between $b_i, b_j \in \mathcal{B}$ and where b_i is closer to the feeder than b_j, then the line is denoted by (i, j). The set of all lines is denoted by the set $\mathcal{L}$, where $|\mathcal{L}| = l$ represents the number of lines in the DN. The power and current flows across line (i, j) are denoted by S_{ij} and I_{ij} respectively. Moreover, the maximum power flow of each $(i, j) \in \mathcal{L}$ is represented by $\overline{S}_{ij}$. The impedance across $(i, j) \in \mathcal{L}$ is denoted by z_{ij} (i.e. $z_{ij} = r_{ij} + \mathbf{j}x_{ij}$, where r_{ij} and x_{ij} are the resistance and reactance across (i, j) respectively). Based on these definitions, we can represent the inter-dependencies among state variables in the DN with the branch flow model presented in reference [14]:

$$\sum_{(j,k)\in\mathcal{L}} S_{ij} = \sum_{(i,j)\in\mathcal{L}} (S_{ij} - z_{ij}|I_{ij}|^2) + S_j, j \in \mathcal{B} \tag{1}$$

$$|V_j|^2 - |V_k|^2 = 2Re(z_{jk}^H S_{jk}) - |z_{jk}|^2|I_{jk}|^2, (j, k) \in \mathcal{L} \tag{2}$$

$$|S_{jk}|^2 = |V_j|^2|I_{jk}|^2, (j, k) \in \mathcal{L} \tag{3}$$

To represent the penetration of the myriad active cyber-physical elements (e.g. EVs and distributed generation) on the smart DN as well as the presence of both the attacker and the EPU, we introduce the following framework: first, we let each $i \subset \mathcal{B}$ host three distinct categories of power generation/consumption capacities. The elements in the first category $\mathcal{E}_C^i$ are controlled solely by the consumer at b_i. Elements belonging to the second category $\mathcal{E}_{EPU}^i$ are power injection and storage components of b_i managed by the EPU, and the elements in the final category $\mathcal{E}_A^i$ are compromised devices in b_i controlled by the adversary. If a bus does not contain any cyber-physical element belonging to a specific category, then its capacity in that category is 0. We further define the set of resources available to the EPU and the attacker as $\mathcal{E}_{EPU} = \{\mathcal{E}_{EPU}^1, \ldots, \mathcal{E}_{EPU}^n\}$ and $\mathcal{E}_A = \{\mathcal{E}_A^1, \ldots, \mathcal{E}_A^n\}$ respectively. Next, we allow the attacker to predict the pre-attack power injection/storage capacities of the EPU on each bus based on past state measurements. For every $b_i \in \mathcal{B}$, the average and minimal EPU injection capacities as predicted by the adversary are $\tilde{d}_i$ and $\underline{d}_i$ respectively. Finally, we allow the EPU to monitor the overall post-attack adversarial power draw p_a^i on every $b_i \in \mathcal{B}$ and define the EPU capacity deployed on b_i post-attack as p_g^i.

3 Design of Stealthy Attacks

The adversary seeks to maximize the deviation of system states from their nominal values through intelligent incremental deployment of its resources in $\mathcal{E}_A$. The incremental nature of the attack increases the difficulty for the EPU to isolate the compromised loads. Moreover, the adversary will utilize the solution concepts in Stackelsberg games [5] to degrade the efficiency of resources controlled by the EPU. Thus, the proposed attack strategy satisfies the stealthy paradigm. After the adversary identifies the specific buses to target, it compromises the cyber-enabled smart loads in these buses using the vulnerabilities in the communication protocol. Next, it implements a distributed actuation strategy for stealthily saturating system limits overtime through coordination of these compromised resources. To avoid detection from the EPU and premature isolation of the attack, the adversary will spread its attacks to a large number of smart loads resides in a multitude of buses in the DN rather than focusing on several specific buses.

3.1 Attack Planning with Stackelberg Games

To maximize the disruption in the DN, the adversary will first leverage the monitoring capability described in Assumption 1 to observe the state measurement signals generated by the load/generation devices located on each bus. These measurements will allow the adversary to infer the availability of defense mechanisms, which in turn enables the adversary to systematically target the vulnerable buses through the application of theoretical constructs based on Stackelsberg games. While the basic principles of Stackelsberg games have been applied on problems from energy management [5] to power system vulnerability analysis [21], this is the first time they are applied to attack planning problems.

In the framework defined by Stackelberg game, a system consists of two agents engaging in competition: 1. A dominant leader that always make the first move, and 2. A reacting follower that observes the strategies adopted by the leader before reacting with a strategy whereby it will minimize its losses with respect to the action performed by the leader. From the perspective of attack planning for power networks, this leader-follower paradigm is analogous to the relationship between the EPU and the adversary. In this case, the EPU takes the leading role by deploying active defense and protection mechanisms that are subjected to inherent limitations (e.g. economical constraints, etc.) against malicious attacks. Meanwhile, the adversary responds by silently observing the state measurement signals, deducing the limitations of the defense mechanisms from it, and using these insights to devise attack strategies that cause maximal disruption. As such, we can formulate the problem for attack target identification as a typical Stackelberg *follower* problem outlined in reference [25].

$$\mathcal{P}_T : \max_{0 \leq a_i \leq 1} \sum_{b_i \in \mathcal{B}} f_i^t(a_i)$$

$$s.t. f_i^t(a_i) = a_i\big(c_i U_a^c(b_i) + (1 - c_i)U_a^u(b_i)\big) - f_i(a_i)$$

In this formulation, the optimization variable a_i represents the probability for the adversary to implement its attack strategy on bus b_i. Moreover, the attacker's cost for deploying its resource on b_i is represented by the function $f_i(a_i)$. To represent the relation between the capacity of defensive mechanisms on b_i and the adversary's willingness to compromise the bus, we further introduce the variable c_i and utility functions $U_a^c(b_i)$ and $U_u^c(b_i)$: c_i is the probability for the bus $b_i \in \mathcal{B}$ to be actively defended by the EPU. $U_a^c(b_i)$ represents the willingness for the attacker to deploy resource on bus b_i if it is actively defended by the EPU, and $U_u^c(b_i)$ represents the willingness for the attacker to deploy resource on bus b_i if it is not defended. Since it benefits the attacker to deploy its resources on undefended buses, the utility for the attacker is higher for undefended buses than otherwise (i.e. $U_a^c(b_i) \leq U_u^c(b_i)$). These utility functions are defined as follows:

$$U_a^c(b_i) = (\min\left[p_{ij}^{lim} - \tilde{p}_{ij}, \check{p}_i, \mathcal{E}_A^i\right] - \tilde{d}_i) * (\tilde{p}_{ij} - \tilde{p}_i) \tag{4}$$

$$U_a^u(b_i) = (\min\left[p_{ij}^{lim} - \tilde{p}_{ij}, \check{p}_i, \mathcal{E}_A^i) - \underline{d}_i\right] * (\tilde{p}_{ij} - \tilde{p}_i) \tag{5}$$

$$c_i = P^i_{d_i \geq \tilde{d}_i} \tag{6}$$

The average real power flow and the upper limit of real power flow across the line $\{i \leftrightarrow j\}$ are denoted by $\tilde{p}_{i,j}$ and $p_{i,j}^{lim}$ respectively. $\check{p}_i$ represents the maximum real power injection on bus b_i before the its voltage magnitude $|V_i|$ drops below the voltage lower limit $\underline{V}_i$ and $\tilde{p}_i$ is the average power injection at b_i. As the intention of the adversary to cause widespread disturbances in the system while protecting its resources in $\mathcal{E}_A$ from premature isolation, the maximal load increment $\check{p}_i$ at each $b_i \in \mathcal{B}$ is designed to maximize the injection of resources without causing voltage constraint violation on b_i. To derive this variable, we first note from the power balance relations presented in Eq. (1) that the voltage drop between two neighboring buses b_i and b_j (i.e. $(i, j) \in \mathcal{E}$) is affected by both the power flow across the line (i, j) and the power loss in the line (i, j). From the studies conducted in reference [7], we have learned that line power loss has minimal impact on voltage drops between neighboring buses in low voltage, resistive DNs such as the IEEE-33 bus, IEEE 123 bus and the Rossi 2056-bus DNs. That is, the maximum voltage deviation caused by ignoring the power loss is 0.0016 p.u. for the aforementioned systems. This finding allows us to drop the power loss term from the expression for the voltage drop relations, and write the voltage drop from the feeder b_0 to any bus b_i in the DN as the following:

$$|V_0|^2 - |V_i|^2 = \sum_{(g,l)\in\mathcal{P}_i} 2Re(S_{gl} z_{gl}^H + z_{gl} S_{gl}^H) \tag{7}$$

where $\mathcal{P}_i$ is the set of lines in the path from b_i to b_0. Since the adversary only has control over resources based on real power loads (based on Assumption 8), the reactive components of the DN (i.e. reactive power flow on the lines and line reactance) has no significant impact on the attack induced voltage changes. Therefore, we can express $\check{p}_i$ with the following relation:

$$\check{p}_i = (|V_i|^2 - \underline{V}_i^2)/(\sum_{(g,l)\in\mathcal{P}_i} 2r_{gl}) \tag{8}$$

where $|V_i|$ is the voltage magnitude at b_i and r_{gl} represents the resistance along the line (g, l). Since both line resistances and voltage magnitude limits are constants, this relation allows every $i \in \mathcal{B}$ to calculate its maximal load increment from its local voltage magnitude $|V_i|$. Furthermore, the active defense probability c_i is defined as the proportion of time EPU capacity at b_i is greater than the average capacity of these resources.

Since $f_i^t(a_i) - f_i(a_i)$ is linear in terms of a_i for every $b_i \in \mathcal{B}$, closed form solution exists for $\mathcal{P}_T$ if the cost functions for resource deployment $f_i(a_i)$ for every b_i is convex and twice continuously differentiable. As the increased probability for attack leads to increased resource requirement which in turn increases the cost for resource deployment, $f_i(a_i)$ can be easily modeled as a convex and twice continuously differentiable function that is defined on the continuous interval $a_i \in [0, 1]$. This allows us to derive the following closed-form solution for each a_i from the first-order condition for optimality:

$$a_i = [f_i^{t-\mathbf{1}}\big(c_i U_a^c(b_i) + (1 - c_i)U_a^u(b_i)\big)]_0^1 \tag{9}$$

where the notation $[.]_0^1$ represents the constraint $a_i \in [0, 1]$. This value is then normalized by $\max_{b_i\in\mathcal{B}}(a_i)$ to guarantee the probability of attack in well-defended DNs. The normalized values a_i^{nom} are then used to form the attack probability vector $a = \{a_i^{nom}, b_i \in \mathcal{B}\}$, which is then used by the adversary to decide which bus to target for compromise. This attack planning method allows the adversary to concentrate its resources on the weaknesses of the EPU defense mechanism, which enables it to launch an attack that is both extremely detrimental to the safe operation of the DN and very hard for the EPU to counter.

3.2 *Attack Actuation via Population Games*

After the adversary identifies the target buses, it will then proceed to compromise the security mechanisms of as many cyber-enabled load-bearing devices (EV, heaters, etc.) located in these buses as possible for the purpose of obtaining direct actuation access of these loads. While an unsophisticated adversary may force all compromised devices to draw the maximum amount of power possible in the hope of causing widespread outages in the DN, the ensuing sudden surge in power flow will alert the EPU of the attack attempt. Moreover, this unsophisticated actuation scheme may lead to unplanned tripping of passive protection devices located in the target buses [8], which will lead to premature identification and isolation of the compromised devices. In this subsection, we present an intelligent attack actuation scheme based on population games that allows the adversary to deploy its resources without causing any sudden surges in power flow which may lead to the isolation of its resources in $\mathcal{E}_A$.

The attack actuation scheme leverages theoretical constructs from Population Game Theory (PGT) to enable stealthy actuation of compromised devices within each target bus. As the adversary aims to implement the attack through the actuation of a large number of small scale appliances located in each bus, the compromised devices in each bus b_i can be compared to agents in PGT and they form the population Q_i where $|Q_i| = q_i$. Under this scheme, each agent will adopt a strategy from an identical and discrete set of strategies $y = \{y_1, \dots y_m\}$, where each strategy corresponds to a level of energy consumption for the agent. Hence, the impact of each strategy choice from the agents is *incremental*. While different agents can have different power consumption ratings and settings and therefore should not be assigned identical strategy sets, we can remedy this issue by establishing a general model for the strategy set that covers most of the settings of typical agents and forcing the agents to adopt the closest feasible strategy. For example, if an agent is unable to support a particular strategy $y_a \in y$, it will adopt a feasible strategy that is closest to y. This solution holds as long as the number of agents in the PGT framework is sufficiently large in comparison to the number of agents that has to adopt the closest feasible strategy. Next, we evaluate the general state of the system with aggregate measures $x = \{x_1, \dots x_m\}$, where each x_k represents the fraction of agents that adopts strategy $y_k \in y$. Furthermore, each agent in b_i may change its adopted strategy from any $y_j \in y$ to any $y_k \in y$ according to its corresponding switching probability $\rho^i_{j,k}(F(x))$, where $F(x)$ is a function of the aggregate measurement variable x.

With this formulation, the PGT paradigm allows for incremental and anonymous changes in the power consumption levels of the compromised devices and this increases the difficulty for the EPU to detect and isolate these agents. Moreover, it allows the adversary to coordinate the attack scheme within every $b_i \in \mathcal{B}$ independently from one another, which enables the adversary to focus on weaknesses specific to each target bus. Based on the aforementioned notations for attack scheme formulation, the attack actuation problem $\mathcal{P}^i_A$ for obtaining the optimal x can then be formulated as follows:

$$\mathcal{P}^{\rangle}_{\mathcal{A}} : \min_{x \in \Delta} \sum_{k=1}^{|y|} \frac{1}{2} q_i x_k^2 y_k \tag{10}$$

$$s.t. \sum_{k=1}^{|y|} q_i y_k x_k = \min[p^{lim}_{ij} - \tilde{p}_{ij}, \check{p}_i, \mathcal{E}^i_A] \tag{11}$$

where $\Delta = \{x | \sum_{j=1 \dots |y|} x_j = 1, x_j \geq 0 \, \forall j = 1 \dots |y|\}$ is the simplex that defines the aggregate variable x in target bus b_i. In this formulation, the linear constraint forces the compromised devices to draw enough power to either saturate the local line or cause voltage outage in neighboring buses. Moreover, the quadratic objective prevents any strategy from becoming dominant, which increases the difficulty for the EPU to identify the compromised resources. Since both the objective function and the constraints for the attack actuation problem $\mathcal{P}^i_A$ are convex, its optimal solution x^* can be easily obtained from any convex solver. Next, we introduce a local actuation framework that guides agents with different constraints and settings toward the optimal aggregate variable x^*. Under this framework, the adversary coordinates the various agents through periodic updates of the cost signal vector $F(x) = \{F_1(x), \dots F_{|y|}(x)\}$. To obtain the expression for $F(x)$, we start from the Lagrangian $\mathcal{L}^i_A$ of the attack formulation problem $\mathcal{P}^i_A$:

$$\mathcal{L}^i_A(x, \nu) = \sum_{k=1}^{|y|} \frac{1}{2} q_i x_k^2 y_k + \nu(\sum_{k=1}^{|y|} q_i y_k x_k - \min[p_{ij}^{lim} - \tilde{p}_{ij}, \check{p}_i, \mathcal{E}^i_A]) \tag{12}$$

The simplex condition on x is intrinsically satisfied by the nature of the distributed strategy selection process performed by the agents. Then, we obtain the optimal Lagrangian dual variable ν^* from solving the dual problem $\max_\nu \mathcal{L}^i_A(x^*, \nu)$ before taking the gradient of $\mathcal{L}^i_A(x, \nu^*)$ to find the following expression for $F(x)$:

$$F_k(x) = q_i x_k y_k + \nu^* q_i y_k, \ \forall k \in \{1, \dots |y|\} \tag{13}$$

After obtaining this signal from the adversary, every agent residing in bus b_i then initiates the local actuation process. The first step of this process is the update of switching probabilities $\rho^i_{j,k}(F(x))$ between every strategy pairs $y_j, \ y_k \in y$:

$$\rho^i_{j,k}(F(x)) = [\frac{F_j(x) - f_k(x)}{q_i x_j}]_+ \tag{14}$$

It is worth noting that $\rho^i_{j,k}(F(x))$ is not equivalent to $\rho^i_{k,j}(F(x))$. With this probability, the agents then periodically perform local strategy revision every τ seconds. To study this dynamic revision process, we represent it with a set of differential equations of the growth/decline rate of any population x_j in the PGT framework:

$$\dot{x}_j = \sum_{k=1}^{|y|} x_k \rho^i_{k,j}(F(x)) - x_i \sum_{k=1}^{|y|} \rho_{j,k}(F(x)) \tag{15}$$

Due to the strong law of large numbers, stochastic effects caused by differences between agents can be safely averaged out. As such, the above state dynamic can be rewritten into the much simpler form

$$\dot{x}_j = \frac{1}{n} \sum_{y_i \in y} F_j(x) - F_i(x) \tag{16}$$

which is none other than the negative of the projection of the cost function $F(x)$ into the simplex Δ. Since $F(x)$ itself is the gradient of $\mathcal{L}_A^i(x, \nu^*)$, the dynamic revision process will gravitate towards lower potential at each revision. According to the constructs from Lyapunov theory of control, which are elaborated in reference [20], this process also drives the system towards convergence to the optimal solution in exponential time. From the formulation of the attack scheme, the adversary would have successfully caused massive voltage drops and outages throughout the DN when the agents converge at the optimal condition. In the following algorithm, we summarize the attack planning and actuation processes outlined in this section.

Attack scheme construction

Target identification:

- For every $b_i \in \mathcal{B}$, the adversary computes its utilities $U_a^c(b_i)$ and $U_a^u(b_i)$ from:
  $U_a^c(b_i) = (\min[p_{i,j}^{lim} - \tilde{p}_{i,j}, \check{p}_i, \mathcal{E}_A^i] - \tilde{d}_i) * (\tilde{p}_{i,j} - \tilde{p}_i)$
  $U_a^u(b_i) = (\min[p_{i,j}^{lim} - \tilde{p}_{i,j}, \check{p}_i, \mathcal{E}_A^i) - \underline{d}_i] * (\tilde{p}_{i,j} - \tilde{p}_i)$.
- The adversary solves the identification problem $\mathcal{P}_T$ for every b_i to obtain the attack probability vector $a = \{a_i^{nom}, i \in \mathcal{B}\}$.
- The adversary compares each a_i^{nom} with a random number generated from $rand(1)$. If $rand(1) < a_i^{nom}$, then the bus b_i is targeted for attack.

Compromised load actuation construction for targeted b_i:

1. The individual agents (i.e. compromised devices) located in b_i updates the adversary monitored aggregate variable x every τ seconds.
2. The adversary updates $F(x)$ by computing $F_j(x_j)$ for every $j \in \{1, \ldots, |y|\}$, then broadcasts it to all agents in b_i.
3. Individual agents updates switching probabilities $\rho_{j,k}^i$ for every strategy pairs $y_j, y_k \in y$ and perform strategy revisions based on the revised switching probabilities. The final strategy adopted by the agents will be the closest feasible approximation of the final outcome of the strategy revision.

3.3 *Case Study*

To evaluate the effectiveness of the DN attack scheme based on Stackelberg game framework and theoretical constructs from PGT, we implement it on a number of

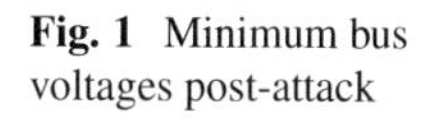

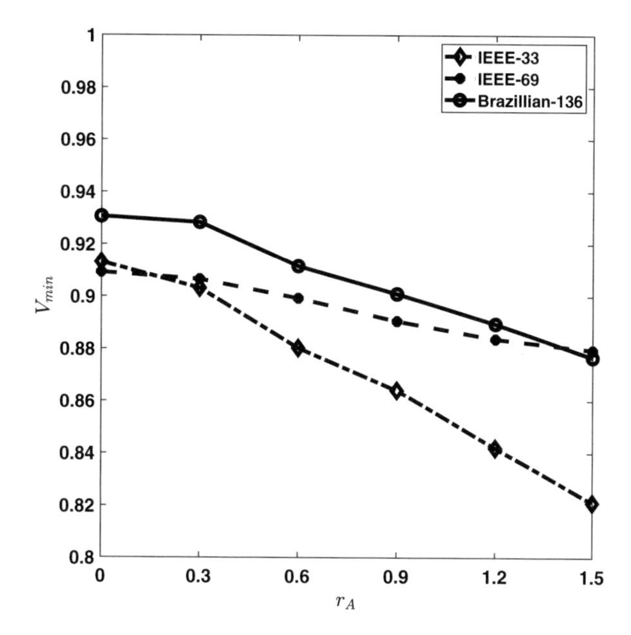

Fig. 1 Minimum bus voltages post-attack

realistic DNs including the IEEE-33 bus [1], IEEE-69 bus [2] and a Brazilian 136-bus [15] DNs. In this case study, we explore the effect of the attack scheme on the minimal bus voltages of these systems and compare its resource draw profile with that of established decentralized state-of-the-art introduced in reference [11]. Based on our findings, we conclude that our attack scheme is capable of causing deeply resonating and hard to detect system failures for a variety of practical systems.

In Fig. 1 we analyze the impact of adversarial attack schemes on the minimum bus voltage of the systems under study. Here, we use the term r_A to represent the ratio of the adversarial resources in $\mathcal{E}$ over the resources in $\mathcal{E}_{EPU}$ that are managed by the EPU. From Fig. 1, we can conclude that the attack scheme's effectiveness in undermining voltage profile of the DN increases notably when more resources are placed under the adversary's disposal. What is more, the effectiveness of adversarial action also depends heavily on the topology and load distribution of the DN: the attack scheme is more effective on the IEEE-33 bus DN than on the IEEE-69 bus and the Brazilian 136 bus DNs for all r_A. Regardless, the attack scheme succeeded in causing low-voltage induced system outage for all tested DNs when the quantity of adversarial resources matches that controlled by the EPU.

Next, we compare the resource draw profiles of the proposed attack scheme with that of the state-of-the-art in Fig. 2. As evidenced from the figure, the state-of-the-art based on theoretical constructs from Sub-Gradient (SG) incurs very noticeable oscillations in its resource draw profile during its convergence while the proposed attack scheme actuates its resources gradually. Since rapid oscillations in resource draw profiles can lead to premature actuations of voltage/power protection devices in the DN as well as attentions from the PMU, the SG algorithm does not satisfy

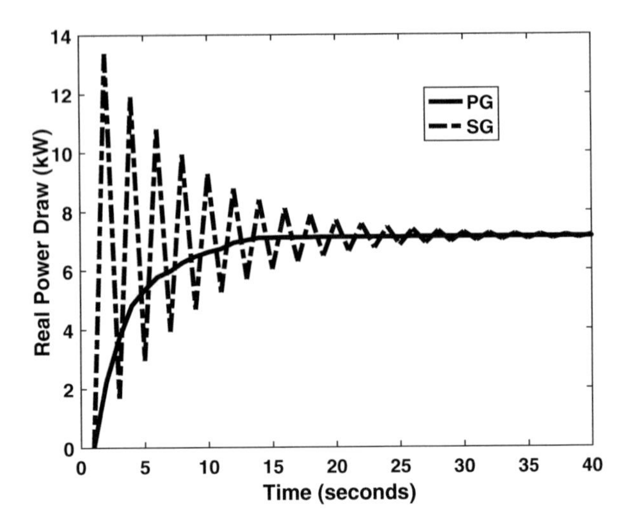

Fig. 2 Comparison of PGT vs SG

paradigm of stealthy attack. Conversely, the proposed attack scheme is stealthy as it allows for the rapid actuation of a large number of resources with no oscillations.

4 Countermeasure Formulation via Dual Updates

After the EPU verifies the existence of an attack from anomalies in power consumption or voltage, it will then deploy all available resources in $\mathcal{E}_{EPU}$ to restore system states such as bus loads and voltages to their nominal values. Moreover, DN defense mechanisms such as storage devices and distributed generators are becoming increasingly economically viable due to government incentives and technological advancements [8]. Therefore, each bus b_i in the DN can host its own power injection capacity $\mathcal{E}^i_{EPU}$ which can be leveraged by the EPU to offset abnormal power drains caused by the adversary. As the stealthy nature of the attack stratagem introduced in Sect. 3 makes it very hard for the EPU to identify the compromised devices, any effective countermeasure of it must be able to function without precise knowledge about the identity of these adversarial agents. Thus, in this section we introduce an iterative approach for defense mechanism actuation based on the principle of monotone operators that allows the EPUs to combat the proposed attack scheme in a systematic and cost effective manner.

4.1 Formulation of Countermeasure

The proposed countermeasure is based on Assumptions 6 and 7 in Sect. 2, which allows the EPU to restore balance in the net power demands of the DN through optimal coordination of its resources located in each bus. As the attack scheme intentionally targets buses with insufficient power injection capacity, the countermeasure compensates for the adversarial power draws by effectively leveraging the surplus capacities located in all buses in the DN. Thus, the EPU countermeasure problem $\mathcal{P}_{EPU}$ can be formulated as follows:

$$\mathcal{P}_{EPU} : \min_{P_g} \sum_{i=1}^{|B|} C_i(P_g^i)$$

$$s.t. \sum_{i=1}^{|\mathcal{B}|} p_g^i = \sum_{i=1}^{|\mathcal{B}|} p_a^i \tag{17}$$

$$\underline{p}_g^i \leq p_g^i \leq \overline{p}_g^i, \forall b_i \in \mathcal{B} \tag{18}$$

where p_g^i is the active power draw from the EPU resource $\mathcal{E}_{EPU}^i$ and p_a^i is the aggregate adversarial power draw on b_i. According to Assumptions 5 and 6 in Sect. 2, the EPU will be able to provide accurate measurements of p_a^i in real-time. The cost function for resource dispatch $C_i(p_g^i)$ is defined as the strictly convex function $(p_g^i)/(1 - |V_i|^2)$. This formulation allows the EPU to prioritize power injection from buses with stressed-out voltage profiles (i.e. buses whose voltage magnitudes are close to the lower limit). Next, the coupling constraint Eq. (17) represents the need for the EPU to restore the balance between power consumption and demand. Furthermore, the final inequality constraint Eq. (18) imposes upper and lower constraints on the generation capacity of the EPU power injection resources located on each bus. While the globally optimal solution of $\mathcal{P}_{EPU}$ can be easily obtained from any convex solver if we solve this problem in a centralized manner, doing so would heavily strain the communication and computation capabilities of the EPU and make the algorithm less scalable. In the rest of this section, we detail a distributed coordination strategy that allows the EPU to offload the computational efforts involved in solving the countermeasure problem to intelligent agents located in each bus $b_i \in \mathcal{B}$.

4.2 Distributed EPU Coordination Strategy

Since the countermeasure problem is strictly convex, its optimal condition is defined by the Karush Kuhn Tucker (KKT) conditions [3]. From the KKT conditions, the optimal solution p_g^{i*} for $\mathcal{P}_{EPU}$ must first satisfy its own constraints Eqs. (17) and (18). Next, the optimal dual variables for the problem λ_1^*, λ_2^* and ν^* have to satisfy the conditions $\lambda_1^* \geq 0$, $\lambda_2^* \geq 0$ and $\nu^* \in \mathbb{R}$, where λ_1^* and λ_2^* are associated with the inequality constraints represented by Eq. (18) and ν^* is associated with the coupling constraint presented in Eq. (17). Furthermore, the primal and dual variables have to satisfy the following equations:

$$\lambda_1^{i*}(\underline{p}_g^i - p_g^{i*}) = 0 \quad (19)$$

$$\lambda_2^{i*}(p_g^{i*} - \overline{p}_g^i) = 0, \forall b_i \in \mathcal{B} \quad (20)$$

which is derived from the complementary slackness property of dual variables and primal constraints. This constraint restricts the λ terms to be 0 if the optimal solution does not violate the aforementioned inequality constraints on EPU power draw. On the other hand, if $\underline{p}_g^i - p_g^{i*} = 0$ or $p_g^{i*} - \overline{p}_g^i$, the λ terms may be non-zero. The final condition from the KKT is attained from the situation of *s*tationarity: $\frac{\partial L}{\partial x_i} = \frac{d}{dt}C_i(p_h^{i*}) + \nu^* - \lambda_1^{i*} + \lambda_2^{i*} = 0$. That is, the gradient of the Lagrangian function of $\mathcal{P}_{EPU}$ with respect to the primal variables p_g^i must be 0 when the system is at its global optimum.

To construct the distributed coordination strategy from the KKT condition, we first extract the following dependencies between ν^* and p_g^i:

$$\text{If } \nu^* \geq -\frac{d}{dt}C_i(\underline{p}_g^i) \text{ then } p_g^{i*} = \underline{p}_g^i \quad (21)$$

$$\text{If } -\frac{d}{dt}C_i(\overline{p}_g^i) \leq \nu^* \leq -\frac{d}{dt}C_i(\underline{p}_g^i) \text{ then } p_g^{i*} = \frac{d}{dt}C_i(-\nu^*) \quad (22)$$

$$\text{If } \nu^* \leq -\frac{d}{dt}C_i(\overline{p}_g^i) \text{ then } p_g^{i*} = \overline{p}_g^i \quad (23)$$

These relations allow us to express the power injection on each bus b_i as the following:

$$p_g^i(\nu) = [\frac{d}{dt}C_i^{-1}(-\nu)]_{\underline{p}_g^i}^{\overline{p}_g^i} \quad (24)$$

Next, we define the difference between the overall power injection from the devices in $\mathcal{E}_{EPU}$ and the overall adversarial power consumption $\sum_{i=1}^{|\mathcal{B}|} p_a^i$ as $d(\nu)$, where

$$d(\nu) = \sum_{i=1}^{|\mathcal{E}_{EPU}|} [\frac{d}{dt}C_i^{-1}(-\nu)]_{\underline{p}_g^i}^{\overline{p}_g^i} - \sum_{i=1}^{|\mathcal{B}|} p_a^i \quad (25)$$

Since the overall power injection from EPU resources must match the cumulative adversarial power draw in the DN for any feasible solution of $\mathcal{P}_{EPU}$, the optimal Lagrangian coefficient ν^* has to satisfy $d(\nu^*) = 0$. As Eq. (25) is a non-decreasing monotonous function with regard to $-\nu$, the optimal solution v^* is the unique solution for $d(\nu) = 0$ [3]. Thus, if we can establish solid upper and lower bounds for ν^* from the power injection capacity ratings of $\mathcal{E}_{EPU}$ and worst possible costs for any bus $b_i \in \mathcal{B}$, we will be able to identify it with a binary search algorithm. To implement the EPU coordination algorithm in a decentralized manner, the EPU will broadcast its guesses for ν^* to the local resources in $\mathcal{E}_{EPU}$ every τ seconds to facilitate updates of active power draws. From theoretical results about binary search algorithms, the convergence rate of this algorithm is $O(\frac{\nu_{max}-\nu_{min}}{\epsilon})$ for an error margin of ϵ (i.e. $d(\nu) \leq |\epsilon|$). The resulting algorithm is detailed as follows:

Countermeasure scheme formulation

Initialization:

- The EPU initializes the upper bound ν_{max} and lower bound ν_{min} of the coordination signal ν as $\max(-\frac{d}{dt}C_i(\underline{p}^i_g))$ and $\min(-\frac{d}{dt}C_i(\overline{p}^i_g))$ respectively. The signal ν is then initialized as $(\nu_{max} + \nu_{min})/2$.

Algorithm:

1. The EPU updates $d(\nu)$ from the present value of ν every τ seconds. If $|d(\nu)|$ is smaller than the tolerance ϵ, then the countermeasure algorithm is terminated.
2. If $d(\nu) > 0$, the EPU updates ν_{min} by setting it to $(\nu_{max}+\nu_{min})/2$.
3. Otherwise, the EPU updates ν_{max} by setting it to $(\nu_{max}+\nu_{min})/2$.
4. Next, the EPU computes $(\nu_{max} + \nu_{min})/2$ from the updated values and assign it to ν.
5. The coordination signal ν is then broadcasted to the $\mathcal{E}^i_{EPU}$ (i.e. the EPU-controlled resources) affiliated with each $b_i \in \mathcal{E}$.
6. Each $\mathcal{E}^i_{EPU}$ uses the coordination signal to update its power injection p^i_g .

4.3 Case Study

In this subsection, we first evaluate the effectiveness of the proposed countermeasure scheme in mitigating the voltage profile of DNs that are compromised by the attack scheme introduced in Sect. 3. Next, we study the dynamic evolution of system states of a IEEE-33 bus DN that is first subjected to the attack scheme for one second then placed under the proposed mitigation scheme.

As illustrated in Fig. 3, the countermeasure we devised is capable of significantly improving the minimum bus voltage magnitudes of the compromised IEEE-33 bus, IEEE-69 bus and Brazilian 136 bus DNs. While the minimum voltages in all test systems display a downward trend when the adversarial resources exceeded half of that of the EPU, it is as expected since the limited EPU capacity will not be

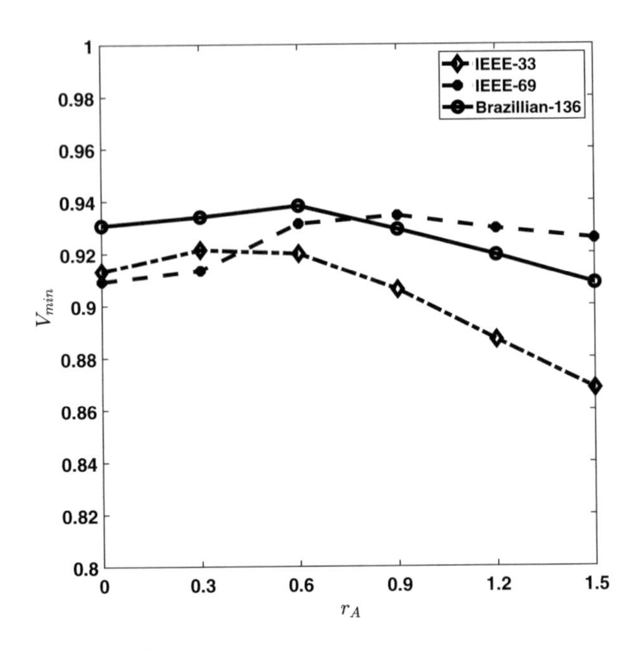

Fig. 3 Minimum bus voltages post-recovery

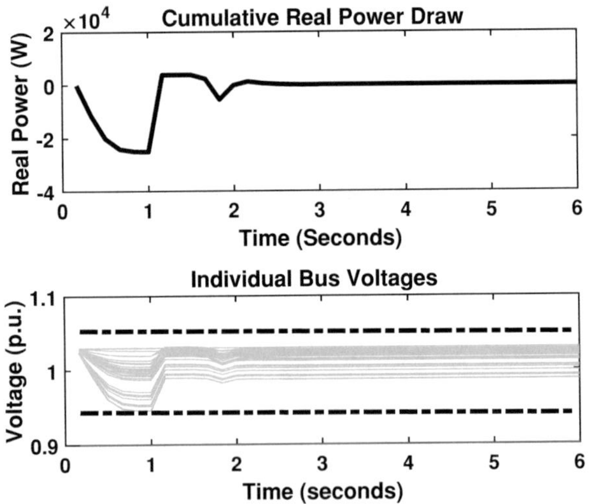

Fig. 4 State evolution in 33-bus DN

able to mitigate the intense power draw enacted by the adversary. Moreover, the effectiveness of the countermeasure in mitigating for minimum voltage also depends heavily on the topology and load profile of the DN.

From Fig. 4, we can see that the cumulative power drawn by compromised devices in the DN increases exponentially prior to the deployment of the mitigation scheme. As illustrated by the second subplot, this drastic change in loading placed considerable stress on the voltage profile of the DN. From Fig. 1 in Sect. 3, we can see that if not for the deployment of the countermeasure the attack scheme would eventually trigger widespread outages in the DN. As it is, the prompt response from the EPU leads to a quick recovery of both load and voltage profiles of the DN.

5 Final Remarks

In this chapter, we have presented a novel and stealthy DN attack scheme based on theoretical constructs from game theory and an intelligent countermeasure based on dual updates. For the attack scheme, we have shown that it is possible for an adversary to effectively and stealthily impair the safe operation of DNs through distributed coordination of a large number of power-consuming small-scale smart appliances. Even so, the performance of the counter-strategy has demonstrated that the EPU is capable of deploying its resources in a distributed manner to counter the worst adverse effects of attack scheme if the attack scheme can be detected in a timely manner. For future work, we intend to study how to effectively counter-attacks on the DN via privately owned, consumer-centric devices by integrating the defensive mechanisms of these devices with those of the EPU. Moreover, we will also study the effects of DN topology and load distribution on the effectiveness of attack and defense strategies of the DN.

References

1. M.E. Baran, F.F. Wu, Network reconfiguration in distribution systems for loss reduction and load balancing. IEEE Trans. Power Deliv. **4**(1), 1401–1407 (1989)
2. M.E. Baran, F.F. Wu, Optimal capacitor placement of radial distribution systems. IEEE Trans. Power Deliv. **4**(2), 725–734 (1989)
3. S.P. Boyd, L. Vandenberghe, *Convex Optimization* (Cambridge University Press, Cambridge, 2011)
4. T.M. Chen, S. Abu-Nimeh, Lessons from Stuxnet. Computer **44**(4), 91–93 (2011)
5. J. Chen, Q. Zhu, A Stackelberg game approach for two-level distributed energy management in smart grids. IEEE Trans. Smart Grid **9**(6), 6554–6565 (2018)
6. K. Gai, M. Qiu, Z. Ming, H. Zhao, L. Qiu, Spoofing jamming attack strategy using optimal power distributions in wireless smart grid networks. IEEE Trans. Smart Grid **8**(5), 431–2439 (2017)
7. L. Gan, S.H. Low, Convex relaxations and linear approximation for optimal power flow in multiphase radial networks, in *Power Systems Computation Conference*, Wroclaw (2014)
8. J.D. Glover, T.J. Overbye, M.S. Sarma, *Power System Analysis and Design*. (Cengage Learning, Boston, 2017)
9. V.C. Gungor, D. Sahin, T. Kocak, S. Ergut, C. Buccella, C. Cecati, G.P. Hancke, Smart grid technologies: communication technologies and standards. IEEE Trans. Ind. Inf. **7**(4), 529–539 (2011)
10. Y. He, G.J. Mendis, J. Wei, Real-time detection of false data injection attacks in smart grid: a deep learning-based intelligent mechanism. IEEE Trans. Smart Grid **8**(5), 21–32 (2017)
11. J. Joo, M.D. Ilic, Multi-layered optimization of demand resources using Lagrangian dual decomposition. IEEE Trans. Smart Grid **4**(4), 2081–2088 (2013)

12. Y. Liu, P. Ning, M.K. Reiter, False data injection attacks against state estimation in electric power grids, in *ACM Conference on Computer and Communications Security* (2009), pp. 21–32
13. Y. Liu, H. Xin, Z. Qu, D. Gan, An attack-resilient cooperative control strategy of multiple distributed generators in distribution networks. IEEE Trans. Smart Grid **7**(6), 2923–2932 (2016)
14. S. Low, Convex relaxation of optimal power flow part I: formulations and equivalence. IEEE Trans. Control Netw. Syst. **1**(1), 15–27 (2014)
15. J.R.S. Mantovani, F. Casari, R.A. Romero, Reconfiguração de sistemas de distribuição radiais utilizando o critério de queda de tensão, Revista Controle e Automação. Sociedade Brasileira de Au- tomática, SBA **11**(3), 150–159 (2000)
16. McAfee White Paper, Global Energy Cyberattacks: Night Dragon. McAfee, 2011
17. U.K. Premaratne, J. Samarabandu, T.S. Sidhu, R. Beresh, J. Tan, An Intrusion detection system for IEC61850 automated substations. IEEE Trans. Power Deliv. **25**(4), 2376–2383 (2010)
18. R. Rodrigo et al., Securing the Internet of Things. Computer **44**(9), 51–58 (2011)
19. E. Ronen, A. Shamir, A.O. Weingarten, C. Oflynn, IoT goes nuclear: creating a ZigBee chain reaction, in *IEEE Symposium on Security and Privacy* (2017)
20. W.H. Sandholm, *Economic Learning and Social Evolution: Population Games and Evolutionary Dynamics* (MIT Press, Cambridge, 2011)
21. D. Shelar, S. Amin, Analyzing vulnerability of electricity distribution networks to DER disruptions, in *2015 American Control Conference (ACC)* (2015)
22. P. Srikantha, D. Kundur, Denial of service attacks and mitigation for stability in cyber-enabled power grid, in *IEEE PES & Innovative Smart Grid Technologies Conference* (2015)
23. P. Srikantha, D. Kundur, Resilient distributed real-time demand response via population games. IEEE Trans. Smart Grid **8**(6), 2532–2543 (2017)
24. C. Ten, K. Yamashita, Z. Yang, A. Vasilakos, A. Ginter. Impact assessment of hypothesized cyberattacks on interconnected bulk power systems. IEEE Trans. Smart Grid **9**(5), 4405–4425 (2018)
25. Z. Yin, D. Korzhyk, C. Kiekintveld, V. Conitzer, M. Tambe, Stackelberg versus Nash in security games: interchangeability, equivalence, and uniqueness. J. Artif. Intell. Res. **41**, 297–327 (2011)

Privacy-Preserving Homomorphic Masking for Smart Grid Data Analytics in the Cloud

Pirathayini Srikantha, Anna Shi, Bokun Zhang, and Deepa Kundur

1 Introduction

The recent cyber-physical revolution in modern societies allows for unprecedented connectivity amongst humans, intelligent devices and complex systems. Devices that range from simple light bulbs to vital monitoring and actuating equipments in critical infrastructures are equipped with intelligence and the ability to communicate. The sheer volume of data generated in this era of the Internet of Things (IoT) offers to all stakeholders tremendous capability to monitor and exact well-informed decisions [1]. One example of this in practice today is the collection and assimilation of smart meter data by Electric Power Utilities (EPUs) [2]. By performing advanced analytics on this data, EPUs can provide to consumers a precise depiction of individual power consumption patterns, comparison of usage with neighbours and even make appropriate recommendations as necessary. Using this feedback, consumers can make informed decisions and alter their daily appliance usage

This work was supported in part by grants from the National Science Foundation and the Natural Sciences and Engineering Research Council of Canada.

P. Srikantha (✉)
Lassonde School of Engineering, York University, Toronto, ON, Canada
e-mail: psrikan@yorku.ca

A. Shi · B. Zhang · D. Kundur
Department of Electrical and Computer Engineering, University of Toronto, Toronto, ON, Canada
e-mail: anna.shi@ece.utoronto.ca; xxmtg.zhang@ece.utoronto.ca; dkundur@ece.utoronto.ca

H. Karimipour et al. (eds.), *Security of Cyber-Physical Systems*,
https://doi.org/10.1007/978-3-030-45541-5_14

patterns for more economical and sustainable electricity consumption. On the other hand, EPUs can also use this data for system maintenance to identify outages and disturbances so that appropriate recovery mechanisms can be activated as necessary.

In order to tap onto the significant potential of the large volumes of data (also referred to as *big data*) continually generated by the IoT, data analytics that enable the computation of statistical measures, perform regression analysis and data mining are vital. These provide deep insights into the underlying structures, dependencies and interactions of participating entities. To manage and perform real-time analytics on these large and complex data sets, efficient storage and computational resources are necessary. As such, a commonly observable recent phenomenon is the migration of large system operators such as the EPUs and banks from traditional in-house data management to cloud solutions. This is mainly due to the attractive pricing options offered by cloud providers and availability of flexible computing resources [3]. However, as the cloud resources are typically shared by the general public, ensuing privacy and security concerns prevent the migrating organizations from taking full advantage of the inherent flexibility and economic viability of the cloud platform [4].

Obfuscation is one approach that holds promise to protect data stored in the cloud. Such algorithms can be divided into two classes; the first represents masking algorithms and the second cryptographic techniques. Most masking algorithms are straightforward bijective transformations that can be leveraged for data analytics. These transformations can be easily deduced by opposing parties which implies that the original data can be uncovered. In contrast, encryption algorithms utilize more sophisticated cryptographic building blocks to protect data from unwanted access and eavesdropping. However, due to the rigid nature of crypto-systems, most encryption techniques do not support data analytics in the cipher domain. Cryptographic systems that support mathematical operations executed on cipher-text are typically referred to as *homomorphic*. Existing homomorphic encryption algorithms in the literature either do not support a wide range of arithmetic operations necessary for data analytics or do not scale well due to significant communication and computational overheads [5].

In this paper, we propose a novel departure from both masking and encryption techniques as we seek to strike a balance between supporting efficient analytics on large data sets and comprehensively addressing privacy/security issues. Specifically, our contribution in this paper is fourfold: we (1) propose a lightweight privacy homomorphism technique that supports a wide range of arithmetic operations; (2) compare the proposed algorithm with state-of-the-art solutions; (3) demonstrate how our proposal can be securely applied to conduct three specific data analytic operations: compute statistical measures, perform regression analysis and conduct data mining operations such as pattern recognition; and (4) apply the proposed technique to practical and realistic smart grid meter data. The reader should note that we do not propose an encryption technique. Instead we assert that the value of individual consumer data in the cloud warrants a more balanced solution favoring efficiency and compatibility with data analytics.

Our paper is organized as follows. Section 2 presents current challenges in the cloud and existing encryption technologies. The system model and notations used in this paper are introduced in Sect. 3. Details on the proposed algorithm are provided in Sect. 4. Secure data analytic computations via the proposed algorithm are demonstrated on practical smart meter data in Sect. 5. Security analysis is presented in Sect. 6 performance comparison with the existing literature is presented in Sect. 7. Finally, we conclude in Sect. 8.

2 Computational and Security Challenges

In this section, we introduce the main motivating factors behind the proposed work and discuss state-of-the-art approaches in the field.

Data generated by cyber-physical devices can reveal highly personal information about consumers and/or organizations. For instance, in the power grid, load profiles obtained from smart meter data can reveal sensitive information that includes the number of occupants in a home at a specific time period and the state of various appliances during the day [6]. Although there exist standards such as *Green Button* [7] in Ontario and parts of the United States that require the removal of personally identifiable meta information from data prior to storage, outlier data set(s) can still provide knowledge of a single or small class of unique consumers with atypical power demand characteristics. Personal information gleaned by adversaries can then be used for a variety of nefarious activities including to perpetrate insidious cyber or physical attacks on targeted consumers. Hence, privacy and security are key aspects that must be preserved in these data sets.

Two main requirements of system operators seeking for a solution to manage large and sensitive data sets are the support for secure storage/processing and a wide range of efficient data analytics. These dual stipulations present a challenging yet interesting research opportunity that we address in this paper. We propose a big data solution that can be readily integrated with distributed computing technologies such as Hadoop that allows system operators to leverage the abundance of computing resources available in the cloud to gain insights from their data while also ensuring privacy and security via a novel lightweight privacy homomorphic obfuscation technique.

2.1 Privacy Issues in the Cloud

The cloud promises a broad spectrum of data management features that can be leveraged by system operators to efficiently manage large data sets. One major deterrent in realizing these extended functionalities is the associated privacy and security challenges. Cloud solutions can be *private* or *public* where the former is composed of dedicated servers hosting the cloud service in a closed private environment while the later consists of servers that are publicly shared.

A private cloud is commonly more costly than its public counterpart and therefore is not typically the most economically viable option. The public cloud, albeit less expensive, is subject to privacy issues stemming from the sharing of physical resources with external public entities. Here, the public cloud can host multiple cloud customers on a single physical server. Virtual machines divide the local computational and storage resources amongst the various tenants of the server. The virtual machine and associated software allocated to each tenant must be patched regularly; otherwise, vulnerabilities can be exploited by malicious tenants to gain unauthorized access to data owned by other tenants. The degree of freedom available to secure the software provisioned by the cloud depends on the service purchased by the customer.

It is well known that three different types of public cloud services are offered: Infrastructure as a Service (IaaS), Platform as a Service (PaaS) and Software as a Service (SaaS) [8]. The granularity of control available to customers for managing the underlying cloud resources decreases across the aforementioned services. The greater the maintenance responsibilities, the greater will be the requirement for additional technical support at the customer's end leading to additional costs. Hence, even though IaaS allows customers to manage resources at the VM level and apply necessary patches if necessary, it requires expensive maintenance. PaaS or SaaS are the more popular choices. With these options, however, cloud providers are not transparent and do not provide system maintenance logs due to the risk of exposing the internal cloud architecture. Hence, cloud tenants have no way of verifying the degree of security of the underlying system.

In addition to the challenges introduced due to multi-tenancy, the physical location of the cloud servers poses another privacy challenge [9]. It is entirely possible for physical servers hosting parts of the cloud platform to reside in foreign locales. Hence, privacy laws native to these jurisdictions apply to any material stored locally. These laws can override laws local to the nation from which the cloud tenant is based in. A law that maybe unconstitutional in one country can be legal in another. Thus, it is entirely possible for governments to gain access to all material stored in servers residing within their territories. As a cloud tenant cannot control the location of storage of their data, significant privacy risks due to these mandates do exist.

2.2 Analytics with Existing Encryption Techniques

In order to overcome the aforementioned issues inherent in cloud solutions in an economically viable manner, one obvious solution is to apply encryption to data prior to storage in the cloud to prevent access by unauthorized parties. This, however, presents a problem with conducting data analytics in an efficient manner. When typical encryption algorithms such as Advanced Encryption Standard (AES) are utilized for securing cloud stored data, encrypted data must be converted back into plaintext prior to applying any mathematical operations necessary for the analytics [10]. If data is decrypted within the cloud environment, then the possibility of privacy or security infringement remains. This is counterproductive as the original purpose of proposing the application of this encryption has been to alleviate this very risk. Another option would be to transport the encrypted data to the local site of the system operator, decrypt the data and then perform the analytics. For this, significant communication overhead will result and the system operator will need extensive computational facilities. This is also ineffective as the very purpose of leveraging the cloud is to minimize resource requirements.

These observations imply that if it is possible to conduct data analytics directly in the cipher domain, then system operators will be able to fully harness available cloud storage and processing resources while overcoming existing privacy/security challenges. For this reason, we consider homomorphic encryption algorithms as these support mathematical operations on cipher-text. Existing popular homomorphic algorithms include Fully Homomorphic Encryption (FHE), Paillier Encryption (PE), Rivest Shamir and Adleman (RSA) to name a few. FHE was first proposed by Gentry in 2009 and supports a wide suite of arithmetic operations on cipher-text [11]. As significant computational overhead is incurred by FHE for most of these operations, scalability of this algorithm to large data sets is a major concern. PE is a lighter version of FHE as it supports a limited number of homomorphic operations such as addition and constant multiplication with significantly better computational performance [12]. An extension of PE is proposed in reference [5] in which native mathematical operations supported by PE are utilized to enable common operations important in data analytics such as the multiplication of two cipher-texts. This proposal however incurs significant communication overhead. RSA is an encryption algorithm proposed in 1978 by Rivest et al. [13] and is still widely in use today. Although RSA inherently supports homomorphic multiplication of cipher-texts, many more mathematical functions such as addition in the encrypted domain are required for secure data analytics.

From the discussion above, it is clear that although stringent security features promote the need for strong encryption, the associated price is computational performance and/or limited/no support for mathematical operations necessary for data analytics. Thus, it is necessary to strike a balance between security and performance. Other techniques presented in the literature that are lightweight and targeted towards practical applications generally involve the addition of noise to polynomials or multivariable equations to construct the cipher-text on which homomorphic operations are conducted. Reference [14] is based on discrete logarithms and factoring. This proposal imposes significant space requirements for homomorphic additions. References [15, 16] support limited number of homomorphic multiplications and unlimited additions on cipher-text. Reference [17] utilizes secret matrices and the addition of random noise for encryption and homomorphic operations.

Our proposal differs significantly from the aforementioned techniques in the literature as our scheme involves significantly more lightweight computations (i.e. dot product multiplication) for secure masking and unmasking. All homomorphic operations required for data analytics are supported by our proposal.

3 System Framework

This section introduces the notations and system model used throughout the paper.

3.1 Notation

In general, plain-text message m is transformed into masked data c using a transformation function ϕ and a key k_e:

$$c = \phi(m, k_e).$$

The original message is recovered from the masked data using an unmasking function ϕ^{-1} and a key k_d:

$$m = \phi^{-1}(c, k_d).$$

If the same key is used for masking and unmasking, this is referred to as a symmetric key technique. Our scheme is analogous to a symmetric masking technique. Thus for notational simplicity, both the masking and unmasking keys are referred to as k where $k = k_e = k_d$. It is assumed that the key k is disclosed only the sender and parties with legitimate access to the information and provides some level of privacy commiserate with the security level of the masking approach and protocol as we discuss later in the paper.

Privacy homomorphism allows a set of mathematical operations $\{f_1(.) \ldots f_n(.)\}$ to be executed in the transformed domain of the masked text. Unmasking the resulting value will preserve in plain-text the operation that was initially applied in the transformed domain as follows:

$$f(m_1, \ldots, m_n) = \phi^{-1}(f(\phi(m_1, k), \ldots, \phi(m_n, k)), k)$$

where m_i is the ith plain-text message to which the operation is applied to. Fundamental homomorphic operations $f(.)$ that must be supported by the proposed scheme are addition and multiplication. We assert that a large class of analytics functions can be derived from these two basic operations.

3.2 Architecture and Components

In this paper, we propose a mechanism for secure light-weight data analytics on large data sets generated in critical infrastructures. Although this proposal can be applied in general to any applications requiring secure analytics, in this paper, we

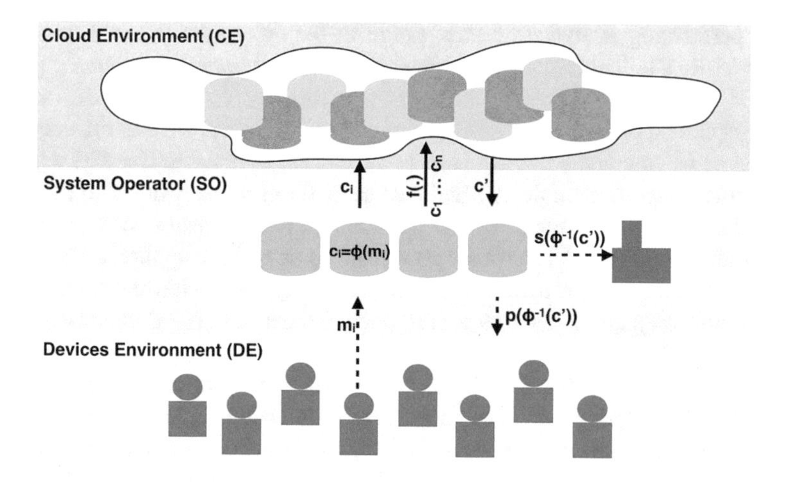

Fig. 1 Information exchange processes between participants

focus on the smart grid setting for demonstration. In the following, system settings and the nature of information exchanged are summarized. In order to aid with this description, an illustration of the associated communication and data transfer processes is included in Fig. 1.

Interactions occur between three main environments. Data is originally generated from the Devices Environment (DE); smart meters located at the residences of electricity consumers and Phasor Measurement Units (PMUs) are examples of measurement devices within the DE. The data from the DE is transferred to the System Operator (SO) environment by the data generating units. Smart meters and PMUs transfer data to the Electric Power Utility (EPU) which is the SO via communication channels available in the Advanced Metering Infrastructure (AMI) [2]. Once this information reaches the EPU, pre-processing maybe applied to this data in order to remove Personally Identifiable Information (PII) in accordance to existing standards such as the Green Button [7]. The resulting meter value represents a numerical data unit $m_i \in \mathbb{R}$. Masking function ϕ is applied to the plain-text message m_i to convert it into c_i, which is then transmitted to the Cloud Environment (CE) that hosts the infrastructure leased by the SO to store and process data generated in the DE; data is masked by the SO prior to transfer to the CE as transmission can occur over insecure communication links.

The CE stores the transmitted masked data within its hosting infrastructure. As the CE is a public entity, it is entirely possible that the physical server within which the data is stored is shared by multiple cloud clients. Moreover, the physical geographical location of the cloud server hosting the data can be a foreign jurisdiction in which local privacy laws are applicable to the data. As the data is stored in secure masked form, it will not be subject to privacy issues stemming from data leakage due to multi-tenancy or foreign data access laws.

For performing analytics on data stored in the cloud, the SO will first send a request to the CE. This request contains information about the operation $f(.)$ to be executed by the CE and the specifics of the data set $c_1 \dots c_n$ on which this operation is to be executed on. The SE may transmit multiple requests such as this depending on the type of data analytics that is to be conducted by the SE. The CE will apply the specified operation on the masked data set and return the result c' to the SO. As c' is still in transformed form, the SO will unmask it in its secure local premises by applying the function ϕ^{-1}. Further operations $s(.)$ or $p(.)$ are applied to the result as necessary and transmitted to either the consumers (e.g. to provide real-time feedback on consumption patterns with respect to one another) or to the grid components as a control signal.

It is important to note that in the CE, consumer data is always in masked form since our model leverages the capabilities available in the cloud without compromising consumer privacy or infrastructure security. All remaining operations necessary for data analytics are performed on the resulting data transmitted by the CE in secure SO premises. This will not require extensive computational or memory resources as the data set that the SE will operating on is now much smaller.

3.3 *Threat Model*

Based on the architecture and entities contained within the system discussed in the above, the following is a summary of the threat model utilized in this paper. The main goal of this paper is to protect the original data transmitted by devices in the DE. Hence, an adversary should not be able to extract the original content from the masked data stored in the CE. The following are specific assumptions made with respect to this:

1. The DE is secure and all communication channels established between the devices and the system operator is contained and not susceptible to intrusion;
2. Masked data transferred between the SO and the CE cannot be altered so that the integrity of the data is preserved;
3. Masked data stored in the CE cannot be altered but maybe susceptible to leakage; and
4. The CE provider is not adversarial.

Assumption 1 is important as data exchanges in the DE occur in the original form. However, as utilities abide by system security requirements as detailed in reference [18], this is a reasonable assumption. Assumption 2 ensures that the masked data is not modified in transit and this can be achieved by incorporating non-intrusive hashing technologies that can detect tampering during these transfers. Assumption 3 identifies the privacy-preservation problem addressed in this paper. The final assumption is practical as any CE provider will strive to preserve self-reputation and therefore will not attempt to corrupt the services provisioned.

The proposed masking technique is not designed to thwart attacks that corrupt the integrity of the data via false data injection and availability of the data. The main focus of this paper is preserving the confidentiality of infrastructure (smart grid) data.

4 Random-Base Privacy Homomorphism

With the preliminaries presented in the previous section, we next introduce the Random-Base (RB) technique for enabling privacy homomorphism on aggregated sensitive data. A homomorphic masking scheme is completely defined by the masking and unmasking functions and supported homomorphic mathematical operators: $\phi, \phi^{-1}, \{f_1(.) \dots f_n(.)\}$; which we present next in detail.

4.1 *Masking Function*

The RB technique is premised on a concept that is similar to the radix representation of numbers. For instance, any real number m can be represented as the following expansion in terms of the *base vector* $\mathbf{s}_b$.

$$m = d_{n-1}s_{n-1} + d_{n-1}s_{n-2} + \dots + d_1 s_1 + d_0 s_0$$

$$m = \mathbf{d}^T \cdot \mathbf{s}_b$$

Unlike standard representation of positional numerical systems, $\mathbf{s}_b = [s_{n-1} \dots s_i \dots s_0]^T$ consists of n randomly selected components where $s_i \in \mathbb{R}/0$. Due to the random nature of individual elements in $\mathbf{s}_b$, this is referred to as the *random base* vector. The coefficients $[d_{n-1} \dots d_0]^T$ in $\mathbf{d}$ represent the expansion of m with respect to the random base vector $\mathbf{s}_b$ and is referred to as the *digit* vector. For a given base $\mathbf{s}_b$, any real-valued m can be specified in terms of the digit vector $\mathbf{d}$. The digit vector for the RB method is computed as outlined in Steps 1–4 of Algorithm 1. This process of obtaining the digit vector can be defined as a mapping R that takes inputs m and $\mathbf{s}_b$ and outputs $\mathbf{d} \in \mathbb{R}^n$ (i.e. $R : (m, \mathbf{s}_b) \rightarrow \mathbf{d}$).

As an example, consider the message $m = 145.96$ and the random base $\mathbf{s}_b = [8.0028\ 1.4189\ 4.2176]^T$ where $n = 3$. The corresponding expansion of m in terms of $\mathbf{s}_b$ is:

$$145.96 = [6.0795\ 34.2904\ 11.5358].\mathbf{s}_b$$

where $\mathbf{d} = [6.0795\ 34.2904\ 11.5358]^T$ is obtained by applying Steps 1–4 in Algorithm 1. The expansion of m in terms of $\mathbf{s}_b$ can be subject to rounding errors due to the floating-point precision supported by the machine performing the

Algorithm 1 Computation of digit vector for RB

Input: s, m
Output: c
1: $len \leftarrow$ dimension of **s**
2: **for all** s_i in **s do**
3: append $m/(len * s_i)$ to d
4: **end for**
5: Generate $n-1$ random numbers $\{r_{n-1} \dots r_0\}$ from $[l, u]$
6: $v_1 \leftarrow [r_{n-1} \ \dots \ r_2 \ r_1 \ r_0]^T$
7: $v_2 \leftarrow [\frac{-s_0}{s_{n-1}} r_0 \ \frac{-s_{n-1}}{s_{n-2}} r_{n-1} \ \dots \ \frac{-s_2}{s_1} r_2 \ \frac{-s_1}{s_0} r_1]^T$
8: $\mathbf{z} \leftarrow v_1 + v_2$
9: $\hat{\mathbf{d}} \leftarrow \mathbf{d} + \mathbf{z}$

computation. However, as discussed in Sect. 7, these inaccuracies are minor and do not detract from the main message conveyed by the data analytics. Next, it is important to note that the expansion of m in terms of **s** is not unique as there exist vectors **z** that satisfy the following relation:

$$0 = \mathbf{z}^T \cdot \mathbf{s}_b \tag{1}$$

We refer to **z** as the *zero-sum* vector. Suppose Steps 1–4 in Algorithm 1 are used to compute the digit vector **d** for m. As $m = (\mathbf{d} + \mathbf{z})^T \cdot \mathbf{s}_b$, the vector $\hat{\mathbf{d}} = \mathbf{d} + \mathbf{z}$ is also a digit vector and this implies that digit vectors are not unique. This one-to-many mapping allows us to introduce significant variation in the representation of m in terms of **d**. For systematically computing **z** of length n (i.e. $n = \dim(\mathbf{s}_b)$), two vectors v_1 and v_2 are defined as follows:

$$v_1 = [r_{n-1} \ \dots r_i \dots \ r_2 \ r_1 \ r_0]^T$$

$$v_2 = [\frac{-s_0}{s_{n-1}} r_0 \ \frac{-s_{n-1}}{s_{n-2}} r_{n-1} \ \dots \ \frac{-s_2}{s_1} r_2 \ \frac{-s_1}{s_0} r_1]^T$$

$$\mathbf{z} = v_1 + v_2$$

where $r_i \in [l, u]$ is a random number selected from an interval in $\mathbb{R}$ defined by the lower bound l and upper bound u. We define the zero-sum vector using v_1 and v_2 as follows:

$$\mathbf{z} = v_1 + v_2$$

It is now straightforward to verify that the **z** defined above satisfies the relation in Eq. (1). This process of obtaining the zero-vector can also be defined as a map $Z : \mathbf{s}_b \to z$.

If we consider $\mathbf{s}_b$ to be the symmetric key and $\hat{\mathbf{d}} = \mathbf{d} + \mathbf{z}$ to be the masked data, even with the variability incorporated into $\hat{\mathbf{d}}$ via the zero vector, an adversary will be able to identify the key $\mathbf{s}_b$ if he or she is aware of n plaintext messages m_i

where $i \in n$ and the corresponding masked values $\hat{\mathbf{d}}_i$ through simple linear algebraic manipulations outlined below:

$$\mathbf{s}_b = \begin{bmatrix} \hat{\mathbf{d}}_1^T \\ \vdots \\ \hat{\mathbf{d}}_n^T \end{bmatrix}^{-1} \begin{bmatrix} m_1 \\ \vdots \\ m_n \end{bmatrix} \tag{2}$$

In order to avoid this issue, the size of the key n can be made very large so that the adversary will need access to more plaintext and masked data pairs to identify the key. However, this will lead to significant resource overheads with the possibility of key identification remaining at large. Another option will be to amplify the original message m_i and add a random offset as follows:

$$\hat{m}_i = am_i + x_i \tag{3}$$

where a is the amplification constant that obscures the range of values taken by the original plaintext message and $x_i \in \mathbb{R}$ is the random offset drawn from the Normal distribution $\mathcal{N}(\mu, \sigma^2)$ defined by the mean μ and variance σ^2 which eliminates the direct correspondence between the original plaintext message and the masked data point. The adversary will not be aware of a, μ and σ^2 which are secret parameters that add the random effect to the original data. Comprehensive security analysis conducted in Sect. 6 highlight the effectiveness of this technique in overcoming the afore-mentioned risk of key exposure. The SO, on the other hand, will be able to readily recover aggregate values via statistical manipulations and known masking parameters.

We define the masking function ϕ that transforms the plaintext message m_i to the masked form c_i as follows:

$$c_i = \phi(m_i, \mathbf{s}_b, a, \mu, \sigma^2) = Z(\mathbf{s}_b) + R(am_i + x_i, \mathbf{s}_b)$$

Thus, the key k is composed of $\{\mathbf{s}_b, a, \mu, \sigma^2\}$ and $c_i \in \mathbb{R}^n$ represents the masked data associated with the random offset x_i drawn from $\mathcal{N}(\mu, \sigma^2)$. As the masked data is essentially a linear combination based on simple dot product operations, statistical operations can be performed homomorphically in the masked domain as outlined next.

4.2 Aggregation Homomorphism and Unmasking Functions

The SO may wish to perform aggregation operations such as computing summation, averages, variances and correlations of various data sets stored in the cloud environment. These fundamental statistical building blocks are derived from addition, subtraction and multiplication operations executed in the masked domain. These operations are relatively straightforward as masking and unmasking involve dot products and other linear operations. The main premise of the proposed masking technique is that these aggregation operations will eliminate the stochastic effects of the random offsets which are highly effective in protecting individual data points. As such, in the following, a detailed overview of how each one of the afore-mentioned statistical operations can be executed in a homomorphic manner in the masked domain along with the corresponding unmasking function that can be employed by the SO to recover the computed data is presented.

The first aggregation operation is the **summation** of plaintext messages in the set $\mathcal{M}$: $f_{add}(\mathcal{M}) = \sum_{i=1}^{|\mathcal{M}|} m_i$. Let C be the set of masked data corresponding to $\mathcal{M}$. The CE will perform the homomorphic aggregation operation in the cloud: $c_a = f_{add}(C) = \sum_{i=1}^{|C|} c_i$. The SO will recover the summation value in plaintext via the unmasking function ϕ_{add}^{-1}:

$$\phi_{add}^{-1}(c_a, \mathbf{s}_b, a, \mu) = \frac{q}{a}[\mathbf{s}_b^T.\frac{c_a}{q} - \mu]$$

where $q = |\mathcal{M}|$. Direct substitution of c_a into the above relation results in the recovery of the summation of the original plaintext messages in $\mathcal{M}$ according to the following:

$$\frac{q}{a}[\mathbf{s}_b^T.\frac{\sum_{i=1}^{q} c_i}{q} - \mu] = \frac{q}{a}[\frac{\sum_{i=1}^{q}(am_i + x_i)}{q} - \mu]$$
$$= \frac{q}{a}[\frac{\sum_{i=1}^{q} am_i}{q} + \frac{\sum_{i=1}^{q} x_i}{q} - \mu] = \sum_{i=1}^{q} m_i$$

The last relation in the above derivation is based on the premise that the addition is performed on a *large data set* which implies that $\lim_{q\to\infty} \sum_{i=1}^{q} x_i/q = \mu$ from the strong law of large numbers. The resulting cancellations will lead to $\sum_{i=1}^{q} m_i$. Thus, the addition operation conducted in the masked domain for big data is homomorphic.

The next operation detailed is the **averaging** of data points in the set $\mathcal{M}$: $\bar{m} = f_{avg}(\mathcal{M}) = \sum_{i=1}^{q} m_i/q$. The CE will perform this operation in the masked domain $\bar{c} = f_{avg}(C) = \frac{\sum_{i=1}^{q} c_i}{q}$. The SO will recover the averaged value $\bar{m}$ using ϕ_{avg}^{-1}:

$$\phi_{avg}^{-1}(\bar{c}, \mathbf{s}_b, a, \mu) = \frac{1}{a}[\mathbf{s}_b^T.\bar{c} - \mu]$$

It can be verified that ϕ_{avg}^{-1} produces $\bar{m}$ in a manner similar to the summation derivation presented in the above.

The next operation considered is the computation of sample **variance** of data points in the set $\mathcal{M}$: $\sigma_M^2 = f_\sigma(\mathcal{M}) = \sum_{i=1}^{q}(m_i - \bar{m})^2/q$. The CE computes the variance in the masked domain: $c_\sigma = f_\sigma^c(C) = \frac{1}{q}\sum_{i=1}^{q}(c_i - \bar{c})(c_i - \bar{c})^T$. It is important to note that c_σ is a $n \times n$ matrix. The SO will unmask c_σ to obtain the plaintext variance σ_M^2 using ϕ_σ^{-1}:

$$\phi_\sigma^{-1}(c_\sigma, \mathbf{s}_b, a, \sigma^2) = \frac{1}{a^2}(\mathbf{s}_b^T c_\sigma \mathbf{s}_b - \sigma^2)$$

It can be verified that ϕ_σ^{-1} results in σ_m^2 by considering the first term in the definition of ϕ_σ^{-1}:

$$\begin{aligned}
\mathbf{s}_b^T c_\sigma \mathbf{s}_b &= \frac{1}{q}\sum_{i=1}^{q} \mathbf{s}_b^T (c_i - \bar{c})(c_i - \bar{c})^T \mathbf{s}_b \\
&= \frac{1}{q}\sum_{i=1}^{q}[am_i + x_i - (a\bar{m} + \mu)]^2 \\
&= \frac{1}{q}\sum_{i=1}^{q}[a(m_i - \bar{m}) + (x_i - \mu)]^2 \\
&= \frac{1}{q}\sum_{i=1}^{q}[a^2(m_i - \bar{m})^2 + 2a(m_i - \bar{m})(x_i - \mu) + (x_i - \mu)^2]
\end{aligned}$$

In the last relation in the above derivation, the second term represents the correlation between the plaintext messages and the random offsets. As the random offsets are independent from the plaintext messages, this term is 0. The third term represents the variance of the random offsets which is σ^2 and this is known to the SO. Hence, it is clear that subtracting σ^2 and dividing the resulting expression by a^2 results in the sample variance σ_M^2 of data set $\mathcal{M}$.

Similarly, the computation of **covariance** amongst two sets of data points $\mathcal{M}^x$ and $\mathcal{M}^y$ in the masked domain C^x and C^y where $|\mathcal{M}^x| = |\mathcal{M}^y| = q$ is: $c_{cov} = f_{cov}(C^x, C^y) = \frac{1}{q}\sum_{i=1}^{q}(c_i^x - \bar{c}^x)(c_i^y - \bar{c}^y)^T$.

$$\phi_{cov}^{-1}(c_{cov}, \mathbf{s}_b, a, \sigma^2) = \frac{1}{a^2}(\mathbf{s}_b^T c_{cov} \mathbf{s}_b - \sigma^2)$$

Once again, the recovery of the covariance of the plaintext data sets can be verified using the notion of independence of the random offsets and the plaintext messages.

Finally, the **correlation** amongst two sets of data points $\mathcal{M}^x$ and $\mathcal{M}^y$ can be computed by the SO using c_{cov}, c_σ^x and c_σ^y which are masked values representing the covariance amongst the two data sets and variances of each data set respectively:

$$\phi_\rho^{-1}(c_{cov}, c_\sigma^x, c_\sigma^y) = \frac{\phi_{cov}^{-1}(c_{cov}, \mathbf{s}_b, a, \sigma^2)}{\sqrt{\phi_\sigma^{-1}(c_\sigma^x, \mathbf{s}_b, a, \sigma^2)}\sqrt{\phi_\sigma^{-1}(c_\sigma^y, \mathbf{s}_b, a, \sigma^2)}}$$

As per the formal definition of the correlation metric.

These statistical operations serve as fundamental building blocks for advanced machine learning and analytics operations discussed in the following section.

5 Data Analytics in the Masked Domain

Data analytics can provide deep insights that are extremely useful for SOs. A few use cases include: enhanced intelligence guiding development of new strategic business ventures (business intelligence), statistical information providing concise description of the data set, regression analysis allowing for prediction or forecasting of future events and data mining where useful patterns in large data sets can be inferred. In this section, we utilize the statistical homomorphic functions supported by the RB technique in the context of smart meter data generated in the power grid to define three common data analytics operations: (1) computation of confidence intervals; (2) classification/clustering; and (3) regression analysis. Although the proposal is demonstrated specifically in this section for smart meter data, it is applicable in general to any application.

The smart meter data set used in the results presented in this section consists of power consumption readings sampled every 15 min for 100 homes over a period of 30 days from load profiles generated using models presented in [19] for North American homes. In this section, the ability to perform data analytics using the proposed RB technique is evaluated and hence the results obtained here can be extrapolated to much larger data sets. Results regarding the efficiency and speed of the proposed masking technique in comparison to conventional homomorphic encryption algorithms for performing computations on large masked data sets is presented later in Sect. 7.

5.1 *Confidence Intervals*

Confidence intervals are utilized commonly to derive the interval in which an unknown parameter of a population such as the *mean* can lie with a particular level of confidence. In the context of smart meter data set, aggregated load profiles provide important information pertaining to overall costs of power demands that allow the EPUs to plan purchasing of power generated from diverse cost-effective sources at an advanced time frame. Computing the confidence interval of the average aggregated power consumption over various seasons and diurnal intervals will allow for proactive planning by the EPU. The computation of confidence intervals for the smart meter data set $\mathcal{M}$ according to the following entails the use of $\bar{c} = f_{avg}(C)$ and $c_\sigma = f_\sigma(C)$ in the masked domain and corresponding unmasking functions $\bar{m} = \phi_{avg}^{-1}(\bar{c}, \mathbf{s}_b, a, \mu)$ and $\sigma_d^2 = \phi_\sigma^{-1}(c_\sigma, \mathbf{s}_b, a, \sigma^2)$:

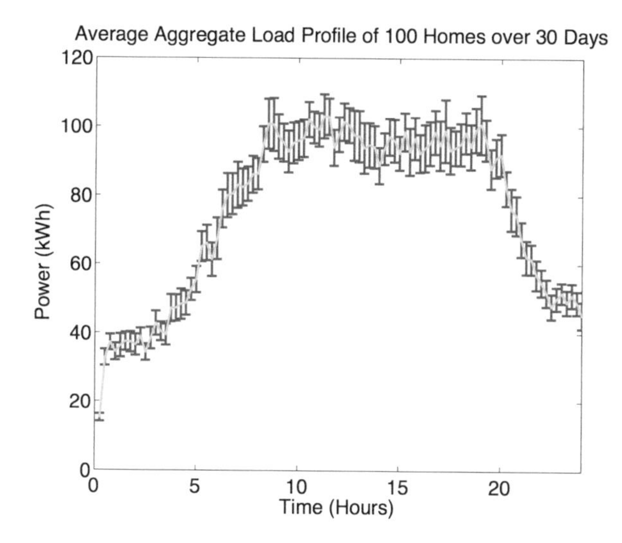

Fig. 2 Confidence interval computed on masked data

$$(\bar{m} - z^* \frac{\sigma_d}{\sqrt{q}}, \bar{m} + z^* \frac{\sigma_d}{\sqrt{q}})$$

where $z^* = 1.96$ for attaining 95% confidence in a Normal distribution.

Figure 2 illustrates the use of the afore-mentioned statistical homomorphic functions and the corresponding unmasking functions in a practical setting consisting of a smart meter dataset. At every sampling period (i.e. every 15 min), the smart meter data accrued for 100 homes is aggregated. Then, the sample mean and variance of the resulting aggregated load profile is computed over a 30 day period. This is used to calculate 95% confidence interval at every sampled period of the aggregated load profiles. Figure 2 is the outcome of computations performed as outlined above in the masked domain. Errors due to floating point accuracy issues are negligible as discussed in Sect. 7. This demonstration shows that we are able to adequately describe the data set by providing confidence intervals in which the data samples are guaranteed to lie with a probability of 95% from homomorphic computations.

5.2 Clustering and Classification

Clustering and classification are integral operations in data mining and pattern recognition applications. Data samples are distinguished from one another via a set of measurable/observable features and these features are used to group data points into various categories or classes. In supervised learning, classes are identified during the training period. Then, classification algorithms are invoked to group new data points into associated classes based on the features of the initial training data set. There are many classification algorithms which are based on probabilistic measures (Bayesian classifiers) or distance metrics.

We show that our proposed homomorphic masking technique allows data mining to be performed directly in the masked domain. Mean and variance measures are used significantly in probabilistic algorithms such as the Bayesian classifiers which can be computed directly in the masked domain. We demonstrate in this section another type of classifier that can be constructed via homomorphic operations called the k-Means algorithm that utilizes the Euclidean distance measure between means of feature classes for clustering purposes [20].

The iterative k-means algorithm presented in the following groups a data point vector m with u features (i.e. $m \in \mathbb{R}^u$) into the cluster k whose mean $\bar{m}_k$ is the closest to this new data point. This measure of closeness is computed using the Euclidean distance:

$$d_i^k = \sqrt{(m - \bar{m}_k)^T (m - \bar{m}_k)}$$

Then, the mean of the cluster is updated to account for this new point. The CO computes the mean of every cluster by applying the averaging homomorphic function f_{avg} to every component of the data vectors forming each cluster. This results in the cluster average $\bar{\mathbf{c}}_k$ which is a $n \times u$ matrix. For simplicity of notation, we present the clustering process of data points consisting of a single feature (i.e. one-dimensional). This can be easily extended to multi-dimensional feature vectors by applying the following computations in a component-wise manner for each feature.

As such, when a new data point m_i is introduced, the SO generates p variations of the message m_i in the masked domain where in each c_i^j a different offset x_i^j drawn from the distribution $\mathcal{N}(\mu, \sigma)$ is applied to the original message. These masked data points form the set C_i. The Euclidean distance of this point from each cluster mean is computed as follows:

$$c_{dist} = f_{dist}(C_i, \bar{c}_k) = \frac{1}{p} \sum_{i=1}^{p} (c_i - \bar{c}_k)(c_i - \bar{c}_k)^T$$

It is evident that the above is identical to the computation of variance as outlined in the previous section. Thus, for unmasking c_{dist}, the function ϕ_{dist}^{-1} is applied to obtain d_i^k:

$$\phi_{dist}^{-1}(c_{dist}, \mathbf{s}_b, \sigma^2) = \sqrt{\frac{1}{a^2}[\mathbf{s}_b^T c_{dist} \mathbf{s}_b - \sigma^2]}$$

This process is repeated until the distances of the new point from all cluster means are computed. Then, the SO assigns the new data point to the cluster k that satisfies the following:

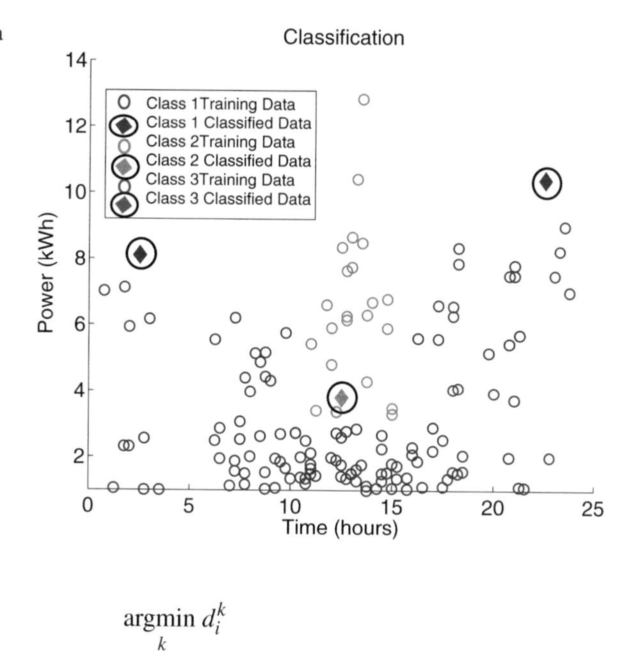

Fig. 3 Classification of data in the masked domain

$$\underset{k}{\operatorname{argmin}}\ d_i^k$$

Then, the corresponding cluster mean is updated by the CE adding the new point m_i to cluster k in the masked domain and evoking f_{avg}.

As an illustration of the above process, the k-means clustering algorithm is applied in the masked domain to the smart meter data set depicted in Fig. 3. Each data point is associated with two features (i.e. $x \in \mathbb{R}^2$) and these are the power demand and the time at which the demand is recorded. Three classes are assigned and points in a subset of this data are manually assigned to one of these classes as depicted in Fig. 3. The trained data points are represented by the circle markers and each class is assigned a different colour label. Three new data points are introduced for classification and these are represented by the filled diamond markers in Fig. 3. The colours assigned to these new data points depict the classes that these points are assigned to via the k-means algorithm operating in the masked domain. Visual inspection ascertains that these labels are indeed correctly assigned. Hence, this example is an illustration of the many complex data mining operations that can be performed on masked data set via our proposed algorithm.

5.3 *Regression Analysis*

Regression analysis can reveal dependencies and relationships between one or more variables in a data set [21]. Suppose that there are p independent variables $x \in \mathbb{R}^p$ on which the measurement $y \in \mathbb{R}$ has some dependencies or correlations. These dependencies can be captured by a curve represented by the function $f(.)$. More specifically, $f(.)$ can be defined to be the curve with the best fit to the available data samples (i.e. measurements y and associate states of the independent variable x). Then, this function can be used to interpolate or extrapolate the behaviour of the dependent variable based on the values taken by x. This is referred to as linear regression when $f(.)$ is linear with the following general form:

$$y = f(x) = \alpha + x \cdot \beta = [\alpha \ \beta] \cdot [1 \ x^T]^T$$

where $\alpha \in \mathbb{R}$ is a bias term and $\beta \in \mathbb{R}^p$ constitutes of the coefficients of the independent variables. Given n samples of y and the values taken by the corresponding independent variables x for each sample, it is possible to compute the parameters $\gamma = [\alpha \ \beta]^T$ that result in the best fit (or least error) by minimizing the error of the fitted data as follows:

$$\min_{\gamma} f(\gamma) = (Y - X\gamma)^T (Y - X\gamma)$$

where $Y = [y_1 \dots y_n]^T$ is the vector consisting of the n available measurements and $X = [[1 \ x_1]^T \dots [1 \ x_n]^T]^T$ is a matrix consisting of the associated values $x_i \in \mathbb{R}^p$ of the independent variables for each measurement sample y_i. This minimization is also known as the least squares problem. A closed-form analytical solution allowing for the computation of γ can be obtained via the first-order condition for optimality $\frac{\partial f}{\partial \gamma} = 0$ which leads to $\gamma = (X^T X)^{-1} X^T Y$.

Linear regression analysis is also possible in the masked domain via homomorphic operations supported by the RB technique as discussed next. Each data point x_i is expanded into p points in the masked domain by the SE in a manner similar to the clustering application. Then, the CE will compute $X^T X$ and $X^T Y$ separately in the masked domain using the homomorphic functions f_σ and f_{cov} for each component of these matrices. These masked values are then transmitted to the SO. Hence, to recover the plain-text value, the SO will apply ϕ_σ^{-1} and ϕ_{cov}^{-1}. Once, these values have been recovered, then the SO will proceed to perform the inversion of $X^T X$ locally and multiply the resulting matrix with $X^T Y$. If the matrix $X^T X$ is invertible, then the inversion can also be performed directly in the CE via methods that involve a set of linear transformations such as Gauss-Jordon elimination which allow the SO to recover the inverted matrix in plain-text.

As such, we have applied the linear regression analysis outlined above to investigate how the total energy consumption of a neighbourhood during peak periods vary with the number of homes in neighbourhood. This analysis can serve to be a useful aid for planning of necessary generation capacity in the neighbourhood to sustain ongoing housing development projects. Figure 4 depicts the results of this analysis. The data measurements are denoted by blue circles and these represent the total energy consumption y of x homes. These data points are obtained by randomly sampling n load profiles generated via smart meter measurements. The parameters obtained via the least squares analysis in the masked domain form the linear red

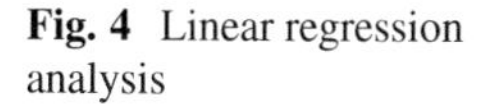

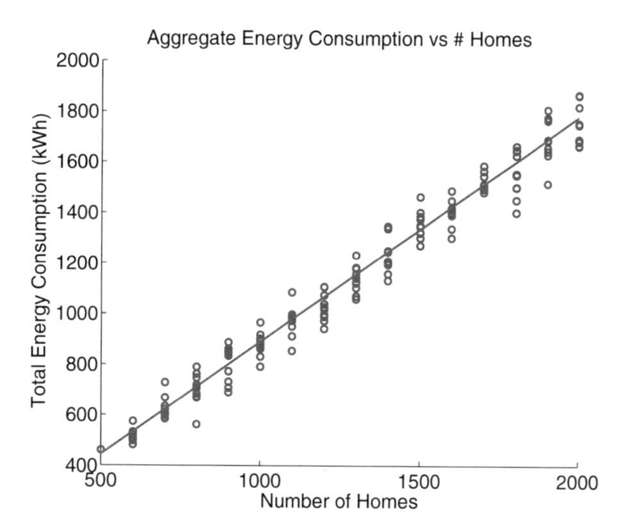

Fig. 4 Linear regression analysis

curve. This line can be extrapolated to predict rise in energy consumption with future increase in homes within the neighbourhood.

5.4 Summary

It is clear from the examples presented earlier of various analytics conducted on large data sets that the homomorphic operations available in the RB technique enables support for a wide range of secure decision-making tools. These provide extremely useful information to SOs that can be used to devise strategies that allow for various operation goals to be met. Moreover, these simplistic computations render easy integration with state-of-the-art database management solutions with minor modifications necessary in the client-server interfaces.

6 Security Analysis

The main goal of the RB masking technique introduced in this paper is to protect individual energy measurements generated by smart meters as these can provide revealing insights into the daily activities of these consumers. We demonstrate in this section that the RB masking technique is indeed successful in obscuring individual energy measurements. We also show that although scaled versions of the masking key parameters can be inferred by gaining access to a large number of plaintext measurements and masked data pairs, these still cannot be utilized to expose individual measurements mainly due to the random offset added to each measurement prior to storage in the CE. As third party entities such as mobile application developers require access to only aggregate values, this exposure is not an issue [7]. However, if the SO wishes to protect the aggregate measures, then it can do so by incorporating hashes for mapping the plaintext and masked data pair. This will eliminate the one-to-one correspondence between the plaintext and masked data pair. As such, in this section security analysis is conducted for two specific scenarios. In the first, brute force search is applied to infer the key $\mathbf{s}_b$ with and without the random offsets when n plaintext and associated masked data pairs are known to the adversary. In the second scenario, we focus on data aggregation and investigate the impact of the size of known plaintext to masked data pairs being aggregated in accurately identifying scaled key parameters and we show that these parameters still do not expose individual consumer measurements.

6.1 Scenario 1: Brute-Force Search

In this case, we consider the key size to be $n = 2$ (i.e. there are two components) for the ease of illustration. It is assumed that the adversary is aware of the plaintext message m and the corresponding masked vector c. First, we consider the case where no amplification or random offset is applied prior to masking. In the brute-force search procedure, the adversary starts with the initial guess of $\mathbf{s}_b^g = (0, 0)$ and then systematically increments the first and second components by $\epsilon = 0.05$. Thus, the total number of key combinations evaluated is 40, 000. The adversary charts the decoded value on the z-axis that results from each one of these guesses by evaluating $c^T.\mathbf{s}_b^g$. When the key is at close proximity to the original key, the decoded guess will be the same as the actual plain-text message. When two plaintext-to-masked data pairs are available, the adversary can simply solve for the key using Eq. (2). This is illustrated in first sub-plot of Fig. 5. Two planes representing the decoded values resulting from the guesses intersect at the actual plaintext value.

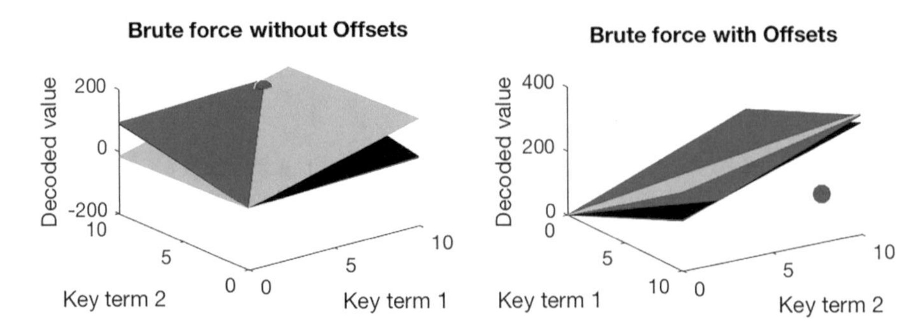

Fig. 5 Brute force search with and without amplification or random offset

When amplification and random offset is applied to the plaintext message prior to masking, the brute force search is not successful in identifying the key as illustrated in second sub-plot in Fig. 5. The original plaintext message is denoted by the red circle in this plot and the decoded guesses are nowhere close to the actual plaintext value mainly due to the random offset applied to the message prior to masking. The amplification embeds a large gap between the actual plaintext message and the decoded guesses. Thus, the amplification and random offset addition process is effective in obscuring individual measurements.

6.2 Scenario 2: Impact of the Number of Known Data Pairs

With the random offset applied to every measurement value stored in the masked domain, it is not possible to identify the original message as the random offset is different for every masked value. However, when aggregation operations are applied in the masked domain, the stochastic effects of the random offsets are eliminated. This principle can be utilized by an adversary to infer attributes of the key. However, the main issue with this is that the adversary will need access to a large number of plaintext and masked data pairs. Suppose that the key size is n, then the adversary will need access to $(n + 1) * N$ data pairs representing $n + 1$ distinct non-zero plaintext messages. Each unique plaintext message must be mapped to N masked vectors. This way, the adversary can execute the following operation to obtain the scaled version of the key $\mathbf{s}_b/a$:

$$\frac{\mathbf{s}_b}{a} = \begin{bmatrix} \lim\limits_{N\to\infty} \frac{\sum_j^N (c_j^1)^T}{N} - \lim\limits_{N\to\infty} \frac{\sum_j^N (c_j^{n+1})^T}{N} \\ \lim\limits_{N\to\infty} \frac{\sum_j^N (c_j^2)^T}{N} - \lim\limits_{N\to\infty} \frac{\sum_j^N (c_j^{n+1})^T}{N} \\ \vdots \\ \lim\limits_{N\to\infty} \frac{\sum_j^N (c_j^n)^T}{N} - \lim\limits_{N\to\infty} \frac{\sum_j^N (c_j^{n+1})^T}{N} \end{bmatrix}^{-1} \begin{bmatrix} m_1 - m_{n+1} \\ m_2 - m_{n+1} \\ \vdots \\ m_n - m_{n+1} \end{bmatrix}$$

The subtraction operation of two averaged masked vectors eliminates the mean μ. The remaining value is the masked version of $a(m_i - m_{n+1})$ without the randomization effect of the offsets in place. With infinite masked data points representing each plaintext, it is possible to exactly recover $\mathbf{s}_b/a$. However, this is not practical. The margin of uncertainty associated with 95% confidence interval is $\pm 1.96\sigma/N$ for finite N. Thus, the uncertainty depends on the variance of the offset σ and the number of known plaintext-mask data pairs available. For example, when $\sigma = 10^3$, N must be 10^3 to obtain a margin of uncertainty of $\approx \pm 2$. Thus, a large number of data pairs for each distinct plaintext is necessary to accurate assimilate the scaled key. In total $(n + 1)N$ points are needed and when the key size is 9, this amounts to 10^4 points for the case where $\sigma = 10^3$. There is still a 5% uncertainty associated with the sample average. Similarly, μ/a can be computed:

$$\frac{\mu}{a} = m_i - \frac{\mathbf{s}_b^T}{a} \lim_{N\to\infty} \frac{\sum_{j=1}^{N} c_j^i}{N}$$

This computation is also subject to the limitations associated with amassing a large number of known data pairs. Moreover, both values recovered are scaled versions of the original key parameters. Thus, the original parameters remain unknown.

6.3 Completely Securing Masked Data

In the analysis presented above, it is clear that individual data points remain obscured even when subject to brute force attacks due to the addition of random offsets. Aggregation operations eliminate the stochastic effects of the random offsets. However, the amplification constant a protects key parameters from being completely exposed. With today's energy data sharing initiatives, allowing a third party entity to compute aggregate measures from consumer data sets while protecting individual measurements is highly desirable [7]. As such, the proposed RB technique is certainly effective in protecting individual data points while allowing access to aggregate metrics with minimal interference to external parties as advocated by current energy data sharing standards and platforms [7]. If the SO wishes to protect aggregate data as well, then, it will need to eliminate the mapping between the plaintext and masked data pairs. For this, the SO can utilize a secure hash for mapping these data pairs so that an adversary will not be aware of the relationship between the plaintext and masked data pairs.

7 Performance of RB Technique

In this section, the RB technique is evaluated in terms of accuracy and performance. Moreover, comparisons of the RB technique against other state-of-the-art techniques in the literature are presented.

7.1 Accuracy Analysis

The symmetric key $\mathbf{s}$ consist of components that can take any real value. Masking involves division by these real-valued components. As these computations are performed on machines that support floating point arithmetic, rounding errors are inevitable. An upper bound on these errors is referred to as the machine epsilon [22]. To reduce these errors, data types that support high precision can be used to

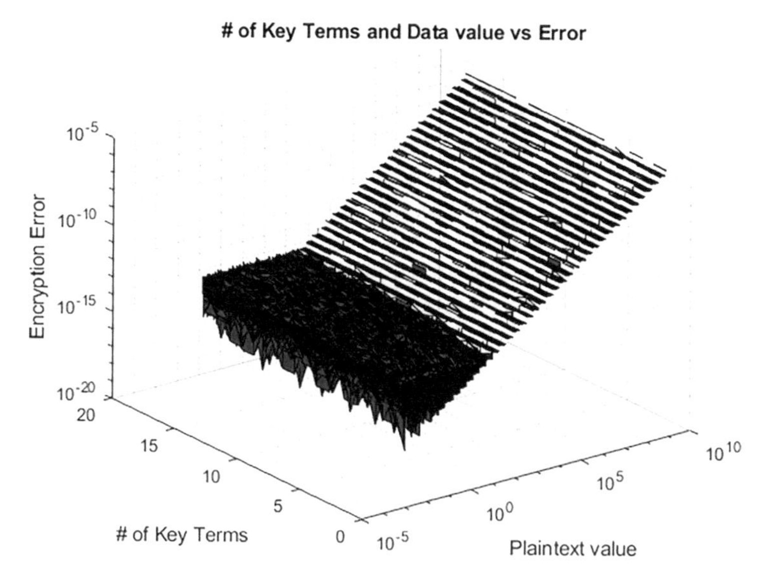

Fig. 6 Masking error analysis

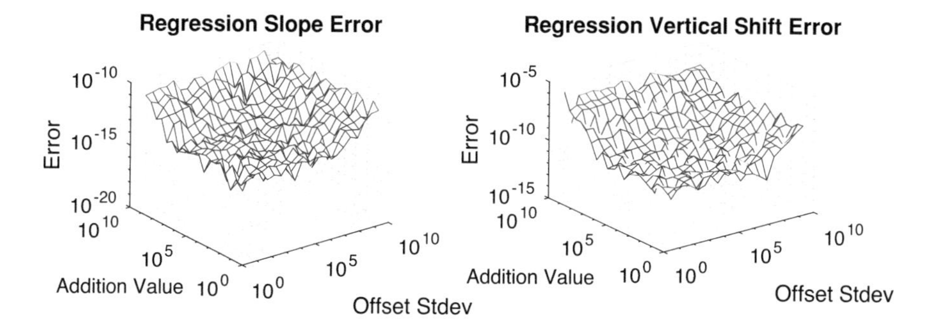

Fig. 7 Regression error analysis

store the transformed data components. For instance, a 64 bit data type that supports double precision has a machine epsilon of 2^{-52}. In Fig. 6, we investigate the impact of n (i.e. number of key terms) and the magnitude of the plaintext value on the error resulting from the masking procedure. It is clear that the error magnitude, which is in the order of 10^-5, is very low especially in the context of aggregation.

Impact of the rounding errors on the data analytics applications presented in this paper is investigated next. Figure 7 illustrates the error resulting in the computation of α (i.e. slope) and β (i.e. vertical shift) parameters. Once again, it is clear that the errors are very minute and this is the case for the other two data analytics applications considered in this paper: confidence interval computation and classification/clustering. Thus, the above accuracy analysis has demonstrated that the errors introduced by rounding in the RB technique is negligible.

7.2 Latencies

Latencies incurred due to masking, unmasking and aggregate homomorphic operations are important performance indicators of masking schemes especially when applied to large data sets. As such, we present the latency measurements obtained for the RB technique and existing masking schemes such as PE and RSA in Table 1. The measurements obtained for the RB technique for which the masking key consists of 14 terms is set as the benchmark and normalized to 1. All other measurements are expressed as factors of the benchmark and this allows for generalization. We do not consider FHE in this performance comparison as it is associated with high latencies [5] and therefore is not directly comparable. Comparisons are made with respect to the supported key sizes. In the RB technique, a key consists of n 'terms'. Each term is defined as a 64 bit data type. Hence, if the key consists of 14 terms, then the key size is $64 * 14 = 896$ bits.

It is clear from Table 1 that the RB technique supports more homomorphic operations in comparison to the other two state-of-the-art techniques. Moreover, for the operations that are supported by all three techniques, the latencies incurred with our proposal are significantly less (in the order of magnitude of 3). Hence, our proposal is suitable for operation on large data sets. These superior performance gains along with effective protection of individual measurement data points in the masked domain makes the RB technique extremely desirable for performing data analytics on sensitive data. Stringent security features available in the PE and RSA schemes result in either noticeable performance overhead or lack of support for homomorphic operations important for most data analytic applications such as the ones presented in Sect. 5.

Table 1 Comparison of operation latencies

	RB 896 bits (14 terms)	RB 1024 bits (16 terms)	PE 1024 bits	RSA 1024 bits
Masking of a message	1	$\times 1.2$	$\times 10^6$	$\times 10^6$
Unmasking of a message	1	$\times 1$	$\times 10^6$	$\times 10^6$
Single addition subtraction	1	$\times 1.2$	$\times 10^6$	N/A
Multiplication	1	$\times 1.2$	N/A	$\times 10^6$
Constant division or multiplication	1	$\times 1.8$	$\times 10^6$ Varies with constant size	$\times 10^6$ Require masking constant

Table 2 Comparison with the state-of-the-art

	Homomorphic operations	Security	Performance
Ref. [23]	Addition only	Paillier	$\times 10^6$ slower
Ref. [24]	Addition only	Paillier	$\times 10^6$ slower
Ref. [25]	Addition only	Masking	$O(n)$
Ref. [26]	Addition only	Masking	Increased comm. overhead
Refs. [27, 28]	N/A	Masking	Battery/DG
Proposal	Statistical operations	Masking	$O(n)$

7.3 Comparison with Prior Art

Existing work in the area of secure energy data analytics can be loosely classified into three categories. In the first class, cryptographic techniques such as Paillier encryption are applied to individual data points which are then aggregated (e.g. [23, 24]). In the second class, noise is added to the data to obscure individual plain-text values which cancel in aggregation operations (e.g. [24–26] and our proposal). In the third class of proposals, physical actuation using batteries or distributed generation (DG) sources residing in consumer premises are utilized to mask actual power demands [27, 28]. Table 2 presents a comparative summary of references [23–28] with our proposal based on three specific metrics.

The first metric examines the *homomorphic operations* that are supported by these techniques. References [23–26] support addition operations only. References [27] and [28] focus on utilizing physical devices to enable consumer privacy only and therefore do not support any homomorphic operations. Our proposal supports a wide variety of statistical operations (e.g. aggregation, average, variance, covariance, correlation, etc.) in the masked domain and thus allows for more comprehensive analysis in comparison to these state-of-the-art proposals.

The second metric considered is the *security* of the masking schemes. As references [23] and [24] apply Paillier encryption to individual data points, these inherit strong cryptographic properties of the Paillier crypto-system and therefore are robust to infringements [12]. References [25, 26] and our proposal add random noise to hide individual data points. This renders recovering the original energy data of individual consumers impossible via bruteforce guesses. When aggregated, the stochastic effects cancel. However, as the resulting value is of aggregated nature, content of individual consumers remain unexposed. References [27] and [28] utilize highly fluctuating physical devices such as DGs whose behaviour depend on external conditions to mask consumer demands. Thus, uncovering individual power demands will also be difficult in these cases.

The third metric considered is the *performance* of the masking techniques. As references [23] and [24] are based on Paillier encryption, the computational overhead is much higher than our proposal (i.e. in the order of 10^6). References [25, 26] and our proposal have similar computational overheads which are much lower than that entailed when using actual cryptographic systems. Specifically, the computational overhead is due to scalar addition and multiplication operations that depend on the size of the key n (i.e. $O(n)$). Significant communication overhead is entailed in the technique proposed by reference [26] as obfuscation values are generated and communicated separately. References [27] and [28] utilize physical devices to diversify energy readings and therefore require no computational resources.

From the above analysis, it is evident that the proposed homomorphic masking technique is light-weight and supports a wide range of homomorphic operations while protecting individual energy usage data generated by consumers. Thus, the proposed technique allows for data analytic and mining operations to be executed in a secure manner.

8 Concluding Remarks

The vast amounts of generated data of our modern society are becoming increasingly cyber-physical. These data sets hold remarkable potential for improving operations within the hierarchies of complex and interconnected systems. Main challenges faced by today's system operators are managing sensitive data in an effective and secure manner. We have discussed in this paper the need to a strike a balance between the stringent security necessary to prevent data compromise and flexibility required for effective data analysis. We have proposed a lightweight masking technique that supports a wide suite of homomorphic operations for highly scalable data analytics while also enabling secure storage of sensitive data in the cloud. This privacy homomorphism technique can be applied by system operators to practical applications as illustrated in this paper while also safely leveraging flexible and economically viable public platforms like the cloud. Significant performance improvements made possible by this lightweight scheme renders it highly scalable and suitable for any big data applications in general. As future work, we will continue to seek secure methods by which privacy can be preserved in a wider range of IoT applications.

References

1. P. Kumar, A. Braeken, A. Gurtov, J. Iinatti, P. Ha, Anonymous secure framework in connected smart home environments. IEEE Trans. Inf. Forensics Secur. **12**(4), 968–979 (2017)
2. S. Jain, N.V. Kumar, A. Paventhan, V.K. Chinnaiyan, V. Arnachalam, M. Pradish, Survey on smart grid technologies - smart metering, IoT and EMS, in *IEEE Students' Conference on Electrical, Electronics and Computer Science*, 2014
3. X. Liu, R.H. Deng, W. Ding, R. Lu, B. Qin, Privacy-preserving outsourced calculation on floating point numbers. IEEE Trans. Inf. Forensics Secur. **11**(11), 2513–2527 (2016)
4. C. Rong, S.T. Nguyen, M.G. Jaatun, Beyond lightning: a survey on security challenges in cloud computing. Comput. Electr. Eng. **39**(1), 47–54 (2013)

5. B. Zhang, P. Srikantha, D. Kundur, Secure management and processing of metered data in the cloud, in *Smart City 360*. Lecture Notes of the Institute for Computer Sciences, Social Informatics and Telecommunications Engineering, 2016, pp. 362–373
6. A.G. Ruzzelli, C. Nicolas, A. Schoofs, G.M.P. O'hare, Real-time recognition and profiling of appliances through a single electricity sensor, in *7th Annual IEEE Communications Society Conference on Sensor, Mesh and Ad Hoc Communications and Networks*, 2010
7. Green Button Data, An Overview of the Green Button Initiative. 25 June 2015 [Online]. Available: http://www.greenbuttondata.org/learn/. Accessed 14 July 2015
8. M. Pham-Hung, P. Srikantha, D. Kundur, A secure cloud architecture for data generated in the energy sector, in *Smart City 360*. Lecture Notes of the Institute for Computer Sciences, Social Informatics and Telecommunications Engineering, 2016, pp. 362–373
9. H. Bohaker, L. Austin, A. Clement, S. Perrin, Seeing through the cloud. *Creative Commons* (2015)
10. P. Rewagad, Y. Pawar, Use of digital signature with Diffie Hellman key exchange and AES encryption algorithm to enhance data security in cloud computing, in *International Conference on Communication Systems and Network Technologies*, 2013
11. C. Gentry, Fully homomorphic encryption using ideal lattices, in *Proceedings of the 41st Annual ACM Symposium on Symposium on Theory of Computing*, 2009
12. P. Paillier, Public-key cryptosystems based on composite degree residuosity classes, in *Advances in Cryptology*. Lecture Notes in Computer Science (Springer, Berlin, Heidelberg, 1999) pp. 223–238
13. R.L. Rivest, L. Adleman, M.L. Dertouzos, On data banks and privacy homomorphism. Found. Secure Comput. **4**(11), 169–180
14. G. Xiang, Z. Cui, The algebra homomorphic encryption scheme based on Fermat's little theorem, in *International Conference on Communication Systems and Network Technologies*, 2012
15. F. Armknecht, A.R. Sadeghi, A new approach for algebraically homomorphic encryption. Cryptology ePrint Archive. Report 2008/422 (2008). http://eprint.iacr.org/2008/422
16. L. Chen, Y. Xu, W. Fang, C. Gao, A new ElGamal-based algebraic homomorphism and its application, in *International Colloquium on Computing, Communication, Control, and Management*, 2008
17. P. Burtyka, O. Makarevich, Symmetric fully homomorphic encryption using decidable matrix equations, in *Proceedings of the 7th International Conference on Security of Information and Networks*, 2014
18. AMI System Security Requirements, *AMI-SEC Task Force*, Department of Energy, Unites States, 2008
19. P. Srikantha, C. Rosenberg, S. Keshav, An analysis of peak demand reductions due to elasticity of domestic appliances, in *Proceedings of the 3rd International Conference on Future Energy Systems Where Energy, Computing and Communication Meet*, 2012
20. C.M. Bishop, *Pattern Recognition and Machine Learning* (Springer, New York, 2006)
21. D. Freedman, Statistical Models: Theory and Practice (Cambridge University Press, Cambridge, 2005)
22. A. Quarteroni, R. Sacco, F. Saleri, *Numerical Mathematics* (Springer, New York, 2000)
23. Z. Guan, G. Si, Achieving privacy-preserving big data aggregation with fault tolerance in smart grid. Digit. Commun. Netw. **3**(4), 242–249 (2017)
24. S. Keoh, Y. Ang, Z. Tang, A lightweight privacy-preserved spatial and temporal aggregation of energy data, in *IEEE International Conference on Information Assurance and Security* (2015), pp. 1–6
25. E. Shi, T. Chan, E. Rieffel, R. Chow, D. Song, Privacy-preserving aggregation of time-series data, in *Proceedings of the Network and Distributed System Security Symposium* (2011), pp. 1–17
26. L. Pang, Y. Wang, A new (t, n) multi-secret sharing scheme based on Shamir's secret sharing. Appl. Math. Comput. **167**(2), 840–848 (2005)

27. J. Gomez-Vilardebo, D. Gunduz, Smart meter privacy for multiple users in the presence of an alternative energy source. IEEE Trans. Inf. Forensics Secur. **10**(1), 132–141 (2014)
28. G. Kalogridis, C. Efthymiou, S. Denic, T. Lewis, R. Cepeda, Privacy for smart meters: towards undetectable appliance load signatures, in *IEEE International Conference on Smart Grid Communication* (2010), pp. 1–5

Towards an Adaptive and Attack-Resilient Communication Infrastructures for Smart Grids

Yifu Wu, Jin Wei, and Bri-Mathias Hodge

1 Introduction

Centralized power generators such as dam, power plant and generating station have used resources like water, nuclear power, coal, oil and natural gas for several decades. As Distributed Energy Resources (DERs) technologies become increasingly popular, power resources also tend to be scattered [1]. DERs, such as renewable energy resources including solar photo voltaic (PV) panels and wind turbines, are accepted by more and more power consumers because they are willing to use the free electricity produced by solar or wind and even sell their excessive electricity back to power grid. The high DER penetrations comprise a larger portion of the generation and consumption fleets, information from these generators becomes much more critical. However, traditional SCADA system is design to work only in substation-level. Therefore, it is necessary to design a new Supervisory Control and Data Acquisition (SCADA) system to maintain power system economic efficiency and reliability.

The extention of SCADA system raises a lot of problems. One problem is how to choose to manage the network used for wide-area communication. Although dedicated network absolutely has best network performance and highest security level, it is too expensive to deploy and maintain for long-time operation. Public network infrastructure provides a more cost-effective solution for wide-area communication. Instead of investing in new communication infrastructures, SCADA system can utilize existing public networks such as the cellular network, cable,

Y. Wu · J. Wei (✉)
Purdue University, West Lafayette, IN, USA
e-mail: wu1584@purdue.edu; kocsis0@purdue.edu

B.-M. Hodge
University of Colorado Boulder, Boulder, CO, USA
e-mail: BriMathias.Hodge@colorado.edu

H. Karimipour et al. (eds.), *Security of Cyber-Physical Systems*,
https://doi.org/10.1007/978-3-030-45541-5_15

Digital Subscriber Line (DSL) and fiber optic internet access to transfer sensing data from DERs to locations in the electricity system where this information can be effectively utilized [2]. However, the utilization of Public network infrastructure significantly raises the complexity of SCADA communication network environment and thus reduces the effectiveness of SCADA network management in detecting the unusual network events. Therefore, the communication networks are more vulnerable to potential cyber attacks, such as Denial of Service (DoS) attacks, which could potentially impede the progress of DER expansion [3]. Thus, how to intelligently select and manage wide-area communication infrastructure for SCADA system extended to distribution level becomes increasing important.

This essential challenge inspires the development of autonomous network systems with adaptive control solutions. Considering the fact that not all power operators are familiar with network programming and configuration [4], middleware has emerged to abstract network topology and infrastructure, with which end users and network programmers can configure the whole network by only calling a local middleware object's attributes and methods [5]. Middleware can be designed to monitor the QoS information of data flow from network infrastructure. Gridstat, an objected-oriented-broker middleware providing QoS information, was designed for smart grid applications [6]. Li et al. suggested to use the distributed probe method to monitor QoS performance of application service [7]. Other researchers use middleware to schedule different application services. For example, Dipippo et al. designed a priority-mapping middleware architecture for static real-time distributed application [8]. Schantz et al. presented how middleware schedules the priority for application services [9]. For middleware implementation, some work has been done by using the co-simulation of VTB or network controller and OPNET [10, 11]. A middleware structure was proposed in [12] to optimize a TCP congestion window of the gateway in Network Simulator 3 (NS-3). For our design, we prefer to use NS-3 because it's an open-source software which is convenient to modify and it has a better running performance than OPNET thanks to that it is written by C++. All these traditional middleware designs focus on serving application service by using middleware to control network resources.

However, since middleware builds a bridge between network infrastructure and application service. we think it can be used to not only manage network resources for application user but also learn the experience and feedback from network application users. Network users do not just configure the network before using network application services. Instead, they can also participate in the network management with middleware by providing their personal experience and historical qualitative and quantitative data from anywhere except network infrastructure. we hope to use middleware to increase the interaction between human and machine rather than just enhancing human's experience. Thus, As to the challenge of wide-area network for SCADA system extended to distribution level, we propose a middleware using user's feedback and historical malicious behavior data to determine switching network media and detect false data injection respectively.

The former function of our middleware is based on Quality of Experience (QoE). QoE design envisions a more humanized, individualized network service providing more comfortable, satisfied experience for users with various requirements [13]. DER applications also aims to enhance the experience of power grid operators. Firstly, conventional QoS criteria only gives a fixed threshold parameter to supervise network service which usually makes sure that whatever degradation network communication causes do not impact the stability of power system. Due to the heterogeneous control structure of DER, however, it is essential to build QoE-based network model to satisfy the demand of different kind of users as far as possible [14]. For instance, different users at different level of power grid may have diverse perception of network service degradation. Network service can be customized by users themselves in order to relax the network QoS criteria or help to solve the shortage of network resources. Secondly, their satisfaction of network service directly influences their experience. For example, Unsatisfied grid operator works inefficiently and is easy to make mistakes. Also Unsatisfied power user is likely to cancel the subscription of power grid. Usually, users' satisfaction is measured in terms of mean opinion score (MOS) [15]. Test volunteers are asked to experience and tell the changes of network QoS. According to these scores, we can figure out what QoS parameter is sensible for users. Further, network service can be improved by referring user's feedback to adjust relative QoS standards. However, each user's perceptual ability may vary a lot and some QoS factors is not easy to recognize. Thus, unconscious QoS factors are often combined with subjective feedback from users to estimate users' QoE [15]. The second function is referring attacker's historical data in order to use attacker's behavior pattern to detect whether there are false data injected in normal data. Traditional false data detection is using state estimation to distinguish good or bad data-gram. However, there still exist some anomalies that are still difficult to be detected. Deep learning techniques are being proven to be efficiency on detecting malicious behavior of False Data Injection (FDI) attacks.

We have conducted some research work as shown in [16, 17]. In this book chapter, we extend our previous work by extensively exploring the QoE and QoS integration for enhancing the resilience to cyber attacks in communication systems for smart grid. Generally speaking, we implement our proposed middleware infrastructure in a co-simulation environment of MATLAB [18] and NS3 [19]. In the next section, we present the background knowledge of our work. Section 3 introduces our proposed middleware architecture. Simulation results and discussion are presented in Sects. 4 and 5, respectively.

2 Background

2.1 SCADA System

Supervisory Control and Data Acquisition (SCADA) is not a complete control system for power grid but a type of software application [20]. Any applications aiming at supervising certain system belongs to the cope of SCADA which is not limited to power system.

In power system, SCADA system usually contains four components: Control center, communication network, Intelligent Electronic Devices (IED) and Remote Terminal Unit (RTU). Control center has two distinguishing functionality: man machine interaction (MMI) and control processing. MMI can be used to display the whole power system in multiple screens with a bunch of compendious texts and graphs. It supplies power operators with real-time or historical information of IEDs as a reference to further control the power grid globally. Control processing is a dedicated and sufficient master station which can automatically handle most data control tasks such as alarm handling, logging/archiving and report generation in the entire system [21]. These functionality ensure the robustness and resiliency of power grid operation. Communication network provides communication media between server and server or server and client in publish-subscribe and event-triggered ways. IED can be considered as sensors for power system and RTU concentrates data from local IEDs then transfer it to control center.

Although SCADA system has worked well to monitor and control power grid in generation and transmission level for several decades, when it comes to distribution level, there is no SCADA application between distribution transformer and power customer where almost 90% of power hitches happen [22]. Thus a extension of SCADA to distribution level of power grid is essential to gather more information from distribution-level power utilities and power customers which is used to enhance the quality of power supply and improve the Quality of Experience (QoE).

On the other hand, traditionally power flow is a one-way delivery from power generator through transmission and sub-transmission line to distribution tier. However, due to the penetration of DERs especially renewable energy resources like solar and wind power, these power generators are introduced to distribution level of power grid which also requires the extension of current SCADA to distribution level. For example, Once SCADA discovers an outage, it will make a rescue strategy based on the current situation of DERs in order to make sure the stability of power grid and ensure the safety of human and power equipment.

What's more, decision making collaboration between human and machine is essential for modern SCADA system. Fixed SCADA control algorithm can not handle contingency problem so that the control decision making of SCADA application should consider human's experience. These experience might be very useful to detect potential threat and failure. Thus collaborative diagnosis is essential to be integrated in SCADA system. on the other hand, human's experience can tell the usefulness of different sensors so that SCADA control center can selectively retrieve relative sensors' data to make decision especially in emergency situation.

2.2 Combining SCADA with AMI

2.2.1 AMI Review

Unlike SCADA system transferring operational signal, Advanced Meter Infrastructure (AMI) handles non-operational information at distribution level. Before the emergence of AMI, power utility companies use uncomplicated meters to record power consumers' usage. These meters are installed in consumers' home or community so that they can only be accessed by meter readers or power consumers themselves manually. Further, consumers' electricity bill will not be ready until these meter readings are gathered and energy consumption charge is calculated. These drawbacks promotes the development of metering system. Then Automated Meter Reading (AMR) replaced manual meter reading. These new meters are able to send meter data hourly or more frequently to utility companies. But it is a unidirectional communication from meters to power utilities. Thus it is impossible to manage meters remotely. Finally, AMI system is designed to provide bidirectional communication between meters, meter data management system (MDMS), consumer information system (CIS). Power utility is able to access power customers' meters. In other word, meters can be occupied or dis-occupied by diverse power suppliers such as brokers, marketer or even generating companies.

Typically, a AMI system contains the following four components: (1) Smart meters: Meters not only record and send data to power utilities via Local Area Network (LAN) but also provide consumers the information about power usage, load profile and electricity bill. (2) Local collection Engine: It is a local data concentrator which reads meter's raw data and helps to deliver to remote server system. (3) Communications network: LAN is a communication network between smart meters and local data aggregator. Wide Area Network is for the communication between data aggregators and remote servers. (4) Meter Data Management System: Meter Data Management System (MDMS) play as a remote server which stores data from smart meters. Then these raw data will be processed by using Validate, Edit and Estimate (VEE) algorithms to provide billing and other service for power consumers. For example, one Ohio's power supplier—FirstEnergy—employs SAP software solution to management their power consumer's billing and other enterprise services.

2.2.2 Integration with AMI

Like we illustrated before, SCADA system needs a extension to distribution level of power grid due to the penetration of DERs. AMI, which focuses on distribution level monitoring, can be used to help SCADA to work in distribution level. And it provides two main benefit as follows.

Firstly, AMI system provide data resource for SCADA system. Considering PV technology is implemented at home-scale or community-scale level, smart meters can be combined with other sensors such as PV invertors, light sensors to provide operational data resource for SCADA application of DER management and power balancing so that smart meters are not only basically record power consumer's usage but also help to sensor home-scale renewable energy generation. For example, according to the information of battery charging, real-time load, power consumption, sun light intensity, SCADA application can predict the future operation of solar panel and determine whether to use the solar energy locally, or transfer to other place requiring power, or store it in battery.

Secondly, AMI system has a prototype of hierarchical data processing for SCADA system. For example, data concentrator can be designed to transform raw sensor data such as voltage and current to cognitive information by using smart analytic algorithms. These information contains the relative features and can be used to predict the power generator's performance, degradation and failure. These information is further delivered to SCADA application in control center which can configure the power system globally.

2.2.3 Communication Challenge and Our Solution

Although AMI provides lots of convenience for distributed SCADA system development, its communication paradigm can not be directly used in SCADA system because traditional SCADA system has a local dedicated network and AMI utilize public network for communication. If SCADA system utilize public network directly, there might be some issues.

Firstly, public communication access media is not ready for tremendous traffic load from distributed SCADA system. Since smart meters will increase drastically in the coming few years, the capacity of communication network and data center processing will face serious challenges. Figure 1 shows the smart meter deployment plan of the FirstEnergy power utility company in Pennsylvania. It is said there will be at least two million smart meters installed [23].

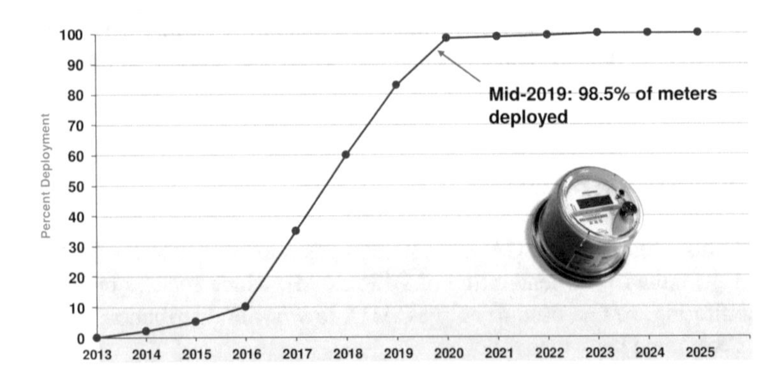

Fig. 1 Communication network vertical structure [23]

Thus, how to get full use of limited network bandwidth to ensure the transmission of huge sensors' data is a big challenge to network infrastructure.

Secondly, security issues of SCADA system is different from other network-based applications. From Fig. 2, it is clear that traditional networks such as office network focus more on the confidentiality of transmitted data rather than the availability and integrity of data.

However, when network is designed to serve SCADA power applications, the availability and integrity tend to be more important than confidentiality because these two security factors directly impact the normally stable operation of large power system especially DER-penetrated smart grid. public network infrastructure is more vulnerable than dedicated network so that the security problems should be considered as first.

To solve these two problems, a new communication architecture, which aims at providing an integration paradigm that can leverage the SCADA techniques and AMI techniques, is proposed to adapt SCADA system for DERs.

For the first issue, enlarging hardware resource such as communication bandwidth, processing units and memory is an effective way to solve the network capacity problem but it is too expensive for deployment. Data conversion and compression is also a efficient application-based solution to solve the data explosion of smart meters. The data collector of AMI system can be modified to process the

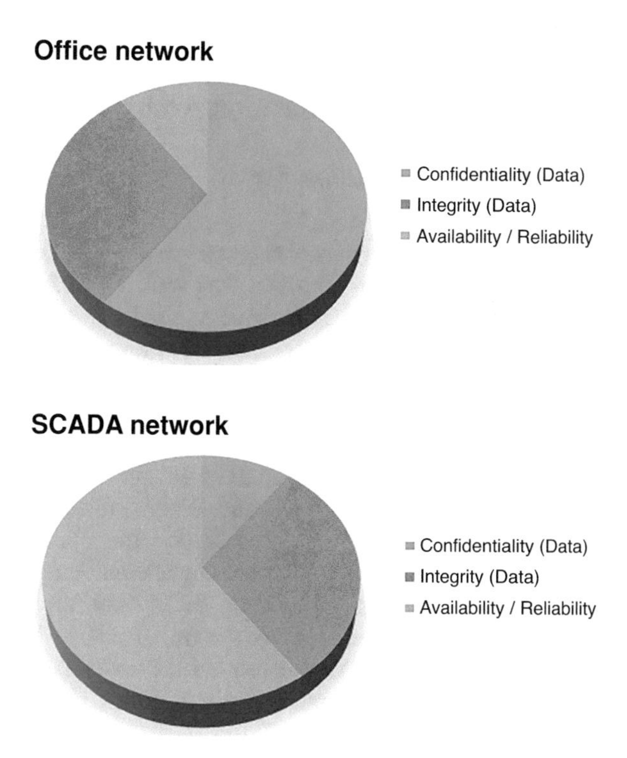

Fig. 2 Communication network vertical structure

data before it is transferred to remote servers. However, it just reduces the traffic load of power application itself but does not consider the influence of other traffic load in public network infrastructure. Thus, it is more reasonable to choose a network-based solution rather than application-based solution. In this book chapter, we propose a mixed wireless network infrastructure to solve the congestion problems of public network especially in urban area. LTE is defined as the main public network carrier to deliver power application messages. And WiFi mesh network is prepared as the backup redundant media for LTE in case congestion occurs. Also we care about the experience of power operators so that we introduce human's feedback as a reference to determine whether it is necessary to use the backup communication media.

As to the second problem, current security power applications mainly use cryptography techniques to protect their data flows. For example, IEC62351 provide some recommendation of cipher suite for IEC 61850. However, smart meters and other distributed sensor of SCADA system gather information from private home facilities and unattended distributed power generation equipment. The integrity of original data can not be guaranteed. In other word, attackers can easily inject false data to SCADA's distributed sensors. In order to solve this problem, we propose a content-aware deep learning method to detect the false data injection on local data concentrators.

In the following section, we will illustrate the detail solution of our proposed communication architecture by aiming at the two problems mentioned before.

3 Proposed Communication Architecture

3.1 Power Application Layer

Power application Layer is not necessary to all network nodes. It merely runs on end host nodes which need to process data such as meters, sensors, data concentrator, and servers. DERs is increasingly being used as a supplement and an alternative to large conventional central power stations. Community-scale distributed energy systems are often coupled with the conventional power distribution system. Currently in this framework sensing devices, such as smart meters and solar panel sensors, samples customers' power usage, load profile and the status of renewable energy such as solar and wind, and send these data to remote servers via multiple network domain such as LAN, internet and WAN. Protection of these data in SCADA system is extremely important. Local concentrator helps them to forward the data to remote servers by converting LAN application data format to universal SCADA application format. MMS based on IEC 61850 has been verified as a excellent application protocol in SCADA system to realize bidirectional communication between sampled devices and substation controller. Also the extension of IEC 61850 is capable of communicating in distribution level in order to support DERs.

The security service of IEC 61850 is provided by IEC62351. To be specific, IEC62351-4 defines the security services for Manufacturing Message Specification (MMS) under IEC 61850 standard, which is the application protocol we plan to use in our design. And MMS is a international standard used to transfer real-time data and control signal between meters, sensors, actuators, and controllers. The A-profile of IEC62351-4 recommends to use cartographic functions to protect ACSE session establishment and payload. Meanwhile, Transport Layer Security (TLS), which has been widely used in applications such as web browsing, is also required to use which is specified in T-profile of IEC62351-4.

As to implementing in our proposed architecture, IEC62351-4 does not give a detailed requirement of security service. For example, the A-profile of IEC62351-4 just suggests to use signature certificates and corresponding private keys but des not specify a exact cipher suite combination, which gives designer a free space to customize the standard to adapt new use cases.

Besides, at application layer, power operator can also input their personal experience feedback and previous attackers' historical data to middleware for data analysis via APIs provided by middleware at control layer.

3.2 Control Layer

In control layer, we propose a real-time distributed middleware to manage data flows of different smart grid application services by referring power operators' experience feedback. Meanwhile, our proposed middleware use attacker's historical data to improve the security of SCADA power application at distribution level. To achieve the balance among computational power consumption, operating efficiency and accuracy, and congested channel bandwidth, middleware instances are only installed in data concentrators and critical gateways. In our work, middleware instances are installed in data concentrators, various gateways and servers which are all red in Fig. 3. All the installed middleware instances have the same structure and functions and the only difference is that the middleware instance installed in data concentrator and servers has Application Program Interfaces (APIs). So power applications can be installed on data concentrator more easily and they can also send control commands to other middleware instances installed in the individual gateway nodes via their local middleware instances in real time. Meanwhile, it is shown in Fig. 3 that power operators can directly use operate on the middleware instances via these APIs.

Figure 4 illustrates the vertical structure of our proposed communication network. Middleware act as a network controller which organize, coordinate and supervise diverse application services in smart grids. The middleware instances in Fig. 4 are allocated between the power application layer and the network infrastructure layer. In Fig. 4, the application module in dashed lines means that only the data concentrators in the network have power application layers and thus only the middleware instances installed in the data concentrators have APIs. As shown

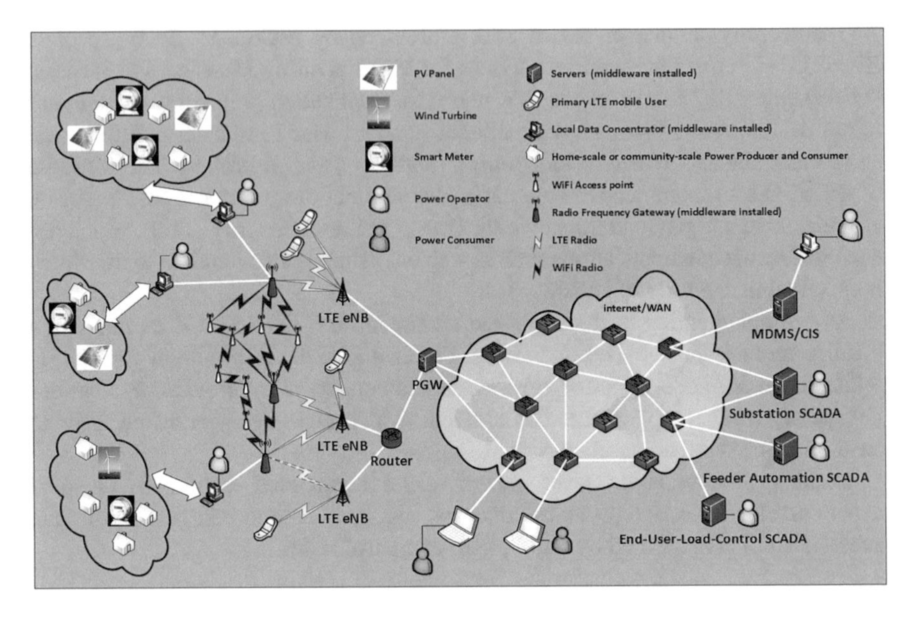

Fig. 3 Communication network horizontal structure

in Fig. 4, middleware instances on data concentrators have four service modules: QoS monitor, QoE evaluation, false data detector and dispatcher. And middleware instances on relay nodes has only QoS monitor and dispatcher. All message must go through dispatcher for message classification. QoE evaluation module is responsible for transforming power operators' qualitative experience to quantitative QoE-based QoS criteria, which will be used by QoS monitor module. QoS monitor observe QoS performance of each data flow at link layer, refer QoE feedback from QoE evaluation module and determine routing strategy in network layer. False data Detection module conducts verifying the realness of data-content.

3.2.1 QoS Monitor

Our proposed mechanism of QoS monitor for one node installed with a middleware instance is illustrated in Fig. 5. As shown in Fig. 5, the QoS monitoring system probes the real-time data-link-layer QoS information of each data flow, evaluates the observed QoS information with our proposed QoS criteria, and stores the QoS information in a buffer for potential usage in the control algorithm module. Our proposed QoS criteria is implemented by using the information provided by QoS standard database that can be updated based on the QoE specification reported by either local users or remote users of end hosts via the QoE evaluation module. If the quality of delivery for a power application's data flow that is represented by the observed QoS information does not pass the QoS criteria, an alert is sent

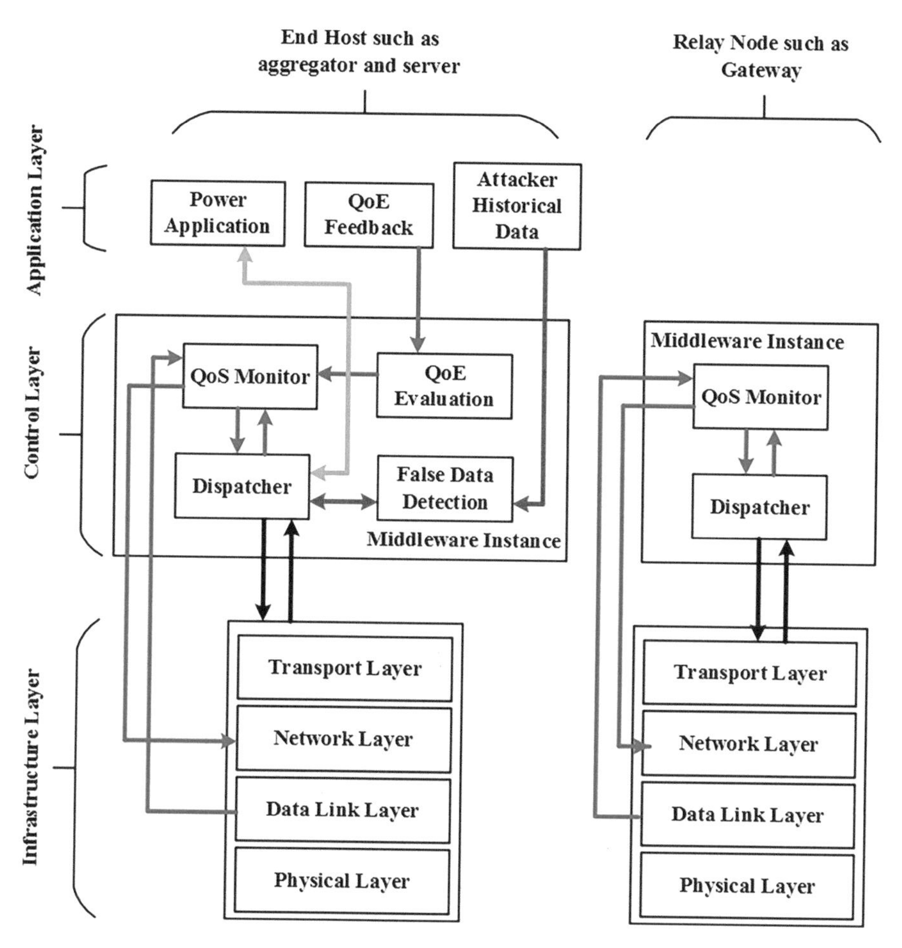

Fig. 4 Communication network vertical structure

out to trigger the congestion control algorithm for the data flow management. Our proposed control algorithm achieves the intelligent data flow management based on the information from different OSI layers including QoS information from data link layer, routing information from network layer, and QoE specification from the application layer.

The authors would like to clarify that we assume that the network system has been initially well designed to fulfill the QoS criteria and the network is in good condition in most time. Moreover, in real world applications, the lowest two layers, the physical layer and link state layer, are integrated together in one network device and one network node can own more than one network devices. For one node, only one middleware instance can be installed, which means that the single middleware instance should be designed to manage all network devices on that node. As shown in Fig. 5, each middleware instance has a QoS standard database that specifies the priorities and requirements of different application services which are

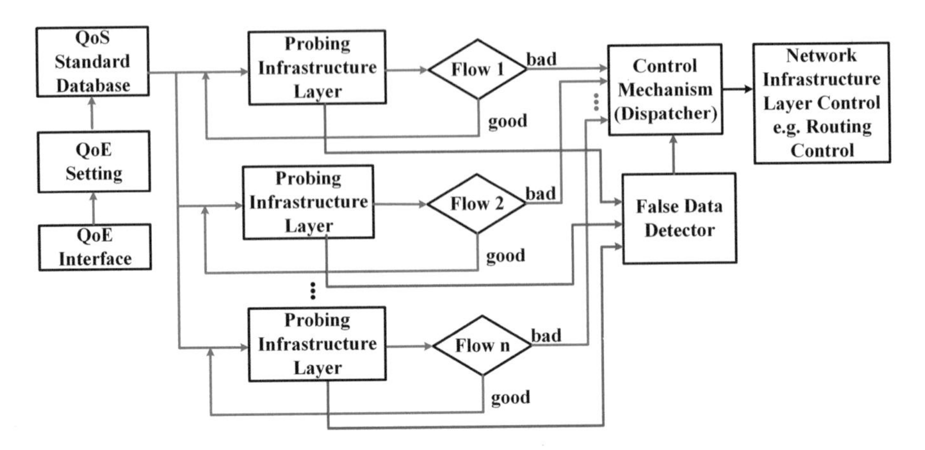

Fig. 5 Mechanism for Middleware Structure

Table 1 Information provided by QoS database

Attribute	Description
i	Interface ID of network devices
n	ID of data flow
$\mathcal{L}_n$	Priority level of data flow n
$\mathcal{S}_n$	Source IP address of data flow n
$\mathcal{D}_n$	Destination IP address of data flow n
p_n^S	Minimum availability rate of data flow n
r_n^S	Minimum data rate of data flow n
d_n^S	Maximum delay of data flow n
j_n^S	Maximum jitter of data flow n
p_n^E	Minimum tolerant availability rate of data flow n
r_n^E	Minimum tolerant data rate of data flow n
d_n^E	Maximum tolerant delay of data flow n
j_n^E	Maximum tolerant jitter of data flow n
w_n^p	Weight of minimum availability rate of data flow n
w_n^r	Weight of data rate of data flow n
w_n^d	Weight of maximum delay of data flow n
w_n^j	Weight of maximum jitter of data flow n

consistent with those in other middleware instances. The information provided by QoS database is detailed in Table 1.

The middleware instance probes the QoS information of different data flows from the data link layer and network layer, such as throughput, latency, jitter, source IP address, and destination IP address in real time, then converts the QoS information to a standard format consistent with QoS criteria stored in its database. The Observed QoS information is listed in Table 2. The authors would like to mention that the similar attributes and the parameters of QoS information are considered in [17]. By using the observed QoS information in Table 2, the quality of the delivery of Data Flow n in Interface i is evaluated as follows [24]:

Table 2 Parameters of observed QoS information

Parameters	Description
$p^o_{n,i}$	Observed availability rate of data flow n transmitted through interface i
$r^o_{n,i}$	Observed throughput of data flow n transmitted through interface i
$d^o_{n,i}$	Observed maximum delay of data flow n transmitted through interface i
$j^o_{n,i}$	Observed maximum jitter of data flow n transmitted through interface i

$$Q_{n,i} = \begin{cases} 1, \text{ if } \Gamma_{n,i} = 1; \\ 0, \text{ Otherwise.} \end{cases} \tag{1}$$

where

$$\Gamma_{n.i} = (Q^p_{n.i} \geq 0) \wedge (Q^r_{n.i} \geq 0) \wedge (Q^d_{n.i} \geq 0) \wedge (Q^j_{n.i} \geq 0),$$

$$\begin{cases} Q^p_{n.i} = \frac{p^S_n - p^o_{n,i}}{p^S_n}, \\ Q^r_{n.i} = \frac{r^o_{n,i} - r^S_n}{r^S_n}, \\ Q^d_{n.i} = \frac{d^S_n - d^o_{n,i}}{d^S_n}, \\ Q^j_{n.i} = \frac{j^S_n - j^o_{n,i}}{j^S_n}, \end{cases} \tag{2}$$

and $p^o_{n,i}$, $r^o_{n,i}$, $d^o_{n,i}$, $j^o_{n,i}$ are observed QoS parameters of data flow n in interface i stated in Table 2. As shown in Fig. 5, if the quality of the delivery $Q_{n.i} = 0$, an alert, that indicates a QoS failure of the corresponding data flow, is sent to the module of congestion control for further data flow management. Furthermore, the evaluation procedures of different flows in different interfaces are executed in parallel at a rate ν.

The QoS monitoring system in the middleware instance utilizes the proposed QoS criteria to evaluate the quality of delivery $Q_{n,i}$ for each data flow n at each interface i in the associated nodes. If $Q_{n,i} = 0$, a local QoS alert is sent out to activate the congestion control algorithm to determine whether it is necessary to reroute relative data flows. Since the default route is the most cost-efficient one, other routes are considered as backup routes. For example, in our work, we predetermine the default wireless route for SCADA data concentrator is LTE network only. And the backup path is LTE plus WiFi.

After receiving one or more QoS alert, the control mechanism in Gateway Node (GN) p will firstly deal with the alert of data flow with higher priority, then evaluates the QoE-based QoS performance $S_{n,i}$ of its network interfaces i. This score actually represents the quantitative analysis of human's satisfactions of a exact network application service. The lower the score is, the better the experience of user is. And it is calculated as follows:

$$S_{n,i} = w_n^p \times E_{n,i}^p + w_n^r \times E_{n,i}^r + w_n^d \times E_{n,i}^d + w_n^j \times E_{n,i}^j, \quad (3)$$

$$\begin{cases} E_{n.i}^p = \frac{|p_n^E - p_{n,i}^o|}{p_n^E}, \\ E_{n.i}^r = \frac{|r_n^E - r_{n,i}^o|}{r_n^E}, \\ E_{n.i}^d = \frac{|d_n^E - d_{n,i}^o|}{d_n^E}, \\ E_{n.i}^j = \frac{|j_n^E - j_{n,i}^o|}{j_n^E}, \end{cases} \quad (4)$$

where w_n^p, w_n^d, w_n^j are weight parameters determined by QoE evaluation module. If $S_{n,i}$ is larger than τ_n, it is undeniable to use backup route to recover the QoS performance of data flow n. Otherwise, it means the end user can tolerant the degradation of QoS and there is no need to reroute data flow n. The routing control is realized by directly configuring the layer-3 routing table predefined by network designer.

3.2.2 QoE Evaluation

In order to obtain the QoE-based QoS score in QoS monitor, it is essential to know how to get aware of the parameter used in that calculation model. Since that model is based on user's experience. Our proposed middleware let authorized user to be able to report the QoE specification of certain application data flow in application layer through a API to middleware, which results in the modification of the corresponding settings in QoS standard database of the local middleware instance. This modification is further synchronized among the middleware instances of the neighboring nodes via network infrastructure layer broadcast. This communication service is routed by dispatcher module which will be illustrated in the next subsection.

Network application operator can use middleware to modify or influence parameters such as p_n, r_n, d_n, j_n, w_n^p, w_n^r, w_n^d and w_n^j. To be specific, These parameters are obtain from Main Opinion Score (MOS) and its label start from 1 to 5 which indicate excellent, good, fair, bad, worse. The MOS evaluation is predetermined by network expert. And the tolerant QoS parameter such as p_n, r_n, d_n, j_n determine its corresponding weight which is proportional to the absolute gradient value of tolerant QoS parameter on MOS curve. In our design, we simplify the MOS label

Table 3 Modeling of the QoE-based QoS weights and their range

	Bad	Fair	Good
Critical availability rate range	$p^o_{n,i} < 99.5\%$	$99.99\% > p^o_{n,i} > 99.5\%$	$p^o_{n,i} > 99.99\%$
Critical availability rate weight	$w^p_n = 1$	$w^p_n = 1000$	$w^p_n = 3000$
Tolerant loss rate range	$p^o_{n,i} > 1\%$	$0.1\% < p^o_{n,i} < 1\%$	$p^o_{n,i} < 0.1\%$
Tolerant loss rate weight	$w^p_n = 100$	$w^p_n = 400$	$w^p_n = 500$
Critical latency range	$d^o_{n,i} > 10$ ms	5 ms $< d^o_{n,i} < 10$ ms	$d^o_{n,i} < 5$ ms
Critical latency weight	$w^d_n = 20$	$w^d_n = 5$	$w^d_n = 2.5$
Tolerant latency range	$d^o_{n,i} > 50$ ms	7 ms $< d^o_{n,i} < 50$ ms	$d^o_{n,i} < 7$ ms
Tolerant latency weight	$w^d_n = 15$	$w^d_n = 3$	$w^d_n = 1$
Throughput weight	N/A	N/A	NA
Jitter Weight	N/A	N/A	NA

and just use good, fair, bad three labels. For example, the bad experience weight is proportional to the absolute gradient value of cut-off point between bad and fair experience. And this value is selected to use when real-time QoS performance is below the cut-off point. In our design, our SCADA application concerns more about latency and availability so that throughput and jitter are ignored in the following simulation. The detailed parameters of tolerant QoS and their corresponding weights are shown in Table 3.

3.2.3 False Data Detection

Since MMS message is transferring between data concentrator and server, IEC62351 can only protect MMS message from network attacks. If meter and sensor's data is falsified before reading, network security defense is not efficient anymore. Thus a new mechanism based on content-aware deep learning is proposed to detect false data injection before sending MMS message. Considering it is hard to understand the pattern of attackers' false data directly, it is more realistic to use machine learning to learn the historical attacker's data and then use the model to find out false data directly. These data can be directly obtain from

A deep belief network (DBN) base classifier is used to formulate the content-aware deep learning mechanism. The classifier uses a set of input data vectors obtained by dividing the input time series data according to a time window. The classifier is trained to detect whether the data is under FDI attacks. DBN is trained through semi-supervised learning with historical time series data. Available historical data are mostly unlabeled data and some labeled data is also available.

As shown in Fig. 6, DBN used in our identification scheme is formed by stacking with conventional Restricted Boltzmann Machines (RBMs) and Gaussian-Bernoulli RBMs (GBRBMs) [25]. Conventional RBMs consist of one visible layer and

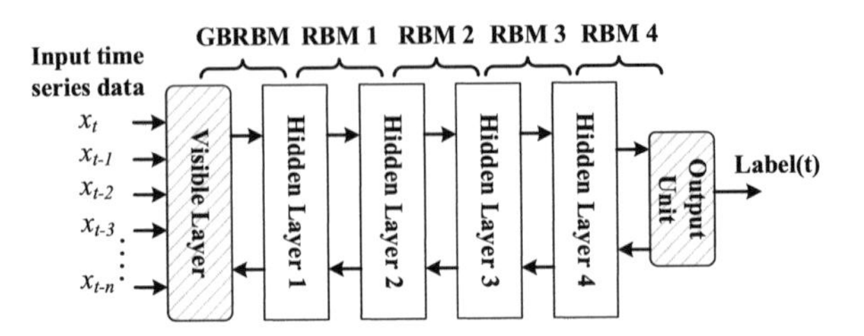

Fig. 6 Architecture of the used DBN as a stack of GBRBM and RBMs. Available time series data are fed into the system with a time window size of n, label is predicted for each time window of data vector from the output unit

one hidden layer of binary units that do not have intra-layer connections [26]. GBRBM is a variation of RBM that has a visible layer with real-valued input units [27, 28]. First GBRBM and RBMs are trained with unlabeled data. By training with unlabeled data,GBRBM and RBMs are able to learn the features embodied by the training data. Then, DBN is further fine tune using back-propagation training method to detect FDI attacks

Energy function of a conventional RBM is given below:

$$E(\mathbf{v}, \mathbf{h}) = -\sum_{i=1}^{n}\sum_{j=1}^{m} w_{ij} h_i v_j - \sum_{j=1}^{m} c_j v_j - \sum_{i=1}^{n} b_i h_i \tag{5}$$

where v_j is the jth element of the vector consisting of input unit values, h_i is the ith element of the vector consisting of hidden unit values, w_{ij} is the ijth element of the weight matrix between the visible and hidden units, while b_i and c_j denote the ith and jth element of the bias vectors for the hidden layer and visible layer, respectively. Note that n and m are the number of hidden units, and number of visible units in the RBM, respectively.

Based on Eq. (5), the activation conditional probability distributions of hidden and visible units, given the adjacent layer unit values, can be calculated by using sigmoid activation function as shown in the followings:

$$\begin{cases} p\left(h_i = 1|\mathbf{v}\right) = sigm\left(b_i + \sum_{j=1}^{m} w_{ij} v_j\right), \\ p\left(v_j = 1|\mathbf{h}\right) = sigm\left(c_j + \sum_{i=1}^{n} w_{ij} h_i\right). \end{cases} \tag{6}$$

Where $sigm\left(x\right) = 1/\left(1 + e^{-x}\right)$ is the sigmoid function.

A conventional RBM can be trained using gradient-based contrastive divergence technique [29]. The update rules for weights and biases are as follows:

$$\begin{cases} w_{ij} = w_{ij} - \rho\left(\left\langle v_j h_i \right\rangle_m - \left\langle v_j h_i \right\rangle_d\right), \\ b_i = b_i - \rho\left(\langle h_i \rangle_m - \langle h_i \rangle_d\right), \\ c_j = c_j - \rho\left(\left\langle v_j \right\rangle_m - \left\langle v_j \right\rangle_d\right), \end{cases} \tag{7}$$

where ρ denotes the learning rate, and $\langle\cdot\rangle_d$ and $\langle\cdot\rangle_m$ are the expectations computed over the data and model distribution, respectively.

Energy function of the Gaussian Bernoulli RBM is defined as follows:

$$E(\mathbf{v}, \mathbf{h}) = -\sum_i^n \sum_j^m \frac{v_j}{\hat{\sigma}_j^2} h_i w_{ij} - \sum_i^n b_i h_i + \sum_j^m \frac{(v_j - c_j)^2}{2\hat{\sigma}_j^2} \tag{8}$$

where $\hat{\sigma}_i$ is the standard deviation of the ith element of the visible units.

Based on Eq. (8), the activation conditional probability distributions of visible units can be calculated as follow:

$$\begin{cases} p\left(h_i = 1|\mathbf{v}\right) = sigm\left(b_i + \sum_{j=1}^m \frac{w_{ij} v_j}{\hat{\sigma}_j^2}\right), \\ p\left(v_j = v|\mathbf{h}\right) = N\left(c_j + \sum_{i=1}^n w_{ij} h_i, \hat{\sigma}_j^2\right), \end{cases} \tag{9}$$

where $\mathcal{N}(\mu, \hat{\sigma}^2)$ is the normal distribution with mean μ and standard deviation $\hat{\sigma}$.

GBRBM are trained by using the following update rules:

$$\begin{cases} w_{ij} = w_{ij} - \rho\left(\left\langle \frac{v_j}{\hat{\sigma}_j^2} h_i \right\rangle_m - \left\langle \frac{v_j}{\hat{\sigma}_j^2} h_i \right\rangle_d\right), \\ b_i = b_i - \rho\left(\langle h_i \rangle_m - \langle h_i \rangle_d\right), \\ c_j = c_j - \rho\left(\left\langle \frac{v_j}{\hat{\sigma}_j^2} \right\rangle_m - \left\langle \frac{v_j}{\hat{\sigma}_j^2} \right\rangle_d\right). \end{cases} \tag{10}$$

The standard deviations $\hat{\sigma}_j$ for GBRBM are updated as follows:

$$\begin{aligned} \Delta z_j =& e^{-z_j} \left\langle \frac{1}{2}(v_j - c_j)^2 - \sum_{i=1}^n w_{ij} h_i v_j \right\rangle_d - \\ & e^{-z_j} \left\langle \frac{1}{2}(v_j - c_j)^2 - \sum_{i=1}^n w_{ij} h_i v_j \right\rangle_m \end{aligned} \tag{11}$$

where $z_j = log\left(\hat{\sigma}_j^2\right)$. Employing Eqs. (7), (10), and (11), GBRBM and conventional RBMs are trained through Gibbs sampling using unlabelled input data.

We fine tune the DBN based classifier by using backpropagation training method [30]. Update rules for back-propagation fine tuning is for hidden layer connections are as follows:

$$\begin{cases} \Delta W_{h,i,j} = -\eta \delta_{h,i} p_{h-1,j} \\ \Delta d_{h,j} = -\eta \delta_{h,j} \end{cases} \tag{12}$$

where $\Delta W_{h,i,j}$ is the update for the i, jth element of the hth hidden layer weights, $\Delta d_{h,j}$ is the update for the jth element of hth hidden layer bias, $p_{h-1,j}$ is the activation probability of jth element of h − 1th hidden layer, η is the learning rate and $\delta_{h,j}$ is calculated as follows:

$$\delta_{h,j} = p_{h,j}(1 - p_{h,j}) \sum_{k}^{K} \delta_{h+1,k} W_{h+1,j,k} \tag{13}$$

where $p_{h,j}$ is the activation probability of jth element of h1th hidden layer, $W_{h+1,j,k}$ is the jkth element of weights corresponds to $h + 1$ hidden layer or the output layer and K is the number of elements in $h + 1$ hidden layer or output layer.

Update rules for back-propagation fine tuning is for output layer connections are as follows:

$$\begin{cases} \Delta W_{o,j} = -\eta \delta_o p_{H,j} \\ \Delta d_o = -\eta \delta_o \end{cases} \tag{14}$$

where $\Delta W_{o,j}$ is the update for the jth element of the output layer weights vector, Δd_o is the update for the bias in output unit, $p_{H,j}$ is the activation probability of jth element of highest hidden layer, η is the learning rate and δ_o is calculated as following:

$$\delta_o = p_o(1 - p_o)(l_o - L) \tag{15}$$

where p_o is the activation probability of output node, l_o is the predicted label, L is the actual label.

3.2.4 Dispatcher

Dispatcher is a application message forwarding and receiving module. In order to distinguish message type, a middleware header with service identity and sequence number is added to the application payload of its local application services like power application and other middleware sub-modules. And the header's encapsulation and decapsulation are done by dispatcher. Service identity indicates the service object and type. Service target is usually set as the MAC address of requesting node installed with middleware instance, which enhances the integrity of middleware payload. Service type is the extended command set of application payload which helps middleware to classify application payload and deliver it to proper application module efficiently. Sequence number is used to avoid processing repeated service request. For example, a middleware instance is updating its QoS database and is going to synchronize this modification with other middleware instance. Initially, it will broadcast to its neighbor nodes. Once the neighboring nodes installed with a middleware instance receive the modification message, their dispatchers first figure out the service type is QoS database updating and sequence number is newer than the former received one. Then they route the message to its indicating service module and broadcast the message to their neighboring nodes again at the same time. If one node with a middleware instance receive a message with same sequence number for the second time, dispatcher will drop the message immediately and stop rebroadcasting it.

3.3 Network Infrastructure Layer

The communication infrastructure layer consists of the first four network layers in the common OSI model [31], which are: the physical layer, data link layer, network layer, and transport layer. Data concentrators collect data from local smart meters and other sensors via LAN such as ZigBee and WiFi. As is known to all, LTE has a better coverage in urban cities rather than rural areas. So it is considered as the premier public network media which provides reliable and secure transmission for SCADA system. As shown in Fig. 3, concentrators continues to transfer the data to remote servers via LTE network to back-haul WAN/internet. Here considering the trade-off between cost and performance of network facilities, public LTE network and dedicated WiFi mesh network are combined together to construct a resilient, cost-efficiency, high-performance network providing QoE-aware tolerant congestion control. Power utility company acts as a Mobile virtual network operator (MVNO) which leasing LTE data service from major public mobile carrier such as Verizon Wireless, AT&T Mobility and T-mobile US. Then they can use public LTE networks to gather DERs information to their data management servers. Since the LTE network is leased from other carrier also having their own consumers, usually these carrier owning LTE infrastructure will prior to serve for their own customers rather than utility companies. Utility companies have lower priority to utilize LTE network so that sometimes the quality of power data transmission service can not be guaranteed if there is a congestion around current LTE eNB. This congestion means LTE inner QoS management service automatically degrade the quality of MVNO's LTE service in order to make sure the QoS of high-priority mobile users. For example, large amount of primary user of major carriers using LTE to access online video service simultaneously will cause the QoS degradation of power utility company's network usage. Thus we add dedicated WiFi mesh network as a detour path and reroute congested power application flows from congested LTE eNB to another LTE eNB via this path. Our proposed middleware efficiently addresses the rerouting problem by frequently checking the QoS performance of each flow in critical gateways such as the radio frequency gateway.

4 Simulation Results

In this section, we demonstrate the performance of our proposed middleware architecture via a real-time co-simulation of NS-3 and MATLAB/Simulink. Our testbed consists of the PV Invertor and other sensors that monitors the operating status of distributed PV systems, data concentrator which collects data such as power, voltage, current, temperature and irradiance from the sensors mentioned above, data transmission network including LAN and WAN and remote servers of utility companies. Data generation and collection are realized by a designed PV system in MATLAB/Simulink. Figure 7 shows the Simulink module used to generate real sensoring data from PV panel such as power of PV generator, mean of voltage, panels' temperature and sun irradiance.

These data will be delivered to a data concentrator node simulated in NS-3. And remote servers also act as packet sink nodes in NS-3 environment. The data concentrator nodes in NS-3 serve as a virtual net device in order to build the connection between MATLAB and NS-3 simulated network. Here, we ignore the LAN simulation between PV invertors and local data concentrator because our core network infrastructure design takes emphasis on the concentrator-side WAN—the LTE and WiFi mixed network.

For NS3 simulation, we use LENA model to simulate LTE network with fully EPC functionality [32]. The specific configuration of LTE is shown in Table 4.

Since power utility have low priority to use LTE AND we want to simulate the congestion for power application data flow in LTE network, we simulate several data flow with diverse QCI index to provide different priority of network service. For example, power applications are assigned to QCI 7, 9 or 10. In other word, their allocated EPS bearer do not ensure bit rate and provide the worst QoS performance guarantee. Conversely, data flows from primary LTE users such as VoLTE and other data services are assigned to lower QCI index which provide both bit rate and other critical QoS guarantee. The WiFi mesh network is simulated by combining wifi net device and Hybrid Wireless Mesh Protocol (HWMP) together which satisfies the requirement of 802.11s. The physical layer of WiFi mesh network is based on 802.11n—5GHz standard and HWMP is selected as the routing protocol in our simulation.

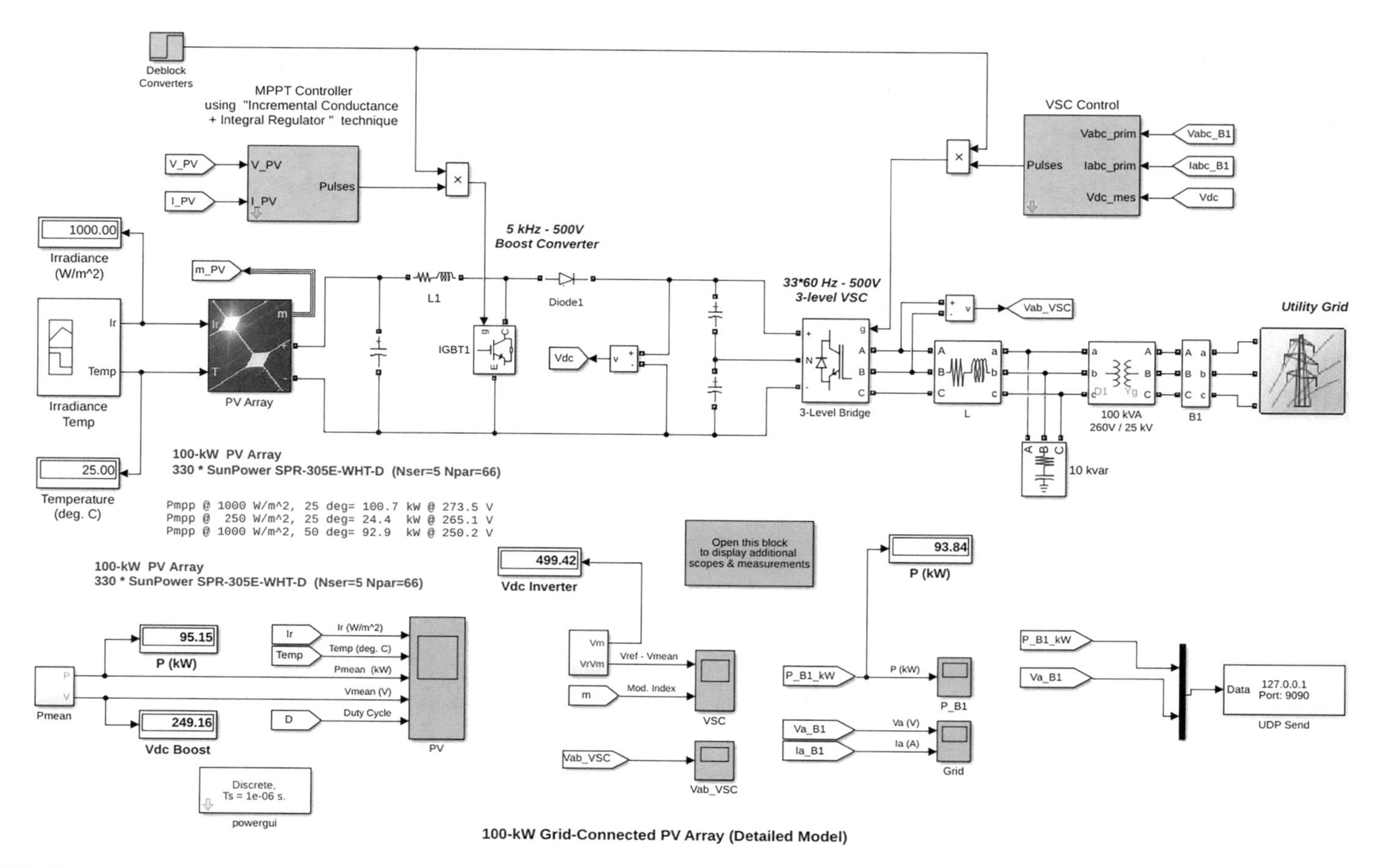

Fig. 7 PV-Array simulation topology based on MATLAB Simulink [33]

Table 4 LTE simulation configuration

Parameters	Value
Uplink EARFCN	18,100 (band 1)
Downlink EARFCN	100 (band 1)
Uplink Bandwidth	10 MHz (50 RBs)
Downlink Bandwidth	10 MHz (50 RBs)
MIMO	2×2
UE tx power	23 dBm
UE noise figure	5 dB
eNB tx power	50 dBm
eNB noise figure	5 dB
Cell Radius	1000 m
CQI reporting period	10 ms

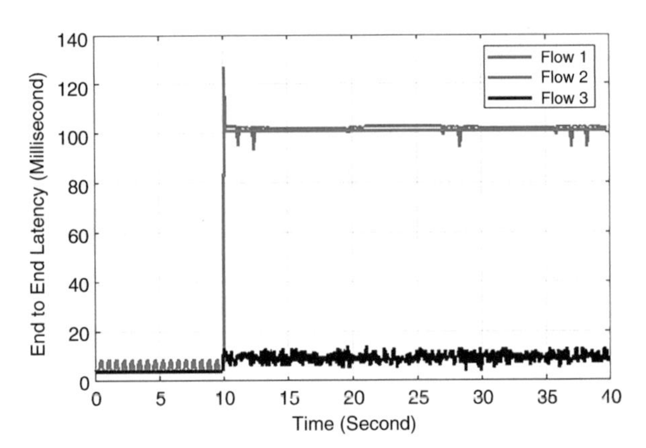

Fig. 8 Latency performance of flow 1, 2 and 3

4.1 *Case Study 1*

For case study one, we mainly focus on the availability protection of power data flow. Since the PV data flow from concentrator to remote servers contains data from dozens of PV panels, the data rate of this flow is much higher than individual data flow from a PV panel. Here, we set three PV data flows sent from data concentrator whose data rate is 4 Mbps. Flow 1 and 3 are TCP-based data-gram and flow 2 is UDP-based data gram. These three data flows originally are routed from different wireless gateway through LTE network to remote servers. In Figs. 8, 9, and 10, before 10 s, it is clear that if there is no congestion in LTE network, three flows all remains a very much low latency less than 10 ms, 100% availability and at least 4 Mbps link-layer throughput.

After 10 s, three flows suffer different congestion and their QoS performance decrease more or less. As shown in Fig. 8, the latency of flow 1 and 2 increases to higher than 100 ms and that of flow 3 just increase a little above 10 ms. In Fig. 8, flow 1 and 2's availability remains high but still have a little bit degradation, but flow 2 do not survive from the congestion because TCP provides reliable transmission by

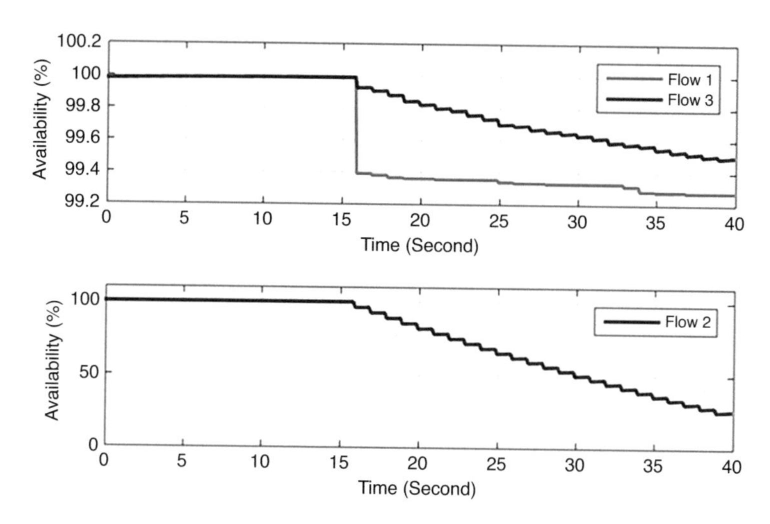

Fig. 9 Availability of flow 1, 2 and 3

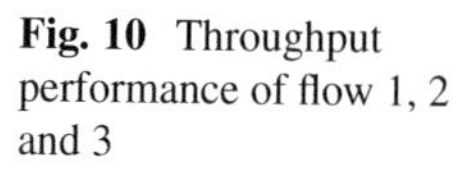
Fig. 10 Throughput performance of flow 1, 2 and 3

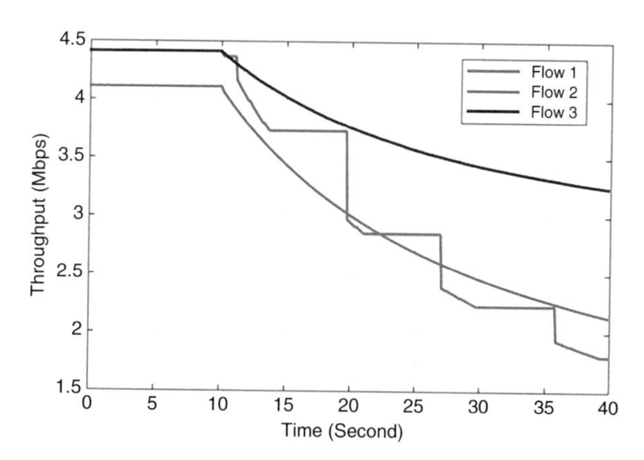

using congestion window and re-transmission to avoid and mask the packet loss and UDP is unreliable transmission. As to the throughput, three flows decrease to different degree and such huge decreases trigger the QoS alert of QoS monitor in middleware instance installed on relative relay nodes. Then the congestion control algorithm is executed to reevaluate the QoE-based QoS score for these flows.

In order to evaluate the QoE-based QoS score, firstly we assume in our simulation the threshold of this score is 5. To be specific, if a flow's score is higher than 5, the flow will be rerouted to WiFi mesh network to recover its normal performance. Otherwise, middleware does nothing and thinks user can accept the performance degradation of served flow. In this case study, for flow 1 and 2, since their availability decreases to less than 99.99%, critical and tolerant users who mainly care about availability will result in different control result. Figure 11 shows the QoE-based QoS score of three scenario—flow 1 with availability-concerned critical user, flow 1

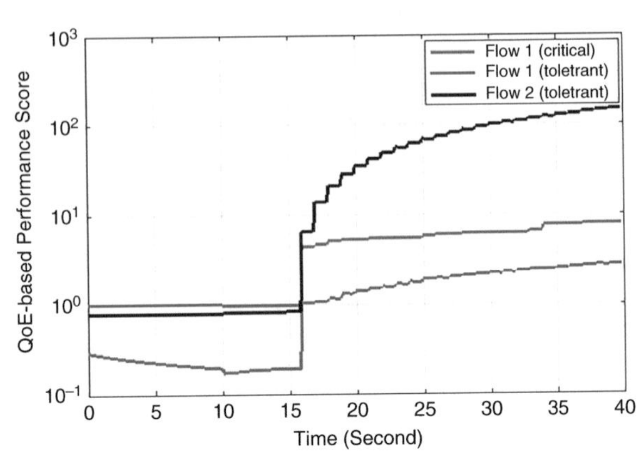

Fig. 11 QoE-based QoS score for Flow 1 and 2 with critical and non-critical users

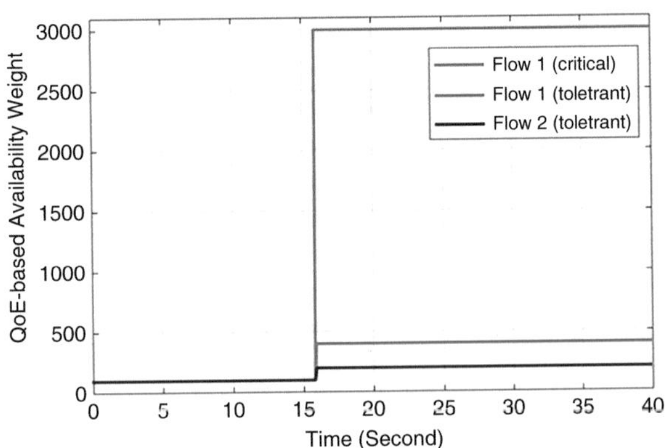

Fig. 12 The value of availability weight W_n^p for flow 1 and 2 with critical and non-critical users

with availability-concerned non-critical user and flow 2 with availability-concerned non-critical user.

It is obvious in Fig. 11 that if user concerns availability critically, flow 1's score will higher than 5 after congestion occurs and this flow will be rerouted. In contrast, if user concerns availability but is tolerant to availability decrease, it will lead the score less than 5 and rerouting control will not be applied. As to flow 2, no matter whether user concerns latency critically or not, flow 2 has a unacceptable availability rate and it will be rerouted by certain middleware instance. From Fig. 12, we can see that the weight for availability changes by time.

Similarly, flow 3 is used to test the result of critical and non-critical users caring about latency. From Fig. 13, it can be seen that if the latency tolerance of user is more than 20 ms, flow 3 is enough to earn a QoE-based QoS score less than 5 because its latency remains lower than 20 ms after congestion happens.

However, if user's tolerant latency value is less than or equal to 10 ms, it will leads to the score higher than 5 and then it will trigger the rerouting control, which is also shown in Fig. 13. The change of latency weights for these two scenarios discussed above are shown in Fig. 14.

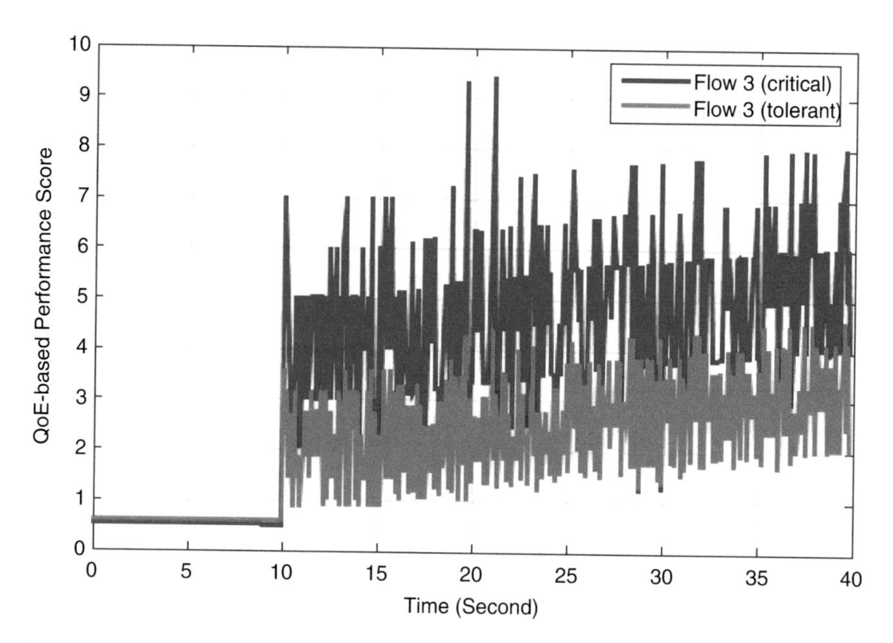

Fig. 13 QoE-based QoS score for Flow 3 with critical and non-critical users

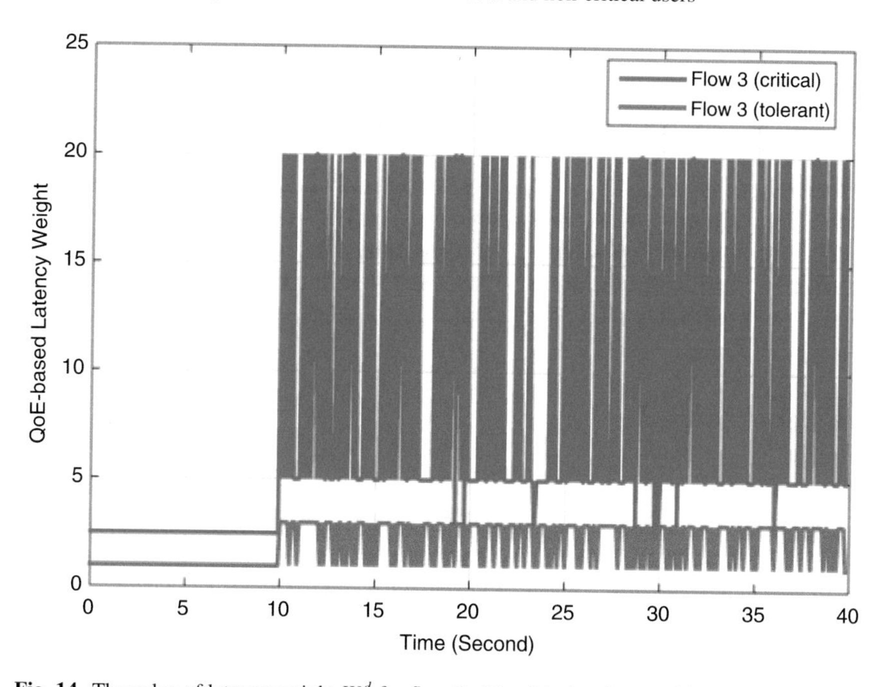

Fig. 14 The value of latency weight W_n^d for flow 3 with critical and non-critical users

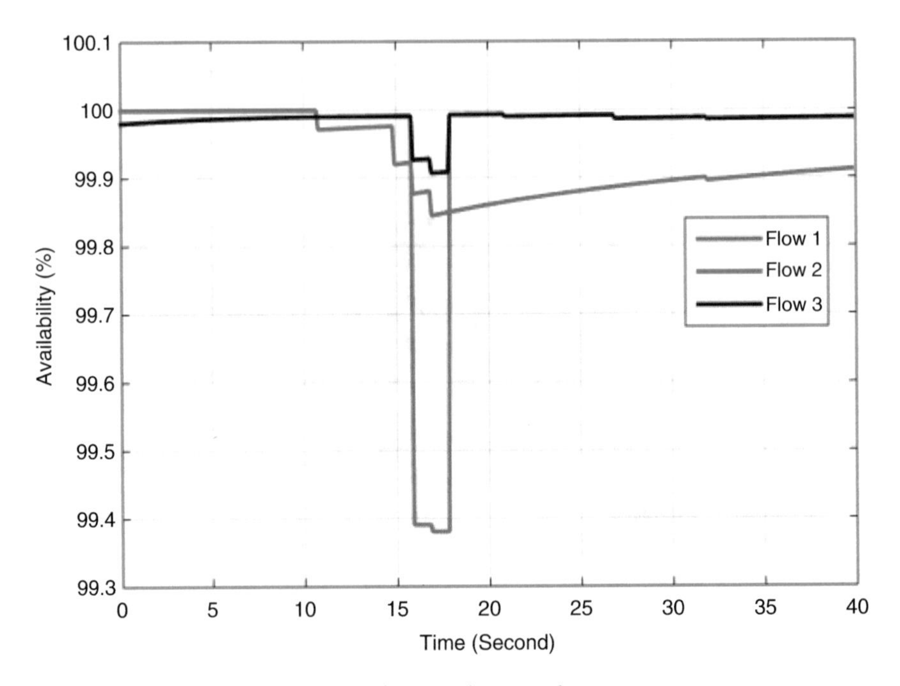

Fig. 15 Availability of flow 1, 2 and 3 after rerouting control

Rerouting of a flow is triggered by counting score higher than 5 for at least 5 times. So after a while of score evaluation, some flows with some kinds of user will be rerouted and the QoS performance of these flows will be recovered soon. For example, from Figs. 15, 16, and 17, it is clear that flow 1 and flow 2 with availability-critical and latency-critical users respectively will be rerouted at 17.8 s and flow 2 with any users will be rerouted definitely after 10.8 s. After the route changes to WiFi mesh network, these flows goes back to normal state with no throughput decrease and perform well in latency and availability aspect.

4.2 Case Study 2

FDI attack is simulated by adding a percentage bias to the power of PV generator. The accuracy of detection of the FDI attack is shown in the Fig. 18 as the percentage of the bias increases. Accuracy remains above 99% and shows a tendency to increase as the bias percentage increase from 1% to 5%.

In order to simulate the effect of noise on the accuracy of FDI attack detection, we add noise with different variances to data with non-attack and 3% bias FDI attack. As shown in Fig. 19 the accuracy remains above 99% and show a tendency to decrease as the noise variance increase from 0 to 4.

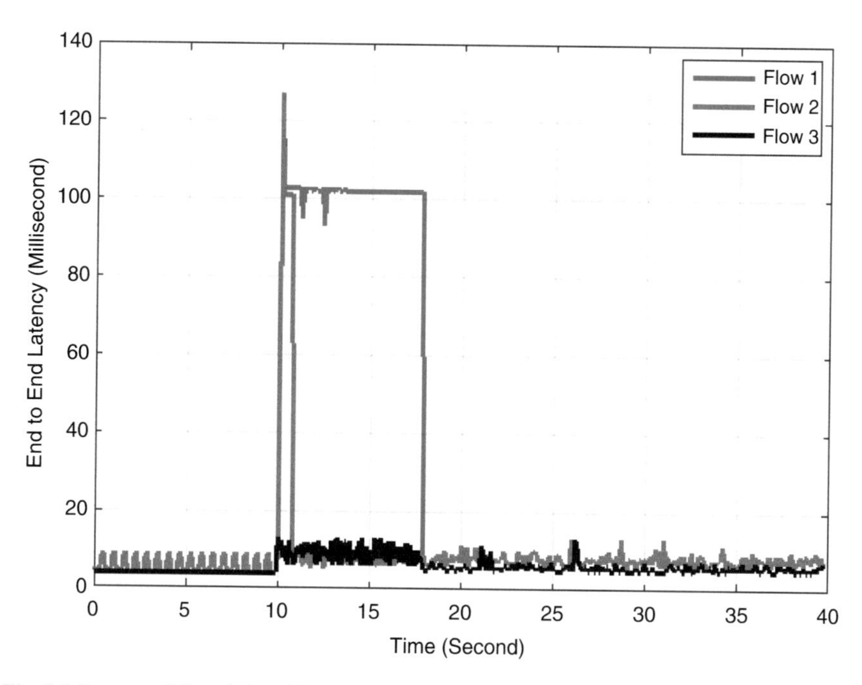

Fig. 16 Latency of flow 1, 2 and 3 after rerouting control

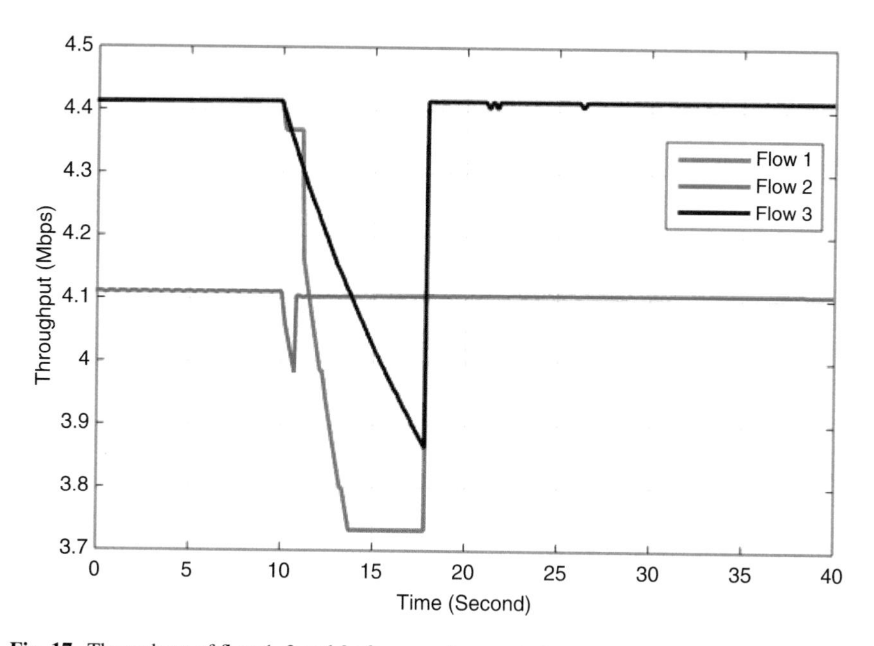

Fig. 17 Throughput of flow 1, 2 and 3 after rerouting control

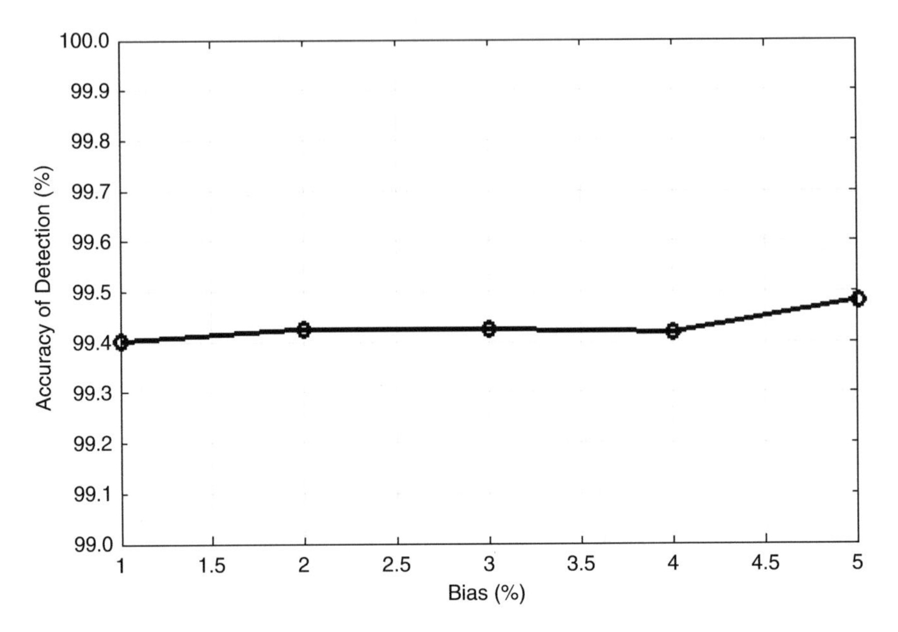

Fig. 18 Accuracy of detection of FDI attack as the percentage bias of FDI attack increases

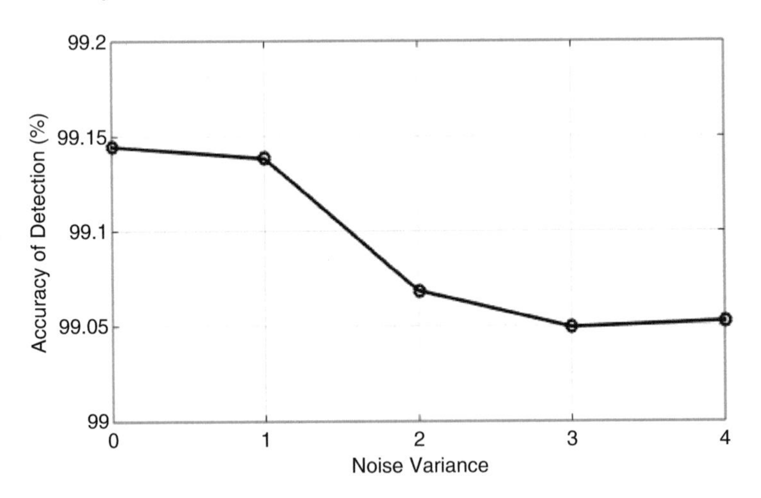

Fig. 19 Accuracy of detection of FDI attack of 3% bias as the variance of noise introduced increases

4.3 Case Study 3

In case study three, we want to apply the authentication part of IEC62351 for MMS message application. From Fig. 20, we can see that at the Association Control Service Elements (ACSE) datagram, AARQ contains a plain-text password "Seven"

```
No.    Time          Source          Destination        Protocol  Length Info
     7 0.010750      10.1.1.1        10.1.1.2           TCP           54 46001 → 102 [ACK] Se
     8 0.012985      10.1.1.1        10.1.1.2           MMS          258 initiate-RequestPDU
     9 0.017484      10.1.1.2        10.1.1.1           TCP           54 102 → 46001 [ACK] Se
    10 0.027743      10.1.1.2        10.1.1.1           MMS          216 initiate-ResponsePDU
▸ Frame 8: 258 bytes on wire (2064 bits), 258 bytes captured (2064 bits)
▸ Point-to-Point Protocol
▸ Internet Protocol Version 4, Src: 10.1.1.1, Dst: 10.1.1.2
▸ Transmission Control Protocol, Src Port: 46001 (46001), Dst Port: 102 (102), Seq: 23, Ack: 23, Le
▸ TPKT, Version: 3, Length: 204
▸ ISO 8073/X.224 COTP Connection-Oriented Transport Protocol
▸ ISO 8327-1 OSI Session Protocol
▸ ISO 8823 OSI Presentation Protocol
▾ ISO 8650-1 OSI Association Control Service
  ▾ aarq
      aSO-context-name: 1.0.9506.2.3 (MMS)
    ▸ called-AP-title: ap-title-form2 (1)
    ▸ called-AE-qualifier: aso-qualifier-form2 (1)
    ▸ calling-AP-title: ap-title-form2 (1)
    ▸ calling-AE-qualifier: aso-qualifier-form2 (1)
      Padding: 3
    ▸ sender-acse-requirements: 80 (authentication)
    ▾ calling-authentication-value: charstring (0)
        charstring: Seven
    ▸ user-information: 1 item
▾ MMS
  ▾ initiate-RequestPDU
      localDetailCalling: 65000
      proposedMaxServOutstandingCalling: 10
      proposedMaxServOutstandingCalled: 10
      proposedDataStructureNestingLevel: 5
    ▾ mmsInitRequestDetail
        proposedVersionNumber: 1
        Padding: 5
      ▸ proposedParameterCBB: f100 (str1, str2, vnam, valt, vlis)
        Padding: 3
      ▸ servicesSupportedCalling: ee1c00000408000079ef18 (status, getNameList, identify, read, wri

0000  00 21 45 00 01 00 00 04  00 00 40 06 00 00 0a 01   .!E..... ..@.....
0010  01 01 0a 01 01 02 b3 b1  00 66 00 00 00 17 00 00   ........ .f......
0020  00 17 80 10 80 00 00 00  00 00 08 0a 00 00 07 dc   ........ ........
0030  00 00 07 d8 00 00 03 00  00 cc 02 f0 80 0d c3 05   ........ ........
0040  06 13 01 00 16 01 02 14  02 00 02 33 02 00 01 34   ........ ...3...4
0050  02 00 01 c1 ad 31 81 aa  a0 03 80 01 01 a2 81 a4   .....1.. ........
0060  81 04 00 00 00 01 82 04  00 00 00 01 a4 23 30 0f   ........ .....#0.
0070  02 01 01 06 04 52 01 00  01 30 04 06 02 51 01 30   .....R.. .0...Q.0
0080  10 02 01 03 06 05 28 ca  22 02 01 30 04 06 02 51   ......(. "..0...Q
0090  01 61 6f 30 6d 02 01 01  a0 6a 60 66 a1 07 06 05   .ao0m... .j`f....
00a0  28 ca 22 02 03 a2 07 06  05 29 01 87 67 01 a3 03   (."..... .)..g...
00b0  02 01 0c a6 06 06 04 29  01 87 67 a7 03 02 01 0c   .......) ..g.....
00c0  aa 02 03 80 ac 07 80 05  53 65 76 65 6e be 33 28   ........ Seven.3(
00d0  31 06 02 01 51 02 01 03  a0 28 a8 26 80 03 00 fd   1...Q... .(.&....
00e0  e8 81 01 0a 82 01 0a 83  01 05 a4 16 80 01 01 81   ........ ........
```

Fig. 20 AARQ partial data-gram with authentication label

used for authentication at responding server which can protect the access to that server to some degree.

5 Conclusions

In this book chapter, we propose a real-time middleware architecture for the application of DER integration. Our proposed middleware effectively utilizes the QoE evaluation from the operators of the power system together with the traditional QoS criteria to determine routing control for congested flows in order to maintain the availability of power data flow. Also, we utilize deep learning method to detect false data injection attack to protect the integrity of data. Furthermore, we realize the initial authentication part of IEC62351 for MMS TO achieve the confidentiality of data acquisition. The simulation results illustrate the efficiency of our proposed architecture. We assert that our work will potentially promote the larger-scale DER integration by providing a resilient and secure communication environment. Future work will examine more detail about the IEC62351 implementation for long-request waiting time power application.

References

1. S. Karnouskos, The cooperative internet of things enabled smart grid, in *Proceedings of the 14th IEEE International Symposium on Consumer Electronics (ISCE2010), June* (2010), pp. 07–10
2. O. Vermesan, P. Friess, P. Guillemin, S. Gusmeroli, H. Sundmaeker, A. Bassi et al., Internet of things strategic research roadmap. Internet Things Glob. Technol. Soc. Trends **1**(2011), 9–52 (2011)
3. J.P. Lopes, N. Hatziargyriou, J. Mutale, P. Djapic, N. Jenkins, Integrating distributed generation into electric power systems: a review of drivers, challenges and opportunities. Electr. Power Syst. Res. **77**(9), 1189–1203 (2007)
4. S. Bandyopadhyay, M. Sengupta, S. Maiti, S. Dutta, A survey of middleware for internet of things, in *Recent Trends in Wireless and Mobile Networks* (Springer, Berlin, 2011), pp. 288–296
5. A.T. Campbell, G. Coulson, M.E. Kounavis, Managing complexity: middleware explained. IT Prof. **1**(5), 22–28 (1999)
6. H. Gjermundrod, D.E. Bakken, C.H. Hauser, A. Bose, GridStat: a flexible QoS-managed data dissemination framework for the power grid. IEEE Trans. Power Deliv. **24**(1), 136–143 (2008)
7. B. Li, K. Nahrstedt, Qualprobes: middleware qos profiling services for configuring adaptive applications, in *IFIP/ACM International Conference on Distributed Systems Platforms and Open Distributed Processing, April* (Springer, Berlin, 2000), pp. 256–272
8. L.C. Dipippo, V.F. Wolfe, L. Esibov, G. Cooper, R. Bethmangalkar, R. Johnston et al., Scheduling and priority mapping for static real-time middleware, in *Challenges in Design and Implementation of Middlewares for Real-Time Systems* (Springer, Boston, 2001), pp. 41–68
9. R.E. Schantz, J.P. Loyall, C. Rodrigues, D.C. Schmidt, Y. Krishnamurthy, I. Pyarali, Flexible and adaptive QoS control for distributed realtime and embedded middleware, in *Proceedings of the ACMIFIPUSENIX 2003 International Conference on Middleware, June* (Springer, New York, 2003), pp. 374–393
10. W. Li, A. Monti, Integrated simulation with VTB and OPNET for networked control and protection in power systems, in *Proceedings of the 2010 Conference on Grand Challenges in Modeling & Simulation, June* (Society for Modeling & Simulation International, Vista, 2010), pp. 386–391
11. M. Bartl, K. Molnar, J. Hosek, Control of network operation generator from opnet modeler environment. Int. J. Comput. Sci. Netw. Secur. **11**(6), 17 (2011)
12. M. Casoni, C.A. Grazia, M. Klapez, N. Patriciello, A congestion control middleware layer with dynamic bandwidth management for satellite communications. Int. J. Satell. Commun. Netw. **34**(6), 739–758 (2016)
13. L.P. Jain, W.J. Scheirer, T.E. Boult, Quality of experience, in *IEEE Multimedia* (2004)
14. L. Zhou, J.J. Rodrigues, L.M. Oliveira, QoE-driven power scheduling in smart grid: architecture, strategy, and methodology. IEEE Commun. Mag. **50**(5), 136–141 (2012)

15. P. Brooks, B. Hestnes, User measures of quality of experience: why being objective and quantitative is important. IEEE Netw. **24**(2), 8–13 (2010)
16. Y. Wu, G.J. Mendis, Y. He, J. Wei, and B.-M. Hodge, An Attack-Resilient Middleware Architecture for Grid Integration of Distributed Energy Resources, IEEE Cyber, Physical and Social Computing (CPSCom), Chengdu, China, December 2016
17. Y. Wu, J. Wei, and B.-M. Hodge, A distributed middleware architecture for attack-resilient communications in smart grids, 2017 IEEE International Conference on Communications (ICC), Paris, France, May 2017
18. MATLAB, 9.0.0.341360($R2016a$) (The MathWorks Inc., Natick, 2016)
19. T.R. Henderson, M. Lacage, G.F. Riley, C. Dowell, J. Kopena, Network simulations with the ns-3 simulator. SIGCOMM Demonstration **14**(14), 527 (2008)
20. B. Berry, A fast introduction to SCADA fundamentals and implementation. DSP Telecom (2009). Retrieved on July, 28
21. A. Daneels, W. Salter, What is SCADA? in *7th International Conference on Accelerator and Large Experimental Physics Control Systems*, (Trieste, Italy, 1999, October), pp. 339–343
22. H. Farhangi, The path of the smart grid. IEEE Power Energy Mag. **8**(1), 18–28 (2009)
23. Pennsylvania Smart Meter (2015). Available via DIALOG. https://www.firstenergycorp.com/content/dam/customer/get-help/files/PASmartMeter/Speakers%20Bureau_Smart%20Meters.pdf. Accessed 15 Dec 2016
24. J. Mirkovic, A. Hussain, S. Fahmy, P. Reiher, R.K. Thomas, Accurately measuring denial of service in simulation and testbed experiments. IEEE Trans. Dependable Secure Comput. **6**(2), 81–95 (2008)
25. G.E. Hinton, S. Osindero, Y.W. Teh, A fast learning algorithm for deep belief nets. Neural Comput. **18**(7), 15271554 (2006)
26. A. Fischer, C. Igel, (2012). An introduction to restricted Boltzmann machines, in *Iberoamerican Congress on Pattern Recognition, September* (Springer, Berlin, 2012), pp. 14–36
27. K.H. Cho, T. Raiko, A. Ilin, Gaussian-bernoulli deep Boltzmann machine, in *The 2013 International Joint Conference on Neural Networks (IJCNN), August* (IEEE, Piscataway, 2013), pp. 1–7
28. K. Cho, A. Ilin, T. Raiko, Improved learning of Gaussian-Bernoulli restricted Boltzmann machines, in *International Conference on Artificial Neural Networks, June* (Springer, Berlin, 2011), pp. 10–17
29. M.A. CarreiraPerpinan, G.E. Hinton, On contrastive divergence learning, in *Aistats*, vol. 10 (2005, January), pp. 33–40
30. J. Li, J.H. Cheng, J.Y. Shi, F. Huang, Brief introduction of back propagation (BP) neural network algorithm and its improvement, in *Advances in Computer Science and Information Engineering* (Springer, Berlin, 2012), pp. 553–558
31. H. Zimmermann, OSI reference model-the ISO model of architecture for open systems interconnection. IEEE Trans. Commun. **28**(4), 425–432 (1980)
32. N. Baldo, The ns3 LTE module by the LENA project. Center Tecnologic de Telecomunicacions de Catalunya (2011)
33. "250-kW Grid-Connected PV Array," [Available] mathworks.com/help/physmod/sps/examples/250-kw-grid-connected-pv-array.html

Printed in the United States
by Baker & Taylor Publisher Services